Student's Solutions

To accompany

Mathematical Excursions

Second Edition

Richard N. Aufmann
Palomar College

Joanne S. Lockwood
New Hampshire Community Technical College

Richard D. Nation
Palomar College

Daniel K. Clegg
Palomar College

Scott Inch
Bloomsburg University

Houghton Mifflin Company   Boston   New York

Publisher: Jack Shira
Senior Sponsoring Editor: Lynn Cox
Development Editor: Lisa Collette
Assistant Editor: Noel Kamm
Editorial Assistant: Laura Ricci
Senior Marketing Manager: Ben Rivera
Marketing Associate: Lisa Lawler

Printed in the U.S.A.

ISBN 10: 0-618-60855-9
ISBN 13: 978-0-618-60855-3

2 3 4 5 6 7 8 9- HES -10 09 08 07

TABLE OF CONTENTS

Chapter 1: Problem Solving

EXERCISE SET 1.1

1. 28. Add 4 to obtain the next number.

3. 45. Add 2 more than the integer added to the previous integer.

5. 64. The numbers are the squares of consecutive integers. $8^2 = 64$.

7. $\dfrac{15}{17}$. Add 2 to the numerator and denominator.

9. -13. Use the pattern of adding 5, then subtracting 10 to obtain the next pair of numbers.

11. Correct.

13. Correct.

15. Incorrect. The resulting number will be 3 times the original number.

17. Doubling the weight of the ball has no effect on the distance it rolls in a given time interval.

19. 150 inches. Distance $= 6t^2$, $t = 5$.

21. A ball rolls 24 inches in 2 seconds and quadruple that distance, 96 inches, in 4 seconds.

23. 1.5 inches is one-fourth of the distance traveled by a ball in 1 second. We know from exercise 21 that doubling the time interval quadruples the distance traveled, so halving the time interval will divide the distance traveled by 4. Thus the ball travels 1.5 inches in 0.5 second.

25. This argument reaches a conclusion based on a specific example, so it is an example of inductive reasoning.

27. The conclusion is a specific case of a general assumption, so this argument is an example of deductive reasoning.

29. The conclusion is a specific case of a general assumption, so this argument is an example of deductive reasoning.
$$1^3 + 5^3 + 3^3 = 1 + 125 + 27$$
$$= 153$$

31. This argument reaches a conclusion based on a specific example, so it is an example of inductive reasoning.

33. Any number less than or equal to -1 or between 0 and 1 will provide a counterexample.

35. Any number less than -1 or between 0 and 1 will provide a counterexample.

37. Any negative number will provide a counterexample.

39. Consider any two odd numbers. Their sum is even, but their product is odd.

41. This procedure does not work if the first and last digits of the number are the same. For example, 313 reversed is 313, and $313 - 313 = 0$, which has a tens digit of 0.

43. Using deductive reasoning:

n	pick a number
$6n$	multiply by 6
$6n + 8$	add 8
$\dfrac{6n + 8}{2} = 3n + 4$	divide by 2
$3n + 4 - 2n = n + 4$	subtract twice the original number
$n + 4 - 4 = n$	subtract 4

45.

	Util	Auto	Tech	Oil
A	Xa	Xa	✓	Xc
T	Xa	Xa	Xc	✓
M	✓	Xb	Xb	Xb
J	Xb	✓	Xb	Xb

47.

	Coin	Stamp	Comic	Baseball
A	Xc	✓	Xc	Xd
C	Xb	Xd	Xa	✓
P	✓	Xc	Xc	Xb
S	Xb	Xc	✓	Xc

49. home, bookstore, supermarket, credit union, home; or home, credit union, supermarket, bookstore, home.

51. N. These are the first letters of the counting numbers: **O**ne, **T**wo, **T**hree, etc. N is the first letter of the next number, which is **N**ine.

53. d is the correct choice. Example 3b found that quadrupling the length of a pendulum doubles its period. Doubling 1 second 4 times gives 16 seconds, which is close to the period of Foucault's pendulum. In order to double the period 4 times, we must quadruple the length of the pendulum 4 times.
$0.25 \times 4 \times 4 \times 4 \times 4 = 64$, so Foucault's pendulum should have a length of approximately 64 meters. Thus D, 67 meters, is the best choice.

EXERCISE SET 1.2

1.
$$
\begin{array}{ccccccc}
1 & 7 & 17 & 31 & 49 & 71 & 97 \\
 & 6 & 10 & 14 & 18 & 22 & 26 \\
 & & 4 & 4 & 4 & 4 & 4
\end{array}
$$

    $26 + 71 = 97$

3.
$$
\begin{array}{ccccccc}
-1 & 4 & 21 & 56 & 115 & 204 & 329 \\
 & 5 & 17 & 35 & 59 & 89 & 125 \\
 & & 12 & 18 & 24 & 30 & 36 \\
 & & & 6 & 6 & 6 & 6
\end{array}
$$

    $125 + 204 = 329$

5.
$$
\begin{array}{ccccccc}
9 & 4 & 3 & 12 & 37 & 84 & 159 \\
-5 & -1 & 9 & 25 & 47 & 75 & \\
 & 4 & 10 & 16 & 22 & 28 & \\
 & & 6 & 6 & 6 & 6 &
\end{array}
$$

    $75 + 84 = 159$

7. Substitute in the appropriate values for n.
   For $n = 1$, $a_1 = \dfrac{1(2(1)+1)}{2} = \dfrac{3}{2}$
   For $n = 2$, $a_2 = \dfrac{2(2(2)+1)}{2} = 5$
   For $n = 3$, $a_3 = \dfrac{3(2(3)+1)}{2} = \dfrac{21}{2}$
   For $n = 4$, $a_4 = \dfrac{4(2(4)+1)}{2} = 18$

For $n = 5$, $a_5 = \dfrac{5(2(5)+1)}{2} = \dfrac{55}{2}$

9. Substitute in the appropriate values for n to obtain 2, 14, 36, 68, 110.

11. Notice that each figure is square with side length n plus an "extra row" of length $n-1$. Thus the nth figure will have $a_n = n^2 + (n-1)$ tiles.

13. Each figure is composed of a horizontal group of n tiles, a horizontal group of $n - 1$ tiles, and a single "extra" tile. Thus the nth figure will have $a_n = n + n - 1 + 1 = 2n$ tiles.

15. a. There are 56 cannonballs in the sixth pyramid and 84 cannonballs in the seventh pyramid.

    b. The eighth pyramid has eight levels of cannonballs. The total number of cannonballs in the eighth pyramid is equal to the sum of the first 8 triangular numbers: $1 + 3 + 6 + 10 + 15 + 21 + 28 + 36 = 120$.

17. a. Five cuts produce six pieces and six cuts produce seven pieces.

    b. The number of pieces is one more than the number of cuts, so $a_n = n+1$.

19. a. Substituting in $n = 5$:
    $$P(5) = \frac{5^3 + 5(5) + 6}{6} = \frac{156}{6} = 26$$

    b. Substituting several values:
    $$P(6) = \frac{6^3 + 5(6) + 6}{6} = 42 < 60.$$
    $$P(7) = \frac{7^3 + 5(7) + 6}{6} = 64 > 60.$$
    Thus the fewest number of straight cuts is 7.

    d. Experimenting:
    For $n > 2$, $5F_n - 2F_{n-2} = F_{n+3}$
    $n = 3 \Rightarrow 5F_3 - 2F_1 = 8 = F_6$
    $n = 4 \Rightarrow 5F_4 - 2F_2 = 13 = F_7$
    $n = 5 \Rightarrow 5F_5 - 2F_3 = 21 = F_8$
    It appears as if this property is valid.

21. Substituting:
$$a_3 = 2 \cdot a_2 - a_1 = 10 - 3 = 7$$
$$a_4 = 2 \cdot a_3 - a_2 = 14 - 5 = 9$$
$$a_5 = 2 \cdot a_4 - a_3 = 18 - 7 = 11$$

23. Substituting:
$$F_{20} = 6765$$
$$F_{30} = 832,040$$
$$F_{40} = 102,334,155$$

25. The drawing shows the nth square number. The question mark should be replaced by n^2.

27. a. The ninth number is
$38.4 + 0.4 = 38.8$ AU

    b. $38.8 - 30.6 = 8.2$ AU. The prediction is not close compared to results obtained for the inner planets.

29. a. $2(76.8) + 0.4 = 154$ AU

    b. Yes

31. a. For $n = 1$, we get $1 + 2(1) + 2 = 5 = F_5$
    For $n = 2$, we get $1 + 2(2) + 3 = 8 = F_6$
    For $n = 3$, we get $2 + 2(3) + 5 = 13 = F_7$.
    Thus, $F_n + 2F_{n+1} + F_{n+2} = F_{n+4}$.

    b. For $n = 1$, we get $1 + 1 + 3 = 5 = F_5$
    For $n = 2$, we get $1 + 2 + 5 = 8 = F_6$
    For $n = 3$, we get $2 + 3 + 8 = 13 = F_7$.
    Thus, $F_n + F_{n+1} + F_{n+3} = F_{n+4}$.

33. Substituting:
$$\frac{F_8}{F_7} = \frac{21}{13} \approx 1.615385$$
$$\frac{F_{10}}{F_9} = \frac{55}{34} \approx 1.617647$$
$$\frac{F_{12}}{F_{11}} = \frac{144}{89} \approx 1.617978$$
$$\frac{F_{14}}{F_{13}} = \frac{377}{233} \approx 1.618026$$
The ratio is always less than Φ.

EXERCISE SET 1.3

1. Let g be the number of first grade girls, and let b be the number of first grade boys. Then $b + g = 364$ and $g = b + 26$. Solving gives $g = 195$, so there are 195 girls.

3. There are 36 1×1 squares, 25 2×2 squares, 16 3×3 squares, 9 4×4 squares, 4 5×5 squares and 1 6×6 square in the figure, making a total of 91 squares.

5. Solving:
$x = $ cost of the shirt
$x - 30 = $ cost of the tie
$$(x - 30) + x = 50$$
$$2x - 30 = 50$$
$$2x = 80$$
$$x = 40$$
The shirt costs $40.

7. There are 14 different routes to get to Fourth Avenue and Gateway Boulevard and 4 different routes to get to Second Avenue and Crest Boulevard. Adding gives that there are 18 different routes altogether.

9. Try solving a simpler problem to find a pattern. If the test had only 2 questions, there would be 4 ways. If the test had 3 questions, there would be 8 ways. Further experimentation shows that for an n question test, there are 2^n ways to answer. Letting $n = 12$, there are $2^{12} = 4,096$ ways

11. a. By adding adjacent pairs, the number of routes from A to B, C, D, E, F, G, and H are 1, 9, 36, 84, 126, 126, and 84 respectively.

    b. The demonstrator is symmetric. F and G are the fifth regions from the left and right, respectively, so there are the same number of paths from A to either one.

13. 8 people shake hands with 7 other people. Multiply 8 and 7 and divide by 2 to eliminate repetitions to obtain 28 handshakes.

15. Let p be the number of pigs and let d be the number of ducks. Then $p + d = 35$ and $4p + 2d = 98$. Solving gives $d = 21$ and $p = 14$, so there are 21 ducks and 14 pigs.

17.

Dimes	Nickels	Pennies
0	0	25
0	1	20
0	2	15
0	3	10
0	4	5
0	5	0
1	0	15
1	1	10
1	2	5
1	3	0
2	0	5
2	1	0

There are 12 ways.

19. The units digits of powers of 4 form the sequence 4, 6, 4, 6, ... Even powers end in 6. Therefore, the units digit of 4^{7022} is 6.

21. The units digits of powers of 3 form the sequence 3, 9, 7, 1, 3, 9, 7, 1, ... Divide 11,707 by 4 to obtain the remainder 3, which corresponds to 7. Therefore the units digit of $3^{11,707}$ is 7.

23. a. Add the numbers in pairs: 1 and 400, 2 and 399, 3 and 398, and so on. There are 200 pair sums equal to 401.
   $200 \times 401 = 80,200$.

   b. Add the numbers in pairs: 1 and 550, 2 and 549, 3 and 548 and so on. There are 275 pair sums equal to 551.
   $275 \times 551 = 151,525$.

   c. Add the numbers in pairs: 2 and 84, 4 and 82, and so on, leaving off the 86. There are 21 sums of 86 plus one additional 86.
   $21 \times 86 + 86 = 1,892$.

25. a. 121, 484, and 676 are the only three-digit perfect square palindromes.

   b. 1331 is the only four-digit perfect cube palindrome.

27. Draw a simpler picture:

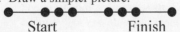

Start   Finish

Note that the first page of the first volume is the second dot on the line, and the last page of the third volume is the seventh dot on the line. This is because when books sit on a shelf, the first pages are on the right side of the book and their last pages are on the left side.
$\frac{1}{8}+\frac{1}{8}+1+\frac{1}{8}+\frac{1}{8}=1\frac{1}{2}$ inches.

29. a. 1.3, 1.5, 1.6 billion

   b. 1995

   c. 2002

   d. 2001 to 2002

31. a. 1994

   b. 2002

   c. The number of ticket buyers was less.

33. Since there is one blue tile in each column, there are n blue tiles on the diagonal that starts in the upper left hand corner. Similarly, there are n blue tiles in the diagonal that starts in the upper right hand corner. The two diagonals have one tile in common, so the actual total number of blue tiles is $2n-1$. Since $2n-1=101$, we can solve to find $n=51$. The total number of tiles is n^2. Substituting the value for n yields 2601.

35. Let b be the number of boys in the family and g be the number of girls. The first two statements imply that the speaker is a girl. Thus, $g-1=b+2$. Solving for b, $b=g-3$. To answer the last question, we must omit the youngest brother, so $b-1=g-4$. There are four more sisters than brothers.

37. The bacteria population doubles every day, so on the 11th day there are half as many bacteria as on the 12th day.

39. Let x be the score that Dana needs on the fourth exam:

$$\frac{82+91+76+x}{4}=85$$
$$249+x=340$$
$$x=91$$

41. a. Place four coins on the left balance pan and the other 4 coins on the right balance pan. The pan that is the higher contains the fake coin. Take the four coins from the higher pan and use the balance scale to compare the weight of two of these coins to the weight of the other two coins. The pan that is the higher contains the fake coin. Take the two coins from the higher pan and use the balance scale to compare the weights. The pan that is the higher contains the fake coin. This procedure enables you to determine the fake coin in 3 weighings.

    b. Place 3 of the coins on one of the balance pans and 3 coins on the other balance pan. If the pans balance, then the fake coin is one of the two remaining coins. You can put each one of these coins on a balance pan and the higher pan contains the fake coin. If the 3 coins on the left do not balance with the 3 coins on the right, then the higher pan contains the fake coin. Pick any 2 of these 3 coins and use the balance scale to compare their weights. If these 2 coins do not balance, then the higher pan contains the fake coin. If these two coins balance, then the 3rd coin (the one that you did not place on the balance pan) is the fake. In either case this procedure enables you to determine the fake coin in 2 weighings.

43. The correct answer is a. Sally likes perfect squares.

45. The correct answer is d. The numbers are all perfect cubes. The missing number is the cube of 4.

47. a. Write an equation. When the people who were born in 1980 are x years old, it will be the year $1980 + x$. We are looking for the year that satisfies $1980 + x = x^2$. Solving gives $x = 45$ and $x = -44$. The solution must be a natural number, so $x = 45$. Therefore when the people born in 1980 are 45 years old, the year will be $45^2 = 2025$.

    b. 2070, because people born in 2070 will be 46 in $2116 = 46^2$.

49. It takes 9 1-digit numbers for pages 1-9, 180 digits for pages 10-99, and 423 digits for pages 100-240. The total is 612.

51. Answers will vary.

CHAPTER 1 REVIEW EXERCISES

1. This argument reaches a conclusion based on a case of a general assumption, so it is an example of deductive reasoning.

2. This argument reaches a conclusion based on specific examples, so it is an example of inductive reasoning.

3. This argument reaches a conclusion based on a specific example, so it is an example of inductive reasoning.

4. This argument reaches a conclusion based on a case of a general assumption, so it is an example of deductive reasoning.

5. Any number from 0 to 1 provides a counterexample. For example, $x = \frac{1}{2}$ provides a counterexample because $\left(\frac{1}{2}\right)^4 = \frac{1}{16}$ and $\frac{1}{16}$ is not greater than $\frac{1}{2}$.

6. $n = 4$ provides a counterexample because $\frac{(4)^3 + 5(4) + 6}{6} = \frac{90}{6} = 15$, which is not even.

7. $x = 1$ provides a counterexample because $(1+4)^2 = (5)^2 = 25$ and $1^2 + 4^2 = 1 + 16 = 17$

8. $a = 1$ and $b = 1$ provides a counterexample because $(1+1)^3 = 2^3 = 8$, but $1^3 + 1^3 = 1 + 1 = 2$.

9. a.
$$
\begin{array}{ccccccc}
-2 & 2 & 12 & 28 & 50 & 78 & \mathbf{112} \\
 & 4 & 10 & 16 & 22 & 28 & 34 \\
 & & 6 & 6 & 6 & 6 & 6
\end{array}
$$

    Add 34 and 78 to obtain 112.

b. -4 -1 14   47   104   191   314   **479**
    3   15   33   57   87   123   165
     12   18   24   30   36   42
       6   6   6   6   6

Add 165 and 314 to obtain 479.

10. a.   5   6   3 -4 -15 -30 -49 **-72**
      1   -3 -7 -11 -15 -19 -23
       -4 -4 -4 -4 -4 -4

Add -23 and -49 to obtain -72.

b.   2   0 -18 -64 -150 -288 -490 **-768**
     -2 -18 -46 -86 -138 -202 -278
      -16 -28 -40 -52 -64 -76
       -12 -12 -12 -12 -12

Add -278 and -490 to obtain -768.

11. Substituting:

$$a_1 = 4(1)^2 - 1 - 2 = 4 - 3 = 1$$

$$a_2 = 4(2)^2 - 2 - 2 = 16 - 4 = 12$$

$$a_3 = 4(3)^2 - 3 - 2 = 36 - 5 = 31$$

$$a_4 = 4(4)^2 - 4 - 2 = 64 - 6 = 58$$

$$a_5 = 4(5)^2 - 5 - 2 = 100 - 7 = 93$$

$$a_{20} = 4(20)^2 - 20 - 2$$
$$= 4(400) - 20 - 2$$
$$= 1600 - 22 = 1578$$

12. Substituting:

$$a_1 = -2(1)^3 + 5(1) = -2 + 5 = 3$$

$$a_2 = -2(2)^3 + 5(2) = -16 + 10 = -6$$

$$a_3 = -2(3)^3 + 5(3) = -54 + 15 = -39$$

$$a_4 = -2(4)^3 + 5(4) = -128 + 20 = -108$$

$$a_5 = -2(5)^3 + 5(5) = -250 + 25 = -225$$

$$a_{25} = -2(25)^3 + 5(25) = -31250 + 125$$
$$= -31125$$

13. Each figure has a horizontal section with $n+1$ tiles, a horizontal section with n tiles, and a vertical section with $n-1$ tiles.
$$a_n = n + 1 + n + n - 1 = 3n$$

14. Each figure is a square with sidelength $n+2$ and n tiles removed.
$$a_n = (n+2)^2 - n = n^2 + 4n + 4 - n$$
$$= n^2 + 3n + 4$$

15. Each figure is a square with sidelength $n+1$ with an attached diagonal with $n+1$ tiles.

$$a_n = (n+1)^2 + (n+1) = n^2 + 3n + 2$$

16. Each figure made up of four sides of length n with a diagonal piece in the middle with length $n-1$.

$$a_n = 4n + (n-1) = 5n - 1$$

17. Let x be the width. Then $5x$ is the length. Since one length already exists, only 3 sides of fencing are needed. The total perimeter is $5x + x + x = 2240$. Solving, we find $x = 320$. The dimensions are 320 ft. by 1600 ft.

18. Solve a simpler problem. If the test has 1 question, there are 3 ways to answer. If the test has 2 questions, there are 9 ways to answer. If the test has 3 questions, there are 27 ways to answer. It appears that for a test with n questions, there are 3^n ways to answer. In this case, $n = 15$, so there are $3^{15} = 14,348,907$ ways to answer the test.

19. If the 11th and 35th are opposite each other, there must be 23 more skyboxes between them (going in each direction). The total is $23 + 1 + 23 + 1 = 48$.

20. On the first trip the rancher takes the rabbit across the river. The rancher returns alone. The rancher takes the dog across the river and returns with the rabbit. The rancher next takes the carrots across the river and returns alone. On the final trip the rancher takes the rabbit across the river.

21. $1400 - $1200 = $200 profit.
$1900 - $1800 = $100 profit.
Total profit = $200 + $100 = $300

22. Multiply 15 and 14 and divide by 2 to eliminate repetitions. 105 handshakes will take place.

23. Answers will vary. Possible answers include: make a list, draw a diagram, make a table, work backwards, solve a simpler similar problem, look for a pattern, write an equation, perform an experiment, guess and check, and use indirect reasoning.

24. Answers will vary. Possible answers include: ensure that the solution is consistent with the facts of the problem, interpret the solution in the context of the problem, and ask yourself whether there are generalizations of the solution that could apply to other problems.

25.

	CS	Chem	Bus	Bio
M	Xa	Xd	Xd	✓
C	Xd	Xb	✓	Xb
R	✓	Xb	Xd	Xb
E	Xd	✓	Xd	Xc

26.

	Bank	Super	Service	Drug
D	Xd	Xd	Xd	✓
P	Xb	✓	Xd	Xb
T	✓	Xd	Xa	Xd
G	Xb	Xc	✓	Xb

27. a. Yes. Answers will vary.

   b. No. The countries of India, Bangladesh, and Myanmar all share borders with each of the other two countries. Thus, at least three colors are needed to color the map.

28. a. If we label the three islands A, B and C from left to right, the following sequence of moves shows a route that starts at North Bay and passes over each bridge once and only once. Starting on the North Bay, cross the bridge to Island A then cross the bridge to the South Bay. Travel to the right on South Bay and cross the bridge back to Island A, then cross the bridge to Island B. From here, cross to the North Bay and travel right on North Bay to cross over to Island C. From here, cross to Island B, then to the South Bay. Travel right on South Bay, crossing to Island C and finally to Island B.

   b. No.

29. Draw the three possible pictures (one with x diagonal from 2, one diagonal from 5 and one diagonal from 10) to find the three possible values for x: 1 square inch, 4 square inches, 25 square inches.

30. a. Adding smaller line segments to each end of the shortest line doubles the total number of line segments. Thus the nth figure has 2^n line segments. For $n = 10$, $a_{10} = 1024$.

   b. $a_{30} = 2^{30} = 1,073,741,824$

31. A = 1, B = 9, D = 0

32. Making a table:

quarters	nickels
4	0
3	5
2	10
1	15
0	20

There are 5 ways.

33. Use a list. WWWLL, WWLWL, WWLLW, WLWLW, WLLWW, WLWWL, LWWWL, LWLWW, LWWLW, LLWWW. There are 10 ways.

34. $7^1 = 7$. $7^2 = 49$. $7^3 = 343$. $7^4 = 2401$. The pattern of the units digits is 7, 9, 3, 1. For 7^{56} divide 56 by 4 to obtain remainder 0. This corresponds to the digit 1.

35. $23^1 = 23$. $23^2 = 529$. $23^3 = 12,167$. $23^4 = 279,841$. The pattern of the unit digits is 3, 9, 7, 1. For 23^{85} divide 85 by 4 to obtain the remainder 1. This corresponds to the digit 3.

36. Pick a number n:

n

$4n$ — multiply by 4

$4n + 12$ — add 12

$\dfrac{4n+12}{2} = 2n+6$ — divide by 2

$2n + 6 - 6 = 2n$ — subtract 6

37. 2000 nickels are equivalent to $100.00. 2004 nickels is $0.20 more than $100 or $100.20.

38. a.   1970 to 1980

    b. $16.2\% - 10.7\% = 5.5\%$

39. a. $\dfrac{112}{7} = 16$ times as many

    b. $\dfrac{427}{7} = 61$ times as many

40. a.   2002

    b.   2004

41. Every multiple of 5 ends in a 5 or a zero. Every palindromic number begins with the same digit it ends with. We cannot begin a number with 0, so the number must end in 5. Thus it must begin with 5. The smallest such number is 5005.

42. Checking all of the two-digit natural numbers shows that there are no narcissistic numbers.

43. a.   10 intersections.

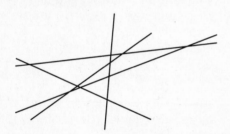

    b.   Yes

44. a.   22. 9^{22} has 21 digits.

    b. $9^9 = 387,420,489$, so $9^{\left(9^9\right)}$ is the product of 387,420,489 nines. At one multiplication per second this would take about 12.3 years. It is probably not a worthwhile project.

CHAPTER 1 TEST

1. This conclusion is based on a specific case of a general assumption, so this argument is an example of deductive reasoning.

2. This argument reaches a conclusion based on a specific example, so it is an example of inductive reasoning.

3. This argument reaches a conclusion based on a specific example, so it is an example of inductive reasoning.

4. This conclusion is based on a specific case of a general assumption, so this argument is an example of deductive reasoning.

5.   −1   0   9   32   75   144   245   **384**
      1   9   23   43   69   101   **139**
        8   14   20   26   32   **38**
         6   6   6   6   **6**

Add 139 and 245 to obtain 384.

6. 1, 1, 2, 3, 5, 8, 13, 21, 34, 55

7. a.   Each figure contains a horizontal group of n tiles, a horizontal group of $n + 1$ tiles, and a vertical group of $n - 1$ tiles.
$$a_n = n + n + 1 + 2n - 1 = 4n.$$

    b.   Each figure contains a horizontal group of $n + 1$ tiles and 2 horizontal groups of n tiles.
$$a_n = n + 1 + n + n = 3n + 1.$$

8. $a_1 = 0$. $a_2 = 1$. $a_3 = -3$. $a_4 = 6$. $a_5 = -10$.
$$a_{105} = (-1)^{105}\left(\frac{105\,(104)}{2}\right)$$
$$= -1(105 \cdot 52) = -5,460$$

9. $50 + 81 = 131$
$81 + 131 = 212$
$131 + 212 = 343$

10. Understand the problem. Devise a plan. Carry out the plan. Review the solution.

11. Making a table:

Half-dollars	Quarters	Dimes
2	0	0
1	0	5
1	2	0
0	4	0
0	2	5
0	0	10

There are 6 ways.

12. Make a list.

    LLWWWW   LWLWWW   LWWLWW
    LWWWLW   LWWWWL   WLWWWL
    WLWWLW   WLWLWW   WLLWWW
    WW LLWW   WWWLWL   WWWWLL
    WWLWWL   WWWLLW   WWLWLW

    There are 15 ways.

13. The units digits form a sequence with 4 terms that repeat, so divide the powers by 4 and look at the remainders. A remainder of 1 corresponds to a 3.

14. Work backwards. Subtract $150 from $326. This is $176. Let x be the amount of money Shelly had before renting the room. Then $x - \frac{x}{3} = \$176$, so $x = \$264$. Adding on $22 and $50 gives $336. Since this is half of her savings (the other half was spent on the plane ticket), double to get $672.

15. 126 ways. Add successive pairs of vertices. The last pair give 70 ways + 56 ways or 126 ways.

16. Mutiply 9 times 8 and divide by 2 to eliminate repetitions. There will be a total of 36 league games.

17.

	5	7	13	15
Rey	Xa	Xc	✓	Xd
Ram	✓	Xc	Xc	Xc
Shak	Xc	Xc	Xd	✓
Sash	Xc	✓	Xc	Xb

18. 606

19. $x = 4$ gives $\frac{(4-4)(4+3)}{(4-4)} = \frac{0}{0}$, which makes the left side of the equation meaningless since division by zero is undefined, but in any case not equal to $4 + 3 = 7$.

20. a. 2002 to 2003

    b. $550{,}000 - 300{,}000 = 250{,}000$

Chapter 2: Sets

EXERCISE SET 2.1

1. {penny, nickel, dime, quarter}

3. {Mercury, Mars}

5. {Reagan, G. H. W. Bush, Clinton, G. W. Bush}

7. The negative integers greater than –6 are –5, –4, –3, –2, –1. Using the roster method, write the set as {–5, –4, –3, –2, –1}.

9. Adding 4 to each side of the equation produces $x = 7$. {7} is the solution set.

11. Solving:

$$x + 4 = 1$$
$$x = -3$$

But –3 is not a counting number so the solution set is empty, $\varnothing$.

In exercises 13 – 19, only one possible answer is given. Your answers may vary from the given answers.

13. the set of days of the week that begin with the letter T

15. the set consisting of the two planets in our solar system that are closest to the sun

17. the set of single digit natural numbers

19. the set of natural numbers less than or equal to 7

21. Because b is an element of the given set, the statement is true.

23. False; although b ∈ {a, b, c}, {b} ∉ {a, b, c}.

25. False; {0} contains 1 element but $\varnothing$ contains no elements.

27. False; "good" is subjective.

29. False; $0 \in I$ but 0 is not an element of the natural numbers.

31. $\{x \mid x \in N \text{ and } x < 13\}$

33. $\{x \mid x \text{ is a multiple of 5 and } 4 < x < 16\}$

35. $\{x \mid x \text{ is the name of a month that has 31 days}\}$

37. $\{x \mid x \text{ is the name of a U.S. state that begins with the letter A}\}$

39. $\{x \mid x \text{ is a season that starts with the letter s}\}$

41. {California, Arizona}

43. {California, Arizona, Florida, Texas}

45. {2000, 2002, 2004}

47. {1997, 1998}

49. {June, October, November}

51. {1985, 1986, 1987, 1989}

53. {1988, 1990, 1991, 1992, 1993, 1995, 1996}

55. 11 since set A has 11 elements

57. 0. The cardinality of the empty set is 0.

59. 4 since 4 states border Minnesota.

61. 16. There are 16 baseball teams in the league.

63. 121

65. Neither. The sets are not equal, nor do they have the same number of elements.

67. Both.

69. Equivalent. The sets are not equal but each has 3 elements.

71. Equivalent. Each set has 2 elements.

73. Not well-defined since the word "good" is not precise.

75. Not well-defined since "tall" is not precise.

77. Well-defined.

79. Well-defined.

81. Not well-defined; "small" is not precise.

83. Not well-defined; "best" is not precise.

85. $A = B$. Replacing n with whole numbers, starting with 0, $A = \{1, 3, 5, ...\}$. Replacing n with natural numbers, starting with one, $B = \{1, 3, 5, ...\}$.

87. $A \neq B$. $A = \{2(1) - 1, 2(2) - 1, 2(3) - 1, ...\}$
$= \{1, 3, 5, ...\}$.
$B = \{\dfrac{1(1+1)}{2}, \dfrac{2(2+1)}{2}, \dfrac{3(3+1)}{2}, ...\}$
$= \{1, 3, 6, ...\}$

89. Answers will vary; however, the set of real numbers between 0 and 1 is one example of a set that cannot be written using the roster method.

EXERCISE SET 2.2

1. The complement of $\{2, 4, 6, 7\}$ contains elements in U but not in the set: $\{0, 1, 3, 5, 8\}$.

3. $\varnothing' = U = \{0, 1, 2, 3, 4, 5, 6, 7, 8\}$

5. $\{x \mid x < 7$ and $x \in N\} = \{1, 2, 3, 4, 5, 6\}$.
$\{1, 2, 3, 4, 5, 6\}' = \{0, 7, 8\}$

7. The set of odd counting numbers less than $8 = \{1, 3, 5, 7\}$. $\{1, 3, 5, 7\}' = \{0, 2, 4, 6, 8\}$

9. $\{a, b, c, d\} \subseteq \{a, b, c, d, e, f, g\}$ since all elements of the first set are contained in the second set.

11. $\{big, small, little\} \not\subseteq \{large, petite, short\}$

13. $I \subseteq Q$

15. $\subseteq$ since the empty set is a subset of every set.

17. $\subseteq$ since every element of the first set is an element of the second set.

19. True; every element of F is an element of D.

21. True; $F \neq D$.

23. True; s is an element of E and $G \neq E$.

25. $G' = \{p, q, r, t\}$. Since q is not an element of D, the statement is false.

27. True; the empty set is a subset of every set.

29. True; $D' = \{q\}$ and $D' \neq E$.

31. False; D does not contain sets.

33. False; D has 4 elements, $2^4 = 16$ subsets and $2^4 - 1 = 15$ proper subsets.

35. False; $F' = \{q, r, s\}$ so F' has $2^3 = 8$ subsets.

37. $2^{16} = 65536$ subsets. 65536 seconds $= 65536 \div 60$ seconds per minute $\approx$ 1092 minutes $\div 60$ minutes per hour $= 18$ hours (to the nearest hour).

39. $\varnothing$, $\{\alpha\}$, $\{\beta\}$, $\{\alpha, \beta\}$

41. $\varnothing$, $\{I\}$, $\{II\}$, $\{III\}$, $\{I, II\}$, $\{I, III\}$, $\{II, III\}$, $\{I, II, III\}$

43. The number of subsets is 2^n where n is the number of elements in the set. $2^2 = 4$.

45. List the elements in the set: $\{8, 10, 12, 14, 16, 18, 20\}$. $2^7 = 128$.

47. $2^{11} = 2,048$

49. There are no negative whole numbers. $2^0 = 1$.

51. a. This is equivalent to finding the number of proper subsets for a set with 4 elements. $2^4 - 1 = 16 - 1 = 15$.

    b. The sets that contain the 1976 dime or the 1992 dime produce duplicate amounts of money.
    There are 8 sets that contain one dime producing 4 sets with the same value. $15 - 4 = 11$. The set containing the nickel and two dimes has the same value as the set containing the quarter. $11 - 1 = 10$ different sums.

    c. Two different sets of coins can have the same value.

53. a. 6 red pieces

b. 2 red squares

c. 4 hexagons

d. 1 large blue triangle

55. a. $2^{10} = 1{,}024$ types of omelets

b. Solve $2^x > 4{,}000$ by guessing and checking.
$2^{11} = 2{,}048$
$2^{12} = 4{,}096$
At least 12 ingredients must be available.

57. a. $\{2\}$ is the set containing 2, not the element 2. $\{1, 2, 3\}$ has only three elements, namely 1, 2, and 3. Because $\{2\}$ is not equal to 1, 2, or 3,
$\{2\} \notin \{1, 2, 3\}$.

b. 1 is not a set, so it cannot be a subset.

c. The given set has the elements 1 and $\{1\}$. Because $1 \neq \{1\}$, there are exactly two elements in $\{1, \{1\}\}$.

59. a. $\{A, B, C\}$, $\{A, B, D\}$, $\{A, B, E\}$,
$\{A, C, D\}$, $\{A, C, E\}$, $\{A, D, E\}$,
$\{B, C, D\}$, $\{B, C, E\}$, $\{B, D, E\}$,
$\{C, D, E\}$, $\{A, B, C, D\}$, $\{A, B, C, E\}$,
$\{A, B, D, E\}$, $\{A, C, D, E\}$,
$\{B, C, D, E\}$, $\{A, B, C, D, E\}$

b. $\{A\}$, $\{B\}$, $\{C\}$, $\{D\}$, $\{E\}$, $\{A,B\}$,
$\{A, C\}$, $\{A, D\}$, $\{A, E\}$, $\{B, C\}$,
$\{B, D\}$, $\{B, E\}$, $\{C, D\}$, $\{C, E\}$, $\{D, E\}$

EXERCISE SET 2.3

1. $A \cup B = \{2, 4, 6\} \cup \{1, 2, 5, 8\}$
$= \{1, 2, 4, 5, 6, 8\}$

3. $A \cap B' = \{2, 4, 6\} \cap \{3, 4, 6, 7\} = \{4, 6\}$

5. $(A \cup B)' = (\{1, 2, 4, 5, 6, 8\})' = \{3, 7\}$

7. $B \cup C = \{1, 2, 5, 8\} \cup \{1, 3, 7\} = \{1, 2, 3, 5, 7, 8\}$. $A \cup (B \cup C) = \{2, 4, 6\} \cup \{1, 2, 3, 5, 7, 8\} = \{1, 2, 3, 4, 5, 6, 7, 8\} = U$

9. $B \cap C = \{1, 2, 5, 8\} \cap \{1, 3, 7\} = \{1\}$.
$A \cap (B \cap C) = \{2, 4, 6\} \cap \{1\} = \emptyset$

11. $B \cap (B \cup C) = \{1, 2, 5, 8\} \cap \{1, 2, 3, 5, 7, 8\} = \{1, 2, 5, 8\} = B$

13. $B \cup B' = \{1, 2, 5, 8\} \cup \{3, 4, 6, 7\} = \{1, 2, 3, 4, 5, 6, 7, 8\} = U$

15. $A \cup C' = \{2, 4, 6\} \cup \{2, 4, 5, 6, 8\} = \{2, 4, 5, 6, 8\}$.
$B \cup A' = \{1, 2, 5, 8\} \cup \{1, 3, 5, 7, 8\} = \{1, 2, 3, 5, 7, 8\}$
$(A \cup C') \cap (B \cup A') = \{2, 4, 5, 6, 8\} \cap \{1, 2, 3, 5, 7, 8\} = \{2, 5, 8\}$

17. $C \cup B' = \{1, 3, 7\} \cup \{3, 4, 6, 7\} = \{1, 3, 4, 6, 7\}$.
$(C \cup B') \cup \emptyset = \{1, 3, 4, 6, 7\} \cup \{\} = \{1, 3, 4, 6, 7\} = (C \cup B')$

19. $A \cup B = \{1, 2, 4, 5, 6, 8\}$
$B \cap C' = \{1, 2, 5, 8\} \cap \{2, 4, 5, 6, 8\} = \{2, 5, 8\}$
$(A \cup B) \cap (B \cap C') = \{1, 2, 4, 5, 6, 8\} \cap \{2, 5, 8\} = \{2, 5, 8\}$

In Exercises 21-27, one possible answer is given. Your answers may vary from the given answers.

21. The set of all elements that are not in L or are in T.

23. The set of all elements that are in A, or are in C, but not in B.

25. The set of all elements that are in T, and are also in J or not in K.

27. The set of all elements that are in both W and V, or are in both W and Z.

29.

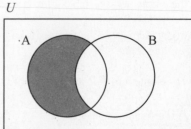

31.

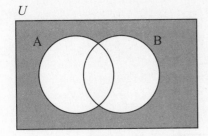

33.

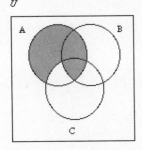

35.

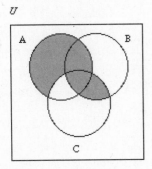

37.

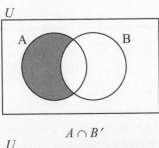

$$A \cap B'$$

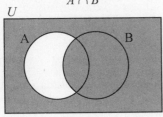

$$A' \cup B$$

Because the sets $A \cap B'$ and $A' \cup B$ are represented by different regions, $(A \cup B') \neq (A' \cup B)$.

39.

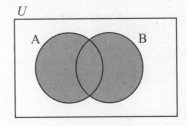

$$A \cup (A' \cap B)$$

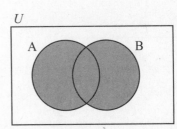

$$A \cup B$$
$$A \cup (A' \cap B) = A \cup B$$

41.

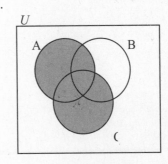

$$(A \cup C) \cap B'$$

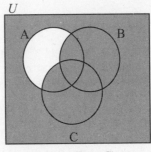

$$A' \cup (B \cup C)$$

$$(A \cup C) \cap B' \neq A' \cup (B \cup C)$$

43.

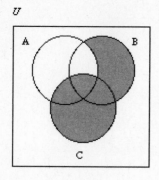

$(A' \cap B) \cup C$

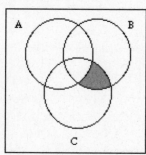

$(A' \cap C) \cap (A' \cap B)$
$(A' \cap B) \cup C \neq (A' \cap C) \cap (A' \cap B)$

45.

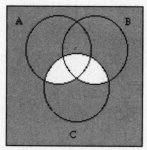

$((A \cup B) \cap C)'$

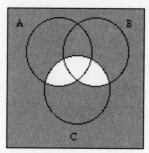

$(A' \cap B) \cup C'$
$((A \cup B) \cap C)' = (A' \cap B) \cup C'$

47. $R \cap G \cap B'$
$B' = \{R, Y, G\}$
$R = \{R, M, W, Y\}$
$G = \{G, Y, W, C\}$
$R \cap G \cap B' = Y$ (yellow)

49. $R' \cap G \cap B$
$R' = \{B, C, G\}$
$G = \{G, Y, W, C\}$
$B = \{B, M, W, C\}$
$R' \cap G \cap B = C$ (cyan)

51. $C' = \{Y, R, M\}$
$M = \{M, B, K, R\}$
$Y = \{Y, G, K, R\}$
$C' \cap M \cap Y = R$ (red)

In Exercises 53 - 61, one possible answer is given.
Your answers may vary from the given answers.

53. $A \cap B'$

55. $(A \cup B)'$

57. $B \cup C$

59. $C \cap (A \cup B)'$

61. $(A \cup B)' \cup (A \cap B \cap C)$

63. a.

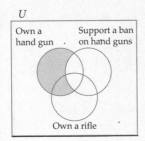

b.

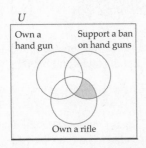

c.

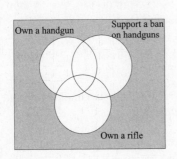

65.

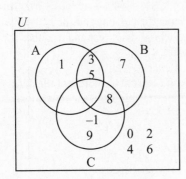

67.

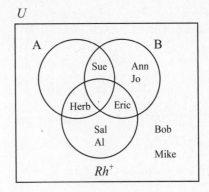

69.

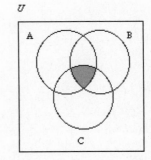

71. $B - A = \{2, 3, 8, 9\} - \{2, 4, 6, 8\}$
    $= \{3, 9\}$

73. $A - B' = \{2, 4, 6, 8\} - \{1, 4, 5, 6, 7\}$
    $= \{2, 8\}$

75. $A' - B' = \{1, 3, 5, 7, 9\} - \{1, 4, 5, 6, 7\}$
    $= \{3, 9\}$

77. Responses will vary.

79.

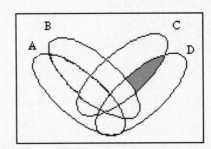

EXERCISE SET 2.4

1. $B \cup C$ = {Math, Physics, Chemistry, Psychology, Drama, French, History} so $n(B \cup C) = 7$

3. $n(B) + n(C) = 5 + 3 = 8$

5. $A \cup B \cup C$ = {English, History, Psychology, Drama, Math, Physics, Chemistry, French} so $n(A \cup B \cup C) = 8$

7. $n(A) + n(B) + n(C) = 4 + 5 + 3 = 12$

9. $n(A \cup B) = n(A) + n(B) - n(A \cap B) = 4 + 5 - 2 = 7$

11. Using the formula:

$$n(J \cup K) = n(J) + n(K) - n(J \cap K)$$
$$310 = 245 + 178 - n(J \cap K)$$
$$310 = 423 - n(J \cap K)$$
$$-113 = -n(J \cap K)$$
$$n(J \cap K) = 113$$

13. Using the formula:

$$n(A \cup B) = n(A) + n(B) - n(A \cap B)$$
$$2250 = 1500 + n(B) - 310$$
$$2250 = 1190 + n(B)$$
$$1060 = n(B)$$

15.

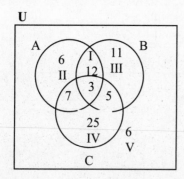

To find $n(I)$: Since $n(A \cap B) = 15$ and 3 has been accounted for, $15 - 3 = 12$ for $n(I)$. To find $n(II)$: $n(A) = 28$. $7 + 3 + 12 = 22$ is accounted for so $28 - 22 = 6$ for $n(II)$. To find $n(III)$: $n(B) = 31$. $5 + 3 + 12 = 20$. $31 - 20 = 11$. To find $n(IV)$: $n(C) = 40$.

$40 - (7 + 3 + 5) = 25$. To find $n(V)$: $n(U) = 75$ so $n(V) = 75 - (6 + 12 + 11 + 7 + 3 + 5 + 25) = 6$.

17. a. S = {investors in stocks} and let B = {investors in bonds}. Since 75 had not invested in either stocks or bonds, $n(S \cup B) = 600 - 75 = 525$.
    $$n(S \cup B) = n(S) + n(B) - n(S \cap B)$$
    $$525 = 380 + 325 - n(S \cap B)$$
    $$525 = 705 - n(S \cap B)$$
    $$-180 = -n(S \cap B)$$
    $n(S \cap B) = 180$ represents the number of investors in both stocks and bonds.

    b. $n(S \text{ only}) = 380 - 180 = 200$

19. Draw a Venn diagram to represent the data: Since 44% responded to both forms, place 44% in the intersection of the two sets. A total of 72% responded to an analgesic, so 72% – 44% = 28% who responded to only the analgesic. Similarly, 59% – 44% = 15% who responded only to the muscle relaxant.

    a.   15% responded to the muscle relaxant but not the analgesic.

    b.   Since the universe must contain 100%, subtract the known values from 100%: 100% – (28% + 44% + 15%) =13%. This represents the percent of athletes who were treated who did not respond to either form of treatment.

21. Draw a Venn diagram to represent the data. Fill in the diagram starting from the innermost region.

    i: 85
    ii: $150 - 85 = 65$
    iii: $135 - 85 = 50$
    iv: $110 - 85 = 25$
    v: $390 - (65 + 85 + 50) = 190$
    vi: $290 - (25 + 85 + 50) = 130$
    vii: $305 - (65 + 85 + 25) = 130$
    viii: $770 - (130 + 25 + 130 + 65 + 85 + 50 + 190) = 95$

    a.   exactly one of these forms of advertising is represented by v, vi, and vii: 190 + 130 + 130 = 450

b. exactly two of these forms are represented
    by ii, iii, and iv:
    $65 + 50 + 25 = 140$

c. PC World and neither of the other two forms
    is represented by vii: 130.

23. Draw a Venn diagram to represent the data. Fill
    in the diagram starting from the innermost
    region.

    i: 52
    ii: $10 - 52 = 88$
    iii: $437 - (52 + 88 + 202) = 95$
    iv: $74 - 52 = 22$
    v: $271 - (22 + 88 + 52) = 109$
    vi: 202
    vii: $497 - (22 + 52 + 95) = 328$
    viii: $1000 - (109 + 88 + 52 + 22 + 202 + 95$
          $+ 328) = 104$

a. this group is represented by v: 109

b. this group is represented by vii: 328

c. this group is represented by viii: 104

25. a. 101

    b. $124 + 82 + 65 + 51 + 48 = 370$

    c. $124 + 82 + 133 + 41 = 380$

    d. $124 + 82 + 101 + 66 = 373$

    e. $124 + 101 = 225$

    f. $124 + 82 + 65 + 101 + 66 + 51 + 41 = 530$

27. Given $n(A) = 47$ and $n(B) = 25$.

    a. If A and B are disjoint sets,
        $n(A \cup B) = n(A) + n(B) =$
        $47 + 25 = 72$.

    b. If $B \subset A$, then $A \cup B = A$ and $n(A) = 47$.

    c. If $B \subset A$, then $A \cap B = B$ and $n(B) = 25$.

    d. If A and B are disjoint sets, then $A \cap B = \varnothing$
        and $n(A \cap B) = 0$.

29. Complete the Venn diagram. Since there are 450
    users of Webcrawler,

$450 - (45 + 41 + 30 + 50 + 60 + 80 + 45) = 99$
gives the total who use only Webcrawler.
Similarly,
$585 - (55 + 50 + 60 + 41 + 34 + 100 + 45) = 200$
gives the total who use only Altavista. To find
Yahoo only users:
$620 - (55 + 50 + 100 + 60 + 80 + 41 + 30) =$
204. To find Lycos only users:
$560 - (50 + 80 + 50 + 60 + 34 + 100 + 45) =$
141.

a. only Altavista: 200

b. To find the number who use exactly three
    search engines, add the numbers given for
    people who use only 3 search engines: $100 +$
    $41 + 50 + 80 = 271$

c. Total number in the regions: $204 + 55 + 200$
    $+ 141 + 50 + 50 + 34 + 45 + 80 + 60 + 100$
    $+ 99 + 30 + 41 + 45 = 1234$. Since 1250
    people were surveyed, this means $1250 -$
    $1234 = 16$ people do not use any of the
    search engines.

EXERCISE SET 2.5

1. a. Comparing:
        $V = \{a, e, i\}$
        $\updownarrow \updownarrow \updownarrow$
        $M = \{3, 6, 9\}$

    b. The possible one-to-one correspondences
        (listed as ordered pairs) are: $\{(a, 6), (i, 3),$
        $(e, 9)\}$,
        $\{(a, 9), (i, 3), (e, 6)\}$,
        $\{(a, 3), (i, 9), (e, 6)\}$,
        $\{(a, 6), (i, 9), (e, 3)\}$,
        $\{(a, 9), (i, 6), (e, 3)\}$ plus the pairing shown
        in part a. produce 6 one-to-one
        correspondences.

3. Write the sets so that one is aligned below the
    other. One possible pairing is shown below.

    $D = \{1, 3, 5, \ldots, 2n - 1, \ldots\}$
    $\updownarrow \updownarrow\updownarrow \quad \updownarrow$
    $M = \{3, 6, 9, \ldots, 3n, \ldots\}$

    Pair $(2n - 1)$ of D with $(3n)$ of M to establish a
    one-to-one correspondence.

5. The general correspondence $(n) \leftrightarrow (7n-5)$ establishes a one-to-one correspondence between the elements of N and the elements of the given set. Thus the cardinality is $\aleph_0$.

7. c

9. c. Any set of the form $\{x \mid a \le x \le b\}$ where a and b are real numbers and $a \ne b$ has cardinality c.

11. Sets with equal cardinality are equivalent. The cardinality of N = cardinality of $I = \aleph_0$ therefore the sets are equivalent.

13. The sets are equivalent since the set of rational numbers and the set of integers have cardinality $\aleph_0$.

15. Let $S = \{10, 20, 30, \ldots, 10n, \ldots\}$. Then S is a proper subset of A. A rule for a one-to-one correspondence between A and S is $(5n) \leftrightarrow (10n)$. Because A can be placed in a one-to-one correspondence with a proper subset of itself, A is an infinite set.

17. Let $R = \left\{ \dfrac{3}{4}, \dfrac{5}{6}, \dfrac{7}{8}, \ldots, \dfrac{2n+1}{2n+2}, \ldots \right\}$. Then R is a proper subset of C. A rule for a one-to-one correspondence between C and R is $\left(\dfrac{2n-1}{2n} \right) \leftrightarrow \left(\dfrac{2n+1}{2n+2} \right)$. Because C can be placed in a one-to-one correspondence with a proper subset of itself, C is an infinite set.

In Exercises 19-25, let $N = \{1, 2, 3, 4, \ldots, n, \ldots\}$. Then a one-to-one correspondence between the given sets and the set of natural numbers N is given by the following general correspondences.

19. $(n+49) \leftrightarrow (n)$

21. $\left(\dfrac{1}{3^{n-1}} \right) \leftrightarrow (n)$

23. $(10^n) \leftrightarrow (n)$

25. $(n^3) \leftrightarrow (n)$

27. a. For any natural number n, the two natural numbers preceding $3n$ are not multiples of 3. Pair these two numbers, $3n-2$ and $3n-1$, with the multiples of 3 given by $6n-3$ and $6n$, respectively. Using the two general correspondences $(6n-3) \leftrightarrow (3n-2)$ and $(6n) \leftrightarrow (3n-1)$, we can establish a one-to-one correspondence between the multiples of 3 (set M) and the set K of all natural numbers that are not multiples of 3.

    b. First find n. $6n-606$, $n=101$. Then $(6n) \leftrightarrow (3n-1)$ means
    $$(6(101)) \leftrightarrow (3(101)-1) = 302.$$

    c. Solve.
    $$3n-1 = 899$$
    $$3n = 900$$
    $$n = 300$$
    Then $(6n) \leftrightarrow (3n-1)$ means
    $$(6(300)) \leftrightarrow (3(300)-1) \text{ and }$$
    $$(1800) \leftrightarrow (899).$$

29. The set of real numbers x such that $0 < x < 1$ is equivalent to the set of all real numbers.

CHAPTER 2 REVIEW EXERCISES

1. $\{0, 1, 2, 3, 4, 5, 6, 7\}$

2. $\{-8, 8\}$. Since the set of integers includes positives and negatives, -8 and 8 satisfy $x^2 = 64$.

3. Solving:
    $$x+3 \le 7$$
    $$x \le 4$$
    $$\{1, 2, 3, 4\}.$$

4. $\{1, 2, 3, 4, 5, 6\}$. Counting numbers begin at 1.

5. $\{x \mid x \in I \text{ and } x > -6\}$

6. $\{x \mid x$ is the name of a month with exactly 30 days$\}$

7. $\{x \mid x$ is the name of a U.S. state that begins with the letter $K\}$

8. $\{x^3 \mid x = 1, 2, 3, 4, 5\}$

9. False. The set contains numbers, not sets. $\{3\} \notin \{1, 2, 3, 4\}$.

10. True. The set of integers includes positive and negative integers.

11. True. The symbol ~ means equivalent and sets with the same number of elements are equivalent.

12. False. The word small is not precise.

13. $A \cap B = \{2, 6, 10\} \cap \{6, 10, 16, 18\} = \{6, 10\}$

14. $A \cup B = \{2, 6, 10\} \cup \{6, 10, 16, 18\} = \{2, 6, 10, 16, 18\}$

15. $A' \cap C = \{8, 12, 14, 16, 18\} \cap \{14, 16\} = \{14, 16\}$

16. $B \cup C' = \{6, 10, 16, 18\} \cup \{2, 6, 8, 10, 12, 18\} = \{2, 6, 8, 10, 12, 16, 18\}$

17. $B \cap C = \{6, 10, 16, 18\} \cap \{14, 16\} = \{16\}$

    $A \cup \{16\} = \{2, 6, 10\} \cup \{16\} = \{2, 6, 10, 16\}$

18. $A \cup C = \{2, 6, 10\} \cup \{14, 16\} = \{2, 6, 10, 14, 16\}$

    $(A \cup C)' = \{8, 12, 18\}$

    $\{8, 12, 18\} \cap \{2, 8, 12, 14\} = \{8, 12\}$

19. $A \cap B' = \{2, 6, 10\} \cap \{2, 8, 12, 14\} = \{2\}$

    $(\{2\})' = \{6, 8, 10, 12, 14, 16, 18\}$

20. $A \cup B \cup C = \{2, 6, 10, 14, 16, 18\}$

    $(A \cup B \cup C)' = \{8, 12\}$

21. All natural numbers are whole numbers, but 0 is not a natural number so $N \subset W$.

22. All integers are real numbers, but $\frac{1}{2}$ is a real number that is not an integer so $I \subset R$.

23. Counting numbers and natural numbers represent the same set of numbers. The set of counting numbers is not a proper subset of the set of natural numbers.

24. The set of real numbers is not a subset of the set of rational numbers.

25. $\emptyset$, $\{I\}$, $\{II\}$, $\{I, II\}$

26. $\emptyset$, $\{s\}$ $\{u\}$, $\{n\}$, $\{s, u\}$, $\{s, n\}$, $\{u, n\}$, $\{s, u, n\}$

27. $\emptyset$, {penny}, {nickel}, {dime}, {quarter}, {penny, nickel}, {penny, dime}, {penny, quarter}, {nickel, dime}, {nickel, quarter}, {dime, quarter}, {penny, nickel, dime}, {penny, nickel, quarter}, {penny, dime, quarter}, {nickel, dime, quarter}, {penny, nickel, dime, quarter}

28. $\emptyset$, {A}, {B}, {C}, {D}, {E}, {A, B}, {A, C}, {A, D}, {A, E}, {B, C}, {B, D}, {B, E}, {C, D}, {C, E}, {D, E}, {A, B, C}, {A, B, D}, {A, B, E}, {A, C, D}, {A, C, E}, {A, D, E}, {B, C, D}, {B, C, E}, {B, D, E}, {C, D, E}, {A, B, C, D}, {A, B, C, E}, {A, B, D, E}, {A, C, D, E}, {B, C, D, E}, {A, B, C, D, E}

29. The number of subsets of a set with n elements is 2^n. The set of four musketeers has 4 elements and $2^4 = 16$ subsets.

30. $n = 26$. $2^{26} = 67,108,864$.

31. The number of letters is 15. $2^{15} = 32,768$

32. $n = 7$. $2^7 = 128$.

33.

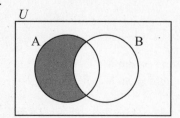

34.

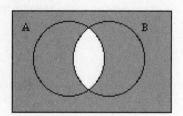

35.

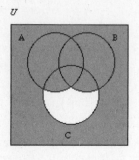

36.

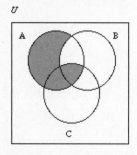

37.

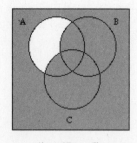

$$A' \cup (B \cup C)$$

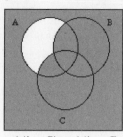

$$(A' \cup B) \cup (A' \cup C)$$

$$A' \cup B \cup C = (A' \cup B) \cup (B' \cup C)$$

38.

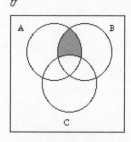

$$(A \cap B) \cap C'$$

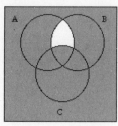

$$(A' \cup B') \cup C$$
$$(A \cap B) \cap C' \neq (A' \cup B') \cup C$$

39.

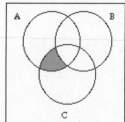

$$A \cap (B' \cap C)$$

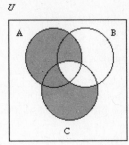

$(A \cup B') \cap (A \cup C)$

$A \cap (B' \cap C) \neq (A \cup B') \cap (A \cup C)$

40.

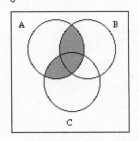

$A \cap (B \cup C)$

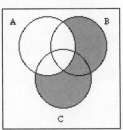

$A' \cap (B \cup C)$

$A \cap (B \cup C) \neq A' \cap (B \cup C)$

41. $(A \cup B)' \cap C$ or $C \cap (A' \cap B')$

42. $(A \cap B) \cup (B \cap C')$

43.

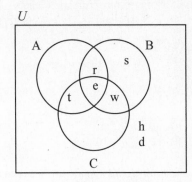

44.

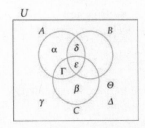

45. Use a Venn diagram to represent the survey results. Total the numbers from each region to find the number of members surveyed: $111 + 97 + 48 + 135 = 391$

46. Use a Venn diagram to represent the survey results: After placing 96 in the intersection of all three types, use the information on customers who like two types of coffee:

116 like espresso and cappuccino – 96 who like all three = 20
136 like espresso and chocolate-flavored – 96 = 40
127 like cappuccino and chocolate-flavored – 96 = 31.
221 like espresso – (20 + 96 + 40) = 65 who like only espresso.
182 like cappuccino – (20 + 96 + 31) = 35 who like only cappuccino.
209 like chocolate-flavored coffee – (40 + 96 + 31) = 42 who like only chocolate-flavored coffee.

a. 42

b. 31

c. 20

d. $65 + 35 + 42 = 142$

47. One possible one-to-one correspondence between $\{1, 3, 6, 10\}$ and $\{1, 2, 3, 4\}$ is given by

$$\{1, 3, 6, 10\}$$
$$\updownarrow \;\updownarrow\; \updownarrow\; \updownarrow$$
$$\{1, 2, 3, 4\}$$

48. $\{x \mid x > 10 \text{ and } x \in N\} =$ $\{11, 12, 13, 14, \ldots, n + 10, \ldots\}$

    Thus a one-to-one correspondence between the sets is given by

$$\{11, 12, 13, 14, \ldots, n+10, \ldots\}$$
$$\updownarrow \;\;\updownarrow\; \;\updownarrow\; \;\updownarrow \qquad \updownarrow$$
$$\{2, \;\; 4, \;\; 6, \;\; 8, \ldots, 2n, \ldots\}$$

49. One possible one-to-one correspondence between the set is given by

$$\{3, \;\; 6, \;\; 9, \;\ldots, \; 3n, \ldots\}$$
$$\updownarrow \;\;\updownarrow\; \;\updownarrow\; \qquad \updownarrow$$
$$\{10, 100, 1000, \ldots, 10^n, \ldots\}$$

50. In the following figure, the line from E that passes through $\overline{AB}$ and $\overline{CD}$ illustrates a method of establishing a one-to-one correspondence between the sets $\{x \mid 0 \le x \le 1\}$ and $\{x \mid 0 \le x \le 4\}$.

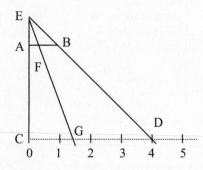

51. A proper subset of A is $S = \{10, 14, 18, \ldots, 4n + 6, \ldots\}$. A one-to-one correspondence between A and S is given by

$$A = \{6, 10, 14, 18, \ldots, 4n + 2, \ldots\}$$
$$\updownarrow \;\updownarrow\; \updownarrow\; \updownarrow \qquad\quad \updownarrow$$
$$S = \{10, 14, 18, 22, \ldots 4n + 6, \ldots\}$$

Because A can be placed in a one-to-one correspondence with a proper subset of itself, A is an infinite set.

52. A proper subset of B is $T = \left\{\dfrac{1}{2}, \dfrac{1}{4}, \dfrac{1}{8}, \dfrac{1}{16}, \ldots, \dfrac{1}{2^n}, \ldots\right\}$. A one-to-one correspondence between B and T is given by:

$$B = \left\{1, \dfrac{1}{2}, \dfrac{1}{4}, \dfrac{1}{8}, \ldots, \dfrac{1}{2^{n-1}}, \ldots\right\}$$
$$\updownarrow \;\updownarrow\; \updownarrow\; \updownarrow \qquad\quad \updownarrow$$
$$T = \left\{\dfrac{1}{2}, \dfrac{1}{4}, \dfrac{1}{8}, \dfrac{1}{16}, \ldots, \dfrac{1}{2^n}, \ldots\right\}$$

Because B can be placed in a one-to-one correspondence with a proper subset of itself, B is an infinite set.

53. 5

54. 10

55. 2

56. 5

57. $\aleph_0$

58. $\aleph_0$

59. c

60. c

61. $\aleph_0$

62. $\aleph_0$

63. $\aleph_0$

64. c

65. c

66. c

67. $\aleph_0$

68. c

CHAPTER 2 TEST

1. $\{3, 5, 7, 8\} \cup \{2, 3, 8, 9, 10\} =$ $\{2, 3, 5, 7, 8, 9, 10\}$

2. $A' \cap B =$ $\{1, 2, 4, 6, 9, 10\} \cap \{2, 3, 8, 9, 10\} =$ $\{2, 9, 10\}$

3. $(A \cap B)' = (\{3, 5, 7, 8\} \cap \{2, 3, 8, 9, 10\})' = \{3, 8\}' = \{1, 2, 4, 5, 6, 7, 9, 10\}$

4. $(A \cup B')' =$
   $(\{3, 5, 7, 8\} \cup \{1, 4, 5, 6, 7\})' =$
   $\{1, 3, 4, 5, 6, 7, 8\}' = \{2, 9, 10\}$

5. $A' \cup (B \cap C') = \{1, 2, 4, 6, 9, 10\} \cup$
   $(\{2, 3, 8, 9\ 10\} \cap \{2, 3, 5, 6, 9, 10\})$
   $= \{1, 2, 4, 6, 9, 10\} \cup \{2, 3, 9, 10\} =$
   $\{1, 2, 3, 4, 6, 9, 10\}$

6. $A \cap (B' \cup C) =$
   $\{3, 5, 7, 8\} \cap (\{1, 4, 5, 6, 7\} \cup \{1, 4, 7, 8\}) =$
   $\{3, 5, 7, 8\} \cap \{1, 4, 5, 6, 7, 8\} = \{5, 7, 8\}$

7. $\{x \mid x \in W \text{ and } x < 7\}$

8. $\{x \mid x \in I \text{ and } -3 \leq x \leq 2\}$

9. a. The set of whole numbers less than 4 = $\{0, 1, 2, 3\}$. $n(\{0, 1, 2, 3\}) = 4$

   b. $\aleph_0$. Any set of the form
      $\{x \mid a \leq x \leq b\}$ where a and b are real
      numbers and $a \neq b$ has cardinality $\aleph_0$.

   c. $\aleph_0$

   d. c

10. a. Equivalent. The set of natural numbers and
       the set of integers have cardinality $\aleph_0$. The
       sets are not equal since integers such as -3
       and 0 are not natural numbers.

    b. Equivalent. Both sets have cardinality $\aleph_0$.
       The sets are not equal because
       $0 \in W$ but 0 is not a positive integer.

11. a. Neither. The cardinality of the set of rational
       numbers is $\aleph_0$ and the cardinality of the set
       of irrational numbers is c.

    b. Equivalent. Both sets have cardinality c.
       The sets are not equal since there are real
       numbers such as $\frac{1}{2}$ and 3 which are not
       irrational numbers.

12. $\varnothing$, $\{a\}$, $\{b\}$, $\{c\}$, $\{d\}$, $\{a, b\}$, $\{a, c\}$, $\{a, d\}$,
    $\{b, c\}$, $\{b, d\}$, $\{c, d\}$, $\{a, b, c,\}$, $\{a, b, d\}$,
    $\{a, c, d\}$, $\{b, c, d\}$, $\{a, b, c, d\}$

13. The number of subsets of a set of n elements is
    2^n. $2^{21} = 2,097,152$.

14. a. False. While $4 \in \{1, 2, 3, 4, 5, 6, 7\}$,
       $\{4\} \notin \{1, 2, 3, 4, 5, 6, 7\}$.

    b. True

    c. False. The symbol $\subset$ means proper subset
       and a set cannot be a proper subset of itself
       since A = A.

    d. True. Both sets have cardinality $\aleph_0$ and so
       are equivalent.

15. a.
    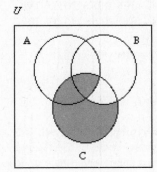

    b.
    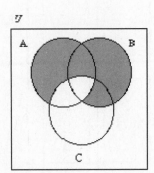

16. Possible answer: $B \cup (A \cap C)$

17. *LeMars*

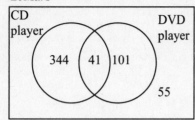

$344 + 41 + 101 + 55 = 541$

18. Draw a Venn diagram to represent the data. Fill in the diagram starting from the innermost region.

    i: 105
    ii: $412 - 105 = 307$
    iii: $280 - (80 + 105 + 64) = 232$
    iv: $185 - 105 = 80$
    v: $724 - (105 + 80 + 307) = 190$
    vi: $545 - (31 + 105 + 307) = 102$
    vii: 64
    viii: $1000 - (105 + 307 + 31 + 80 + 232 + 102 + 64) = 79$

    a. this group is represented by v: 232

    b. this group is represented by vi: 102

    c. this group is represented by i, ii, iii, iv, v, and vi: $105 + 307 + 31 + 80 + 232 + 102 = 857$

    d. this group is represented by viii: 79

19. A possible correspondence:

    $\{5, 10, 15, 20, 25, \ldots 5n, \ldots\}$
    ↕ ↕ ↕ ↕   ↕
    $\{0, 1, 2, 3, 4, \ldots n-1, \ldots\}$
    $(5n) \leftrightarrow (n-1)$

20. A possible correspondence:

    $\{3, 6, 9, 12, \ldots 3n, \ldots\}$
    ↕ ↕ ↕ ↕   ↕
    $\{6, 12, 18, 24, \ldots 6n, \ldots\}$
    $(3n) \leftrightarrow (6n)$

Chapter 3: Logic

EXERCISE SET 3.1

1. West Virginia is east of the Mississippi, so this sentence is false and it is a statement.

3. You may not know the area code for Storm Lake, Iowa; however, you do know that it is either 512 or it is not 512. The sentence is either true or it is false, and it is not both true and false, so it is a statement.

5. The sentence "Have a fun trip," is a command, so it is not a statement.

7. All hexagons have exactly six sides, so this sentence is false and it is a statement.

9. The sentence is an opinion. It can be true for one person and false for another person. Since it can be both true and false, it is not a statement.

11. One component is "The principal will attend the class on Tuesday." The other component is "The principal will attend the class on Wednesday."

13. One component is "A triangle is an acute triangle." The other component is "A triangle has three acute angles."

15. One component is "I ordered a salad." The other component is "I ordered a cola."

17. One component is "$5 + 2 > 6$" The other component is "$5 + 2 = 6$."

19. The Giants did not lose the game.

21. The game went into overtime.

23. $w \rightarrow t$; conditional

25. $s \rightarrow r$; conditional

27. $l \leftrightarrow a$; biconditional

29. $d \rightarrow f$; conditional

31. $m \vee c$; disjunction

33. The tour goes to Italy and the tour does not go to Spain.

35. If we go to Venice, then we will not go to Florence.

37. We will go to Florence if and only if we do not go to Venice.

39. All cats have claws.

41. Some classic movies were not first produced in black and white.

43. Some of the numbers were even numbers.

45. Some irrational numbers can be written as terminating decimals.

47. Some cars do not run on gasoline.

49. Some items are not on sale.

51. $3 > 1$ is true, so the statement is true.

53. $(-1)^{50} = 1$ is true and $(-1)^{99} = -1$ is true, so the statement is true.

55. $-5 > -11$, so the statement is true.

57. "2 is an even number" is a true statement, so the statement is true.

59. 2 is an even prime number, so the statement is true.

61. π is a real number that is irrational, so the statement is true.

63. Since every integer x can be written in the form $\dfrac{x}{1}$, the statement is true.

65. $p \rightarrow q$, where p represents "you can count your money" and q represents "you don't have a billion dollars."

67. $p \rightarrow q$, where p represents "you do not learn from history" and q represents "you are condemned to repeat it."

69. $p \rightarrow q$, where p represents "people concentrated on the really important things in life" and q represents "there'd be a shortage of fishing poles."

71. $p \leftrightarrow q$, where p represents "an angle is a right angle" and q represents "its measure is 90°."

73. $p \rightarrow q$, where p represents "two sides of a triangle are equal in length" and q represents "the angles opposite those sides are congruent."

75. $p \rightarrow q$, where p represents "it is a square" and s represents "it is a rectangle."

EXERCISE SET 3.2

1. True.

p	q	r	p	$\vee$	$(\sim q$	$\vee$	$r)$
F	T	T	T	T	F	T	T
			1	5	2	4	3

3. False.

p	q	r	$(p$	$\wedge$	$q)$	$\vee$	$(\sim p$	$\wedge$	$\sim q)$
F	T	T	F	F	T	F	T	F	F
			1	6	2	7	3	5	4

5. False.

p	q	r	$[\sim$	$(p$	$\wedge$	$\sim q)$	$\vee$	$r]$	$\wedge$	$(p$	$\wedge$	$\sim r)$
F	T	T	T	F	F	F	T	T	F	F	F	F
			8	1	7	2	9	3	10	4	6	5

7. False.

p	q	r	$[(p$	$\wedge$	$\sim q)$	$\vee$	$\sim r]$	$\wedge$	$(q$	$\wedge$	$r)$
F	T	T	F	F	F	F	F	F	T	T	T
			1	7	2	8	3	9	4	6	5

9. False.

p	q	r	$[(p$	$\wedge$	$q)$	$\wedge$	$r]$	$\vee$	$[p$	$\vee$	$(q$	$\wedge$	$\sim r)]$
F	T	T	F	F	T	F	T	F	F	F	T	F	F
			1	7	2	8	3	11	4	10	5	9	6

11. a. If p is false, then $p \wedge (q \vee r)$ must be a false statement.

   b. For a conjunctive statement to be true, it is necessary that all components of the statement be true. Because it is given that one of the components (p) is false, $p \wedge (q \vee r)$ must be a false statement.

13. The truth table is given below:

p	q	~p	∨	q
T	T	F	T	T
T	F	F	F	F
F	T	T	T	T
F	F	T	T	F
		1	3	2

15. The truth table is given below:

p	q	p	∧	~q
T	T	T	F	F
T	F	T	T	T
F	T	F	F	F
F	F	F	F	T
		1	3	2

17. The truth table is given below:

p	q	(p	∧	~q	)	∨	[~	(p	∧	q)]
T	T	T	F	F		F	F	T	T	T
T	F	T	T	T		T	T	T	F	F
F	T	F	F	F		T	T	F	F	T
F	F	F	F	T		T	T	F	F	F
		1	7	2		8	6	3	5	4

19. The truth table is given below:

p	q	r	~	(p	∨	q)	∧	(~r	∨	q)
T	T	T	F	T	T	T	F	F	T	T
T	T	F	F	T	T	T	F	T	T	T
T	F	T	F	T	T	F	F	F	F	F
T	F	F	F	T	T	F	F	T	T	F
F	T	T	F	F	T	T	F	F	T	T
F	T	F	F	F	T	T	F	T	T	T
F	F	T	T	F	F	F	F	F	F	F
F	F	F	T	F	F	F	T	T	T	F
			7	1	6	2	8	3	5	4

21. The truth table is given below:

p	q	r	(p	∧	~r)	∨	[~q	∨	(p	∧	r)]
T	T	T	T	F	F	T	F	T	T	T	T
T	T	F	T	T	T	T	F	F	T	F	F
T	F	T	T	F	F	T	T	T	T	T	T
T	F	F	T	T	T	T	T	T	T	F	F
F	T	T	F	F	F	F	F	F	F	F	T
F	T	F	F	F	T	F	F	F	F	F	F
F	F	T	F	F	F	T	T	T	F	F	T
F	F	F	F	F	T	T	T	T	F	F	F
			1	8	2	9	3	7	4	6	5

23. The truth table is given below:

p	q	r	[(p	∧	q)	∨	(r	∧	~p)]	∧	(r	∨	~q)
T	T	T	T	T	T	T	T	F	F	T	T	T	F
T	T	F	T	T	T	T	F	F	F	F	F	F	F
T	F	T	T	F	F	F	T	F	F	F	T	T	T
T	F	F	T	F	F	F	F	F	F	F	F	T	T
F	T	T	F	F	T	T	T	T	T	T	T	T	F
F	T	F	F	F	T	F	F	F	T	F	F	F	F
F	F	T	F	F	F	T	T	T	T	T	T	T	T
F	F	F	F	F	F	F	F	F	T	F	F	T	T
			1	7	2	9	3	8	4	11	5	10	6

25. The truth table is given below:

p	q	r	q	∨	[~r	∨	(p	∧	r)]
T	T	T	T	T	F	T	T	T	T
T	T	F	T	T	T	T	T	F	F
T	F	T	F	T	F	T	T	T	T
T	F	F	F	T	T	T	T	F	F
F	T	T	T	T	F	F	F	F	T
F	T	F	T	T	T	T	F	F	F
F	F	T	F	F	F	F	F	F	T
F	F	F	F	T	T	T	F	F	F
			1	7	2	6	3	5	4

27. The truth table is given below:

p	q	r	(~q	∧	r)	∨	[p	∧	(q	∧	~r)]
T	T	T	F	F	T	F	T	F	T	F	F
T	T	F	F	F	F	T	T	T	T	T	T
T	F	T	T	T	T	T	T	F	F	F	F
T	F	F	T	F	F	F	T	F	F	F	T
F	T	T	F	F	T	F	F	F	T	F	F
F	T	F	F	F	F	F	F	F	T	T	T
F	F	T	T	T	T	T	F	F	F	F	F
F	F	F	T	F	F	F	F	F	F	F	T
			1	6	2	9	3	8	4	7	5

29. The truth tables are given below:

p	r	p	∨	(p	∧	r)
T	T	T	T	T	T	T
T	F	T	T	T	F	F
F	T	F	F	F	F	T
F	F	F	F	F	F	F
		1	5	2	4	3

p
T
T
F
F

31. The truth tables are given below:

p	q	r	p	∧	(q	∨	r)
T	T	T	T	T	T	T	T
T	T	F	T	T	T	T	F
T	F	T	T	T	F	T	T
T	F	F	T	F	F	F	F
F	T	T	F	F	T	T	T
F	T	F	F	F	T	T	F
F	F	T	F	F	F	T	T
F	F	F	F	F	F	F	F
			1	5	2	4	3

(p	∧	q)	∨	(p	∧	r)
T	T	T	T	T	T	T
T	T	T	T	T	F	F
T	F	F	T	T	T	T
T	F	F	F	T	F	F
F	F	T	F	F	F	T
F	F	T	F	F	F	F
F	F	F	F	F	F	T
F	F	F	F	F	F	F
1	5	2	7	3	6	4

33. The truth tables are given below:

p	q	p	∨	(q	∧	~p)
T	T	T	T	T	F	F
T	F	T	T	F	F	F
F	T	F	T	T	T	T
F	F	F	F	F	F	T
		1	5	2	4	3

p	∨	q
T	T	T
T	T	F
F	T	T
F	F	F
1	3	2

35. The truth tables are given below:

p	q	r	[(p	∧	q)	∧	r]	∨	[p	∧	(q	∧	~r)]
T	T	T	T	T	T	T	T	T	T	F	T	F	F
T	T	F	T	T	T	F	F	T	T	T	T	T	T
T	F	T	T	F	F	F	T	F	T	F	F	F	F
T	F	F	T	F	F	F	F	F	T	F	F	F	T
F	T	T	F	F	T	F	T	F	F	F	T	T	F
F	T	F	F	F	T	F	F	F	F	F	T	T	T
F	F	T	F	F	F	F	T	F	F	F	F	F	F
F	F	F	F	F	F	F	F	F	F	F	F	F	T
			1	7	2	8	3	11	4	10	5	9	6

(p	∧	q)
T	T	T
T	T	T
T	F	F
T	F	F
F	F	T
F	F	T
F	F	F
F	F	F
1	3	2

37. Let *p* represent the statement "It rained." Let *q* represent the statement "It snowed." In symbolic form, the original sentence is ~(p ∨ q). One of De Morgan's laws states that this is equivalent to ~p ∧ ~q. Thus an equivalent sentence is "It did not rain and it did not snow."

39. Let *p* represent the statement "She visited France." Let *q* represent the statement "She visited Italy." In symbolic form, the original sentence is ~p ∧ ~q. One of De Morgan's laws states that this is equivalent to ~(p ∨ q). Thus an equivalent sentence is "She did not visit either France or Italy."

41. Let *p* represent the statement "She received a promotion." Let *q* represent the statement "She received a raise." In symbolic form, the original sentence is ~(p ∨ q). One of De Morgan's laws states that this is equivalent to ~p ∧ ~q. Thus an equivalent sentence is "She did not receive a promotion and she did not receive a raise."

43. The statement is always true, so it is a tautology.

p	p	∨	~p
T	T	T	F
F	F	T	T
	1	3	2

45. The statement is always true, so it is a tautology.

p	q	(p	∨	q)	∨	(~p	∨	q)
T	T	T	T	T	T	F	T	T
T	F	T	T	F	T	F	F	F
F	T	F	T	T	T	T	T	T
F	F	F	F	F	T	T	T	F
		1	5	2	7	3	6	4

47. The statement is always true, so it is a tautology.

p	q	r	(~p	∨	q)	∨	(~q	∨	r)
T	T	T	F	T	T	T	F	T	T
T	T	F	F	T	T	T	F	F	F
T	F	T	F	F	F	T	T	T	T
T	F	F	F	F	F	T	T	T	F
F	T	T	T	T	T	T	F	T	T
F	T	F	T	T	T	T	F	F	F
F	F	T	T	T	F	T	T	T	T
F	F	F	T	T	F	T	T	T	F
			1	5	2	7	3	6	4

49. The statement is always false, so it is a self-contradiction.

r	~r	∧	r
T	F	F	T
F	T	F	F
	1	3	2

51. The statement is always false, so it is a self-contradiction.

p	q	p	∧	(~p	∧	q)
T	T	T	F	F	F	T
T	F	T	F	F	F	F
F	T	F	F	T	T	T
F	F	F	F	T	F	F
		1	5	2	4	3

53. The statement is not always false, so it is not a self-contradiction.

p	q	[p	∧	(~p	∨	q)]	∨	q
T	T	T	T	F	T	T	T	T
T	F	T	F	F	F	F	F	F
F	T	F	F	T	T	T	T	T
F	F	F	F	T	T	F	F	F
		1	6	2	5	3	7	4

55. The symbol ≤ means "less than *or* equal to."

57. There are 5 simple statements, *p*, *q*, *r*, *s*, and *t*, so $2^5 = 32$ rows are needed.

59. The truth table is given below:

p	q	r	s	[(p	∧	~q)	∨	(q	∧	~r)]	∧	(r	∨	~s)
T	T	T	T	T	F	F	F	T	F	F	F	T	T	F
T	T	T	F	T	F	F	F	T	F	F	F	T	T	T
T	T	F	T	T	F	F	T	T	T	T	F	F	F	F
T	T	F	F	T	F	F	T	T	T	T	T	F	T	T
T	F	T	T	T	T	T	T	F	F	F	T	T	T	F
T	F	T	F	T	T	T	T	F	F	F	T	T	T	T
T	F	F	T	T	T	T	T	F	F	T	F	F	F	F
T	F	F	F	T	T	T	T	F	F	T	T	F	T	T
F	T	T	T	F	F	F	F	T	F	F	F	T	T	F
F	T	T	F	F	F	F	F	T	F	F	F	T	T	T
F	T	F	T	F	F	F	T	T	T	T	F	F	F	F
F	T	F	F	F	F	F	T	T	T	T	T	F	T	T
F	F	T	T	F	F	T	F	F	F	F	F	T	T	F
F	F	T	F	F	F	T	F	F	F	F	F	T	T	T
F	F	F	T	F	F	T	F	F	F	T	F	F	F	F
F	F	F	F	F	F	T	F	F	F	T	F	F	T	T
				1	7	2	9	3	8	4	11	5	10	6

EXERCISE SET 3.3

1. *Antecedent:* I had the money
   Consequent: I would buy the painting

3. *Antecedent:* they had a guard dog
   Consequent: no one would trespass on their property

5. *Antecedent:* I change my major
    Consequent: I must reapply for admission

7. Because the consequent is true when the antecedent is true, this is a true statement.

9. Because the antecedent is false, this is a true statement.

11. Because the antecedent is false, this is a true statement.

13. Consider the case $x = -6$: $|-6| = 6$ is true, but $-6 = 6$ is false. Because the consequent is false when the antecedent is true, this is a false statement.

15. The truth table is given below:

p	q	(p	∧	~q)	→	[~	(p	∧	q)
T	T	T	F	F	T	F	T	T	T
T	F	T	T	T	T	T	T	F	F
F	T	F	F	F	T	T	F	F	T
F	F	F	F	T	T	T	F	F	F
		1	5	2	8	7	3	6	4

17. The truth table is given below:

p	q	[(p	→	q)	∧	p]	→	q
T	T	T	T	T	T	T	T	T
T	F	T	F	F	F	T	T	F
F	T	F	T	T	F	F	T	T
F	F	F	T	F	F	F	T	F
		1	5	2	6	3	7	4

19. The truth table is given below:

p	q	r	[r	∧	(~p	∨	q)]	→	(r	∨	~q)
T	T	T	T	T	F	T	T	T	T	T	F
T	T	F	F	F	F	T	T	T	F	F	F
T	F	T	T	F	F	F	F	T	T	T	T
T	F	F	F	F	F	F	F	T	F	T	T
F	T	T	T	T	T	T	T	T	T	T	F
F	T	F	F	F	T	T	T	T	F	F	F
F	F	T	T	T	T	T	F	T	T	T	T
F	F	F	F	F	T	T	F	T	F	T	T
			1	7	2	6	3	9	4	8	5

21. The truth table is given below:

p	q	r	[(p	→	q)	∨	(r	∧	~p)]	→	(r	∨	~q)
T	T	T	T	T	T	T	T	F	F	T	T	T	F
T	T	F	T	T	T	T	F	F	F	F	F	F	F
T	F	T	T	F	F	F	T	F	F	T	T	T	T
T	F	F	T	F	F	F	F	F	F	T	F	T	T
F	T	T	F	T	T	T	T	T	T	T	T	T	F
F	T	F	F	T	T	T	F	F	T	F	F	F	F
F	F	T	F	T	F	T	T	T	T	T	T	T	T
F	F	F	F	T	F	T	F	F	T	T	F	T	T
			1	7	2	9	3	8	4	11	5	10	6

23. The truth table is given below:

p	q	r	~	(p	→	~r)	∧	~q]	→	r
T	T	T	T	T	F	F	F	F	T	T
T	T	F	F	T	T	T	F	F	T	F
T	F	T	T	T	F	F	T	T	T	T
T	F	F	F	T	T	T	F	T	T	F
F	T	T	F	F	T	F	F	F	T	T
F	T	F	F	F	T	T	F	F	T	F
F	F	T	F	F	T	F	F	T	T	T
F	F	F	F	F	T	T	F	T	T	F
			6	1	5	2	7	3	8	4

25. She cannot sing or she would be perfect for the part.

27. Either x is not an irrational number or x is not a terminating decimal.

29. The fog must lift or our flight will be canceled.

31. They offered me the contract and I didn't accept.

33. Pigs have wings and they still can't fly.

35. She traveled to Italy and she didn't visit her relatives.

37. If $x = -3$, the first component is true and the second component is false. Thus this is a false statement.

39. Both components are true when $x \neq 0$ and false when $x = 0$. Both components have the same truth value for any value of x, so this is a true statement.

41. The number $\frac{2}{3}$ is a rational number. Its decimal form is $0.\overline{6}$, which does not terminate. The first component is true while the second component is false. Thus this is a false statement.

43. Both components are always false, so this statement is true.

45. Both components are true when triangle ABC is equilateral and false when triangle ABC is not equilateral. Thus this statement is true.

47. $p \rightarrow v$

49. $t \rightarrow \sim v$

51. $(\sim t \wedge p) \rightarrow v$

53. The statements are not equivalent.

p	r	p	→	~r
T	T	T	F	F
T	F	T	T	T
F	T	F	T	F
F	F	F	T	T
		1	3	2

r	∨	~p
T	T	F
F	F	F
T	T	T
F	T	T
1	3	2

55. The statements are not equivalent.

p	r	~p	→	(p	∨	r)
T	T	F	T	T	T	T
T	F	F	T	T	T	F
F	T	T	T	F	T	T
F	F	T	F	F	F	F
		1	5	2	4	3

r
T
F
T
F
1

57. The statements are equivalent.

p	q	r	p	→	(q	∨	r)
T	T	T	T	T	T	T	T
T	T	F	T	T	T	T	F
T	F	T	T	T	F	T	T
T	F	F	T	F	F	F	F
F	T	T	F	T	T	T	T
F	T	F	F	T	T	T	F
F	F	T	F	T	F	T	T
F	F	F	F	T	F	F	F
			1	5	2	4	3

(p	→	q)	∨	(p	→	r)
T	T	T	T	T	T	T
T	T	T	T	T	F	F
T	F	F	T	T	T	T
T	F	F	F	T	F	F
F	T	T	T	F	T	T
F	T	T	T	F	T	F
F	T	F	T	F	T	T
F	T	F	T	F	T	F
1	5	2	7	3	6	4

59. If a number is a rational number, then it is a real number.

61. If a number is a repeating decimal, then it is a rational number.

63. If an animal is a Sauropod, then it is herbivorous.

EXERCISE SET 3.4

1. If we take the aerobics class, then we will be in good shape for the ski trip.

3. If the number is an odd prime number, then it is greater than 2.

5. If he has the talent to play a keyboard, then he can join the band.

7. If I was able to prepare for the test, then I had the textbook.

9. If you ran the Boston marathon, then you are in excellent shape.

11. a. If I quit this job, then I am rich.

    b. If I were not rich, then I would not quit this job.

    c. If I would not quit this job, then I would not be rich.

13. a. If we are not able to attend the party, then she did not return soon.

    b. If she returns soon, then we will be able to attend the party.

    c. If we are able to attend the party, then she returned soon.

15. a. If a figure is a quadrilateral, then it is a parallelogram.

    b. If a figure is not a parallelogram, then it is not a quadrilateral.

    c. If a figure is not a quadrilateral, then it is not a parallelogram.

17. a. If I am able to get current information about astronomy, then I have access to the Internet.

    b. If I do not have access to the Internet, then I will not be able to get current information about astronomy.

    c. If I am not able to get current information about astronomy, then I don't have access to the Internet.

19. a. If we don't have enough money for dinner, then we took a taxi.

    b. If we did not take a taxi, then we will have enough money for dinner.

    c. If we have enough money for dinner, then we did not take a taxi.

21. a. If she can extend her vacation for at least two days, then she will visit Kauai.

    b. If she does not visit Kauai, then she could not extend her vacation for at least two days.

    c. If she cannot extend her vacation for at least two days, then she will not visit Kauai.

23. a. If two lines are parallel, then the two lines are perpendicular to a given line.

    b. If two lines are not perpendicular to a given line, then the two lines are not parallel.

    c. If two lines are not parallel, then the two lines are not both perpendicular to a given line.

25. The second statement is the converse of the first. The statements are not equivalent.

27. The second statement is the contrapositive of the first. The statements are equivalent.

29. The second statement is the inverse of the first. The statements are not equivalent.

31. If $x = 7$, then $3x - 7 \neq 11$. The original statement is true.

33. If $|a| = 3$, then $a = 3$. The original statement is false.

35. If $a + b = 25$, then $\sqrt{a+b} = 5$. The original statement is true.

37. The contrapositive of $p \to q$ is $\sim q \to \sim p$. The inverse of $\sim q \to \sim p$ is $\sim(\sim q) \to \sim(\sim p)$ or $q \to p$. The converse of $q \to p$ is $p \to q$.

For exercise 39, a sample answer is given. Your answer may vary from the given answer.

39. a. "If a figure is a triangle, then it is a 3-sided polygon." The converse, "If a figure is a 3-sided polygon, then it is a triangle" is true.

    b. "If it is an apple, then it is a fruit." The converse, "If it is a fruit, then it is an apple" is false.

41. If you can dream it, then you can do it.

43. If I were a dancer, then I would not be a singer.

45. A conditional statement and its contrapositive are equivalent. They always have the same truth value.

47. Suppose the Hatter is the only person telling the truth. Then the statements of the Dodo and the March Hare must be false. Therefore, negating the Dodo's statement gives "The Hatter does not lie," which is true. Negating the March Hare's statement gives "Either the Dodo or the Hatter tells lies," which is also true. The Hatter's statement is also true. Since all three statements are true, the Hatter is telling the truth.

EXERCISE SET 3.5

1. In symbolic form:
$$r \to c$$
$$\underline{\quad r \quad}$$
$$\therefore c$$

3. In symbolic form:
$$g \to s$$
$$\underline{\quad \sim g \quad}$$
$$\therefore \sim s$$

5. In symbolic form:
$$s \to i$$
$$\underline{\quad s \quad}$$
$$\therefore i$$

7. In symbolic form:
$$\sim p \to \sim a$$
$$\underline{\quad a \quad}$$
$$\therefore p$$

9.

p	q	First premise $p \vee \sim q$	Second premise $\sim q$	Conclusion p
T	T	T	F	T
T	F	T	T	T
F	T	F	F	F
F	F	T	T	F

The premises are true in rows 2 and 4. Because the conclusion in row 4 is false and the premises are both true, we know the argument is invalid.

11.

p	q	First premise $p \to \sim q$	Second premise $\sim q$	Conclusion p
T	T	F	F	T
T	F	T	T	T
F	T	T	F	F
F	F	T	T	F

The premises are true in rows 2 and 4. Because the conclusion in row 4 is false and the premises are both true, we know the argument is invalid.

13.

p	q	First premise $\sim p \to \sim q$	Second premise $\sim p$	Conclusion $\sim q$
T	T	T	F	F
T	F	T	F	T
F	T	F	T	F
F	F	T	T	T

The premises are true in row 4. Because the conclusion is true and the premises are both true, the argument is valid.

15.

p	q	First premise $(p \to q)$ $\land$ $(\sim p \to q)$	Second premise q	Conclusion p
T	T	T	T	T
T	F	F	F	T
F	T	T	T	F
F	F	F	F	F

The premises are true in rows 1 and 3. Because the conclusion in row 3 is false and the premises are both true, we know the argument is invalid.

17.

p	q	First premise $(p \land \sim q)$ $\lor$ $(p \to q)$	Second premise $q \lor p$	Conclusion $\sim p \land q$
T	T	T	T	F
T	F	T	T	F
F	T	T	T	T
F	F	T	F	F

The premises are true in rows 1, 2, and 3. Because the conclusions in rows 1 and 2 are false and the premises are true, we know the argument is invalid.

19.

p	q	r	First premise $(p \land \sim q)$ $\lor$ $(p \lor r)$	Second premise r	Conclusion $p \lor q$
T	T	T	T	T	T
T	T	F	T	F	T
T	F	T	T	T	T
T	F	F	T	F	T
F	T	T	T	T	T
F	T	F	F	F	T
F	F	T	T	T	F
F	F	F	F	F	F

The premises are true in rows 1, 3, 5, and 7. Because the conclusion in row 7 is false and the premises are both true, we know the argument is invalid.

21.

p	q	r	First premise $(p \leftrightarrow q)$	Second premise $p \to r$	Conclusion $\sim r \to \sim p$
T	T	T	T	T	T
T	T	F	T	F	F
T	F	T	F	T	T
T	F	F	F	F	F
F	T	T	F	T	T
F	T	F	F	T	T
F	F	T	T	T	T
F	F	F	T	T	T

The premises are true in rows 1, 7 and 8. Because the conclusions are true and the premises are true, the argument is valid.

23.

p	q	r	First premise $(p \land \sim q)$	Second premise $p \leftrightarrow r$	Conclusion $q \lor r$
T	T	T	F	T	T
T	T	F	F	F	T
T	F	T	T	T	T
T	F	F	T	F	F
F	T	T	F	F	T
F	T	F	F	T	T
F	F	T	F	F	T
F	F	F	F	T	F

The premises are true in row 3. Because the conclusion is true and the premises are true, the argument is valid.

25. In symbolic form:

$h \rightarrow r$

$\sim h$

$\therefore \sim r$

h	r	First premise $h \rightarrow r$	Second premise $\sim h$	Conclusion $\sim r$
T	T	T	F	F
T	F	F	F	T
F	T	T	T	F
F	F	T	T	T

The premises are true in rows 3 and 4. Because the conclusion in row 3 is false and the premises are both true, we know the argument is invalid.

27. In symbolic form:

$\sim b \rightarrow d$

$b \vee d$

$\therefore b$

b	d	First premise $\sim b \rightarrow d$	Second premise $b \vee d$	Conclusion b
T	T	T	T	T
T	F	T	T	T
F	T	T	T	F
F	F	F	F	F

The premises are true in rows 1, 2, and 3. Because the conclusion in row 3 is false and the premises are both true, we know the argument is invalid.

29. In symbolic form:

$c \rightarrow t$

t

$\therefore c$

c	t	First premise $c \rightarrow t$	Second premise t	Conclusion c
T	T	T	T	T
T	F	F	F	T
F	T	T	T	F
F	F	T	F	F

The premises are true in rows 1 and 3. Because the conclusion in row 3 is false and the premises are both true, we know the argument is invalid.

31. Label the statements

    f: You take Art 151 in the fall.

    s: You will be eligible to take Art 152 in the spring.

In symbolic form:

$f \rightarrow s$

$\sim s$

$\therefore \sim f$

The argument is valid using modus tollens.

33. Label the statements

    n: I had a nickel for every logic problem I have solved.

    r: I would be rich.

In symbolic form:

$n \rightarrow r$

$\sim n$

$\therefore \sim r$

The argument is invalid using the fallacy of the inverse.

35. Label the statements

    s: We serve salmon.

    v: Vicky will join us for lunch.

    m: Marilyn will join us for lunch.

In symbolic form:

$s \rightarrow v$

$v \rightarrow \sim m$

$\therefore s \rightarrow \sim m$

The argument is valid using the law of syllogism.

37. Label the statements

    c: My cat is left alone in the apartment.

    s: She claws the sofa.

In symbolic form:

$c \rightarrow s$

c

$\therefore s$

The argument is valid using modus ponens.

39. Label the statements

    n: Rita buys a new car.

    c: Rita goes on a cruise.

In symbolic form:

$n \rightarrow \sim c$

c

$\therefore \sim n$

The argument is valid using modus tollens.

41. Applying the law of syllogism to Premise 1 and Premise 2 produces

$\sim p \to r$	Premise 1
$r \to t$	Premise 2
$\therefore \sim p \to t$	Law of Syllogism

The above conclusion $\sim p \to t$ can be written as $\sim t \to p$, using the contrapositive form. Combining the conclusion above with Premise 3 gives

$\sim t \to p$	Equivalent form of conclusion
$\sim t$	Premise 3
$\therefore p$	Modus Ponens

This sequence of valid arguments has produced the desired conclusion. Thus the original argument is valid.

43. In symbolic form the argument is

$s \to \sim r$	Premise 1
$\sim r \to c$	Premise 2
$\sim t \to \sim c$	Premise 3
$\therefore s \to t$	Conclusion

Applying the law of syllogism to Premise 1 and Premise 2 produces:

$s \to \sim r$	Premise 1
$\sim r \to c$	Premise 2
$\therefore s \to c$	Law of Syllogism

Premise 3 can be written as $c \to t$ using the contrapositive form. Combining Premise 3 with the conclusion from above gives

$s \to c$	Conclusion from above
$c \to t$	Equivalent form of Premise 3
$\therefore s \to t$	Law of Syllogism

This sequence of valid arguments has produced the desired conclusion. Thus the original argument is valid.

45. In symbolic form the argument is

$\sim o \to \sim f$	Premise 1
$c \to \sim o$	Premise 2
$\therefore f \to \sim c$	Conclusion

Premise 1 can be written as $f \to o$ using the contrapositive form. Premise 2 can be written as $o \to \sim c$ using the contrapositive form. Applying the equivalent forms and the law of syllogism to Premise 1 and Premise 2 produces

$f \to o$	Equivalent form of Premise 1
$o \to \sim c$	Equivalent form of Premise 2
$\therefore f \to \sim c$	Law of Syllogism

This sequence of valid arguments has produced the desired conclusion. Thus the original argument is valid.

47. In symbolic form the argument is

$\sim (p \land \sim q)$	Premise 1
p	Premise 2
$\therefore ?$	

De Morgan's law gives $\sim p \lor q$ as an equivalent form of Premise 1. Combining this form with premise 2 creates a disjunctive syllogism:

$\sim p \lor q$	Equivalent form of Premise 1
p	Premise 2
$\therefore q$	Disjunctive Syllogism

The conclusion is q.

49. Label the statements
    t: It is a theropod.
    h: It is herbivorous.
    s: It is a sauropod.
In symbolic form the argument is:

$t \to \sim h$	Premise 1
$\sim h \to \sim s$	Premise 2
s	Premise 3
$\therefore ?$	

Applying the law of syllogisms to Premise 1 and

Premise 2 gives

$t \rightarrow \sim h$ Premise 1

$\sim h \rightarrow \sim s$ Premise 2

$\therefore t \rightarrow \sim s$ Law of Syllogism

Write the conclusion from above in its equivalent contrapositive form, $s \rightarrow \sim t$. Use the equivalent form of the conclusion from above and Premise 3 to give

$s \rightarrow \sim t$ Equivalent form of conclusion

s Premise 3

$\therefore \sim t$ Modus ponens

Therefore, it is not a theropod.

51. Label the statements
   b: A person is a baby.
   i: A person is illogical.
   c: A person can manage a crocodile.
   d: A person is despised.
In symbolic form the argument is:

$b \rightarrow i$ Premise 1

$c \rightarrow \sim d$ Premise 2

$i \rightarrow d$ Premise 3

$\therefore b \rightarrow \sim c$ Conclusion

Applying the law of syllogism to Premise 1 and Premise 3 produces

$b \rightarrow i$ Premise 1

$i \rightarrow d$ Premise 3

$\therefore b \rightarrow d$ Law of Syllogism

Premise 2 can be written as $d \rightarrow \sim c$ using the contrapositive form. Applying the law of syllogism to the conclusion from above and the equivalent form of Premise 2 produces

$b \rightarrow d$ Conclusion from above

$d \rightarrow \sim c$ Equivalent form of Premise 2

$\therefore b \rightarrow \sim c$ Law of Syllogism

This sequence of valid arguments has produced the desired conclusion. Thus the original argument is valid.

EXERCISE SET 3.6

1. The argument is valid.

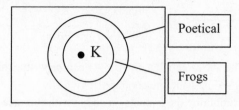

3. The argument is valid.

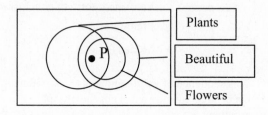

5. The argument is valid.

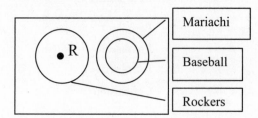

7. The argument is valid.

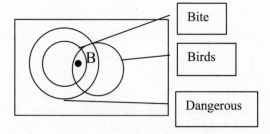

9. The argument is invalid.

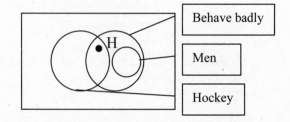

11. The argument is invalid.

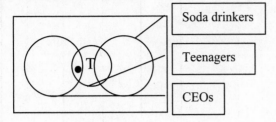

13. The argument is invalid.

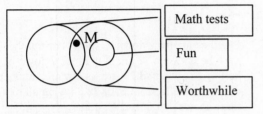

15. The argument is valid.

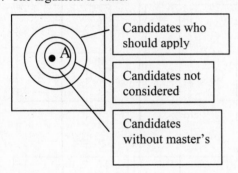

17. The argument is valid.

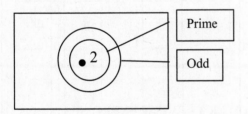

19. The argument is invalid.

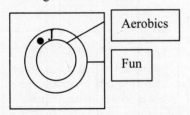

21. All Reuben sandwiches need mustard.

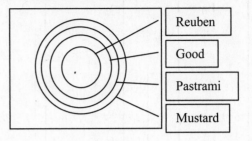

23. 1001 ends with a 5.

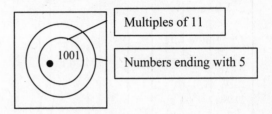

25. Some horses are grey.

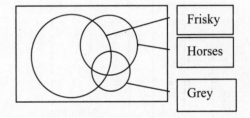

27. a. Invalid

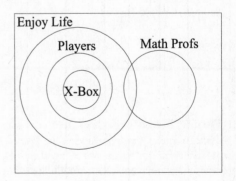

b.   Invalid

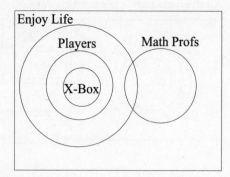

c.   Invalid

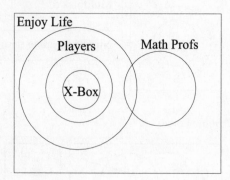

d.   Invalid

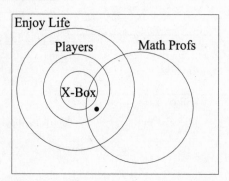

e.   Valid

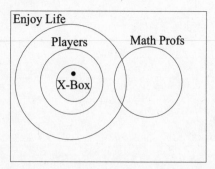

f.   Valid

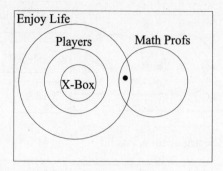

29.   Use the clue for 1 down to start. The number must be between 21 and 29. Since 2 down is half of 1 down, and 2 down is a whole number, 1 down can be 22, 24, 26, or 28. Trying all of these shows that only 28 works.

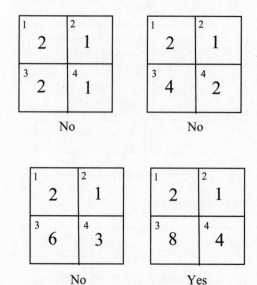

CHAPTER 3 REVIEW EXERCISES

1.   The sentence "How much is a ticket to London?" is a question, not a declarative sentence. Thus it is not a statement.

2.   You may not know if 91 is a prime number or not; however, you do know that it is a whole number larger than 1, so it is either a prime number or it is not a prime number. The sentence is either true or it is false, and it is not both true and false, so it is a statement.

3. We do not know whether the given sentence is true or false, but we know that the sentence is either true or false and that it is not both true and false. Thus the sentence is a statement.

4. Since the square of any real number is either positive or zero, this is a true sentence. Thus the sentence is a statement.

5. The sentence "Lock the car" is a command. It is not a statement.

6. The sentence "Clark Kent is Superman" is either true or false, so it is a statement.

7. $m \wedge b$; conjunction

8. $d \rightarrow e$; conditional

9. $g \leftrightarrow d$; biconditional

10. $t \rightarrow s$; conditional

11. No dogs bite.

12. Some desserts at the Cove restaurant are not good.

13. Some winners do not receive a prize.

14. All cameras use film.

15. Some of the students received an A.

16. Nobody enjoyed the story.

17. $5 > 2$ is true, so the statement is true.

18. $3 \neq 5$ is true and "7 is a prime number" is true, so the statement is true.

19. $4 < 7$ is true, so the statement is true.

20. $-3 < -1$ is true, so the statement is true.

21. Since all repeating decimals can be written in the form $\dfrac{p}{q}$, $q \neq 0$, the statement is true.

22. The number 0 is neither positive nor negative, so the statement is true.

23. False.

p	q	$(p$	$\wedge$	$q)$	$\vee$	$(\sim p$	$\vee$	$q)$
T	F	T	F	F	F	F	F	F
		1	5	2	7	3	6	4

24. False.

p	q	$(p$	$\rightarrow$	$\sim q)$	$\leftrightarrow$	$\sim$	$(p$	$\vee$	$q)$
T	F	T	T	T	F	F	T	T	F
		1	5	2	8	7	3	6	4

25. True.

p	q	r	$(p$	$\wedge$	$\sim q)$	$\wedge$	$(\sim r$	$\vee$	$q)$
T	F	F	T	T	T	T	T	T	F
			1	5	2	7	3	6	4

26. True.

p	q	r	$(r$	$\wedge$	$\sim p)$	$\vee$	$[(p$	$\vee$	$\sim q)$	$\leftrightarrow$	$(q$	$\rightarrow$	$r)]$
T	F	F	F	F	F	T	T	T	T	T	F	T	F
			1	10	2	11	3	7	4	9	5	8	6

27. True.

p	q	r	$[p$	$\wedge$	$(r$	$\rightarrow$	$q)]$	$\rightarrow$	$(q$	$\vee$	$\sim r)$
T	F	F	T	T	F	T	F	T	F	T	T
			1	7	2	6	3	9	4	8	5

28. False.

p	q	r	(~q	∨	~r)	→	[(p	↔	~r)	∧	q]
T	F	F	T	T	T	F	T	T	T	F	F
			1	8	2	9	3	6	4	7	5

29.

p	q	(~p	→	q)	∨	(~q	∧	p)]
T	T	F	T	T	T	F	F	T
T	F	F	T	F	T	T	T	T
F	T	T	T	T	F	F	F	F
F	F	T	F	F	F	T	F	F
		1	5	2	7	3	6	4

31.

p	q	~	(p	∨	~q)	∧	(q	→	p)]
T	T	F	T	T	F	F	T	T	T
T	F	F	T	T	T	F	F	T	T
F	T	F	F	F	F	F	T	T	F
F	F	F	F	T	T	F	F	T	F
		6	1	5	2	8	3	7	4

30.

p	q	~p	↔	(q	∨	p)
T	T	F	F	T	T	T
T	F	F	F	F	T	T
F	T	T	T	T	T	F
F	F	T	F	F	F	F
		1	5	2	4	3

32.

p	q	(p	↔	q)	∨	(~q	∧	p)]
T	T	T	T	T	T	F	F	T
T	F	T	F	F	T	T	T	T
F	T	F	F	T	F	F	F	F
F	F	F	T	F	T	T	F	F
		1	5	2	7	3	6	4

33.

p	q	r	(r	↔	~q)	∨	(p	→	q)
T	T	T	T	F	F	T	T	T	T
T	T	F	F	T	F	T	T	T	T
T	F	T	T	T	T	T	T	F	F
T	F	F	F	F	T	F	T	F	F
F	T	T	T	F	F	T	F	T	T
F	T	F	F	T	F	T	F	T	T
F	F	T	T	T	T	T	F	T	F
F	F	F	F	F	T	T	F	T	F
			1	5	2	7	3	6	4

34.

p	q	r	(~r	∨	~q)	∧	(q	→	p)
T	T	T	F	F	F	F	T	T	T
T	T	F	T	T	F	T	T	T	T
T	F	T	F	T	T	T	F	T	T
T	F	F	T	T	T	T	F	T	T
F	T	T	F	F	F	F	T	F	F
F	T	F	T	T	F	F	T	F	F
F	F	T	F	T	T	T	F	T	F
F	F	F	T	T	T	T	F	T	F
			1	5	2	7	3	6	4

35.

p	q	r	[p	↔	(q	→	~r)]	∧	~q
T	T	T	T	F	T	F	F	F	F
T	T	F	T	T	T	T	T	F	F
T	F	T	T	T	F	T	F	T	T
T	F	F	T	T	F	T	T	T	T
F	T	T	F	T	T	F	F	F	F
F	T	F	F	F	T	T	T	F	F
F	F	T	F	F	F	T	F	F	T
F	F	F	F	F	F	T	T	F	T
			1	6	2	5	3	7	4

36.

p	q	r	~	(p	∧	q)	→	(~q	∨	~r)
T	T	T	F	T	T	T	T	F	F	F
T	T	F	F	T	T	T	T	F	T	T
T	F	T	T	T	F	F	T	T	T	F
T	F	F	T	T	F	F	T	T	T	T
F	T	T	T	F	F	T	F	F	F	F
F	T	F	T	F	F	T	T	F	T	T
F	F	T	T	F	F	F	T	T	T	F
F	F	F	T	F	F	F	T	T	T	T
			6	1	5	2	8	3	7	4

37. Let p represent the statement "Bob failed the English proficiency test." Let q represent the statement "He registered for a speech course." In symbolic form, the original sentence is $\sim(p \wedge q)$. One of De Morgan's laws states that this is equivalent to $\sim p \vee \sim q$. Thus an equivalent sentence is "Bob passed the English proficiency test or he did not register for a speech course."

38. Let p represent the statement "Ellen went to work this morning." Let q represent the statement "She took her medication." In symbolic form, the original sentence is $\sim p \wedge \sim q$. One of De Morgan's laws states that this is equivalent to $\sim(p \vee q)$. Thus an equivalent sentence is "It is not true that Ellen went to work this morning or she took her medication."

39. Let p represent the statement "Wendy will not go to the store this afternoon." Let q represent the statement "She will be able to prepare her fettuccine al pesto recipe." In symbolic form, the original sentence is $\sim p \vee \sim q$. One of De Morgan's laws states that this is equivalent to $\sim(p \wedge q)$. Thus an equivalent sentence is "It is not the case that Wendy will not go to the store this afternoon and she will be able to prepare her fettuccine al pesto recipe."

40. Let p represent the statement "Gina did not enjoy the movie." Let q represent the statement "She did enjoy the party." In symbolic form, the original sentence is $\sim p \wedge \sim q$. One of De Morgan's laws states that this is equivalent to $\sim(p \vee q)$. Thus an equivalent sentence is "It is not the case that Gina did not enjoy the movie or she enjoyed the party."

41.

p	q	~p	→	~q
T	T	F	T	F
T	F	F	T	T
F	T	T	F	F
F	F	T	T	T
		1	3	2

p		∨	~q
T	T	T	F
T	T	T	T
F	F	F	F
F	F	T	T
		1	3	2

42.

p	q	~p	∨	q
T	T	F	T	T
T	F	F	F	F
F	T	T	T	T
F	F	T	T	F
		1	3	2

~	(p	∧	~q)
T	T	F	F
F	T	T	T
T	F	F	F
T	F	F	T
4	1	3	2

43.

p	q	p	∨	(q	∧	~p)
T	T	T	T	T	F	F
T	F	T	T	F	F	F
F	T	F	T	T	T	T
F	F	F	F	F	F	T
		1	7	2	6	3

p	q	p	∨	q
T	T	T	T	T
T	F	T	T	F
F	T	F	T	T
F	F	F	F	F
		4	8	5

44.

p	q	p	↔	q
T	T	T	T	T
T	F	T	F	F
F	T	F	F	T
F	F	F	T	F
		1	7	2

p	q	(p	∧	q)	∨	(~p	∧	~q)
T	T	T	T	T	T	F	F	F
T	F	T	F	F	F	F	F	T
F	T	F	F	T	F	T	F	F
F	F	F	F	F	T	T	T	T
		3	8	4	10	5	9	6

45. The statement is always false, so it is a self-contradiction.

p	q	p	∧	(q	∧	~p)
T	T	T	F	T	F	F
T	F	T	F	F	F	F
F	T	F	F	T	T	T
F	F	F	F	F	F	T
		1	5	2	4	3

46. The statement is always true, so it is a tautology.

p	q	(p	∧	q)	∨	(p	→	~q)
T	T	T	T	T	T	T	F	F
T	F	T	F	F	T	T	T	T
F	T	F	F	T	T	F	T	F
F	F	F	F	F	T	F	T	T
		1	5	2	7	3	6	4

47. The statement is always true, so it is a tautology.

p	q	[~(p	→	q)]	←	(p	∧	~q)
T	T	F	T	T	T	T	T	F	F
T	F	T	T	F	F	T	T	T	T
F	T	F	F	T	T	T	F	F	F
F	F	F	F	T	F	T	F	F	T
		6	1	5	2	8	3	7	4

48. The statement is always true, so it is a tautology.

p	q	p	∨	(p	→	q)
T	T	T	T	T	T	T
T	F	T	T	T	F	F
F	T	F	T	F	T	T
F	F	F	T	F	T	F
		1	5	2	4	3

49. *Antecedent:* he has talent
Consequent: he will succeed

50. *Antecedent:* I had a credential
Consequent: I could get the job

51. *Antecedent:* I join the fitness club
Consequent: I will follow the exercise program

52. *Antecedent:* I will attend
Consequent: It is free

53. She is not tall or she would be on the volleyball team.

54. He cannot stay awake or he would finish the report.

55. Rob is ill or he would start.

56. Sharon will not be promoted or she closes the deal.

57. I get my paycheck and I do not purchase a ticket.

58. The tomatoes will get big and you did not provide them with plenty of water.

59. You entered Cleggmore University and you did not have a high score on the SAT exam.

60. Ryan enrolled at a university and he did not enroll at Yale.

61. When $x = 3$ and $y = -3$, the second statement is true and the first statement is false. Thus this is a false statement.

62. Both components are true when $x > y$ and false when $x \le y$. Both components have the same truth value for any values of x and y, so this is a true statement.

63. Because the consequent is true when the antecedent is true, this is a true statement.

64. When $x = 5$ and $y = 2$, the antecedent is true and the consequent is false. So this is a false statement.

65. When $x = -1$, the antecedent is true and the consequent is false. So this is a false statement.

66. When $x = 2$ and $y = -2$, the antecedent is true and the consequent is false. So this is a false statement.

67. If a real number has a nonrepeating, nonterminating decimal form, then the real number is irrational.

68. If you are a politician, then you are well known.

69. If I can sell my condominium, then I can buy the house.

70. If a number is divisible by 9, then the number is divisible by 3.

71. a. *Converse:* If $x > 3$, then $x + 4 > 7$.

    b. *Inverse:* If $x + 4 \le 7$, then $x \le 3$.

    c. *Contrapositive:* If $x \le 3$, then $x + 4 \le 7$.

72. a. *Converse:* If the recipe can be prepared in less than 20 minutes, then the recipe is in this book.

    b. *Inverse:* If the recipe is not in this book, then the recipe cannot be prepared in less than 20 minutes.

    c. *Contrapositive:* If the recipe cannot be prepared in less than 20 minutes, then the recipe is not in this book.

73. a. *Converse:* If $(a + b)$ is divisible by 3, then a and b are both divisible by 3.

    b. *Inverse:* If a and b are not both divisible by 3, then $(a + b)$ is not divisible by 3.

    c. *Contrapositive:* If $(a + b)$ is not divisible by 3, then a and b are not both divisible by 3.

74. a. *Converse:* If they come, then you built it.

    b. *Inverse:* If you do not build it, then they will not come.

    c. *Contrapositive:* If they do not come, then you did not build it.

75. a. *Converse:* If it has exactly two parallel sides, then it is a trapezoid.

    b. *Inverse:* If it is not a trapezoid, then it does not have exactly two parallel sides.

    c. *Contrapositive:* If it does not have exactly two parallel sides, then it is not a trapezoid.

76. a. *Converse:* If they returned, then they liked it.

    b. *Inverse:* If they do not like it, then they will not return.

    c. *Contrapositive:* If they do not return, then they did not like it.

77. The contrapositive of $p \rightarrow q$ is $\sim q \rightarrow \sim p$. The inverse of $\sim q \rightarrow \sim p$ is $q \rightarrow p$, the converse of the original statement.

78. The inverse of $p \rightarrow q$ is $\sim p \rightarrow \sim q$. The contrapositive of $\sim p \rightarrow \sim q$ is $q \rightarrow p$. The converse of $q \rightarrow p$ is $p \rightarrow q$, the original statement.

79. The original statement is "If x is an odd prime number, then $x > 2$."

80. The negation of $p \wedge \sim q$ is $\sim p \vee q$, which is equivalent to $p \rightarrow q$. The original statement is "If the senator attends the meeting, then she will vote on the motion."

81. The original statement is "If their manager contacts me, then I will purchase some of their products."

82. The original statement is "If I can rollerblade, then Ginny can rollerblade."

83.

p	q	First premise $(p \wedge \sim q)$ $\wedge$ $(\sim p \rightarrow q)$	Second premise p	Conclusion $\sim q$
T	T	F	T	F
T	F	T	T	T
F	T	F	F	F
F	F	F	F	T

The premises are true in row 2. Because the conclusion is true and the premises are both true, the argument is valid.

84.

p	q	First premise $p \rightarrow \sim q$	Second premise q	Conclusion $\sim p$
T	T	F	T	F
T	F	T	F	F
F	T	T	T	T
F	F	T	F	T

The premises are true in row 3. Because the conclusion is true and the premises are both true, the argument is valid.

85.

p	q	r	First premise r	Second premise $p \rightarrow \sim r$	Third premise $\sim p \rightarrow q$	Conclusion $p \wedge q$
T	T	T	T	F	T	T
T	T	F	F	T	T	T
T	F	T	T	F	T	F
T	F	F	F	T	T	F
F	T	T	T	T	T	F
F	T	F	F	T	T	F
F	F	T	T	T	F	F
F	F	F	F	T	F	F

The premises are true in row 5. Because the conclusion is false and the premises are true, we know the argument is invalid.

86.

p	q	r	First premise $(p \vee \sim r) \rightarrow (q \wedge r)$	Second premise $r \wedge p$	Conclusion $p \vee q$
T	T	T	T	T	T
T	T	F	F	F	T
T	F	T	F	T	T
T	F	F	F	F	T
F	T	T	T	F	T
F	T	F	F	F	T
F	F	T	T	F	F
F	F	F	F	F	F

The premises are true in row 1. Because each conclusion is true and the premises are true, the argument is valid.

87. Label the statements.
    f: We will serve fish for lunch.
    c: We will serve chicken for lunch.
In symbolic form the argument is:

$f \vee c$

$\sim f$
_____

$\therefore c$

The argument is valid using the disjunctive syllogism.

88. Label the statements.
    c: Mike is a CEO.
    d: He can afford to make a donation.
    s: He loves to ski.
In symbolic form the argument is:

$c \rightarrow d$

$d \rightarrow s$
_____

$\therefore \sim s \rightarrow \sim c$

The argument is valid using the law of syllogism, since $\sim s \rightarrow \sim c$ is equivalent to $c \rightarrow s$.

89. Label the statements.
    w: We wish to win the lottery.
    b: We must buy a lottery ticket.
In symbolic form the argument is:

$w \rightarrow b$

$\sim w$
_____

$\therefore \sim b$

The argument is invalid using the fallacy of the inverse.

90. Label the statements.
    m: Robert can charge it on his MasterCard.
    v: Robert can charge it on his Visa.
In symbolic form the argument is:

$m \vee v$

$\sim m$
_____

$\therefore v$

The argument is valid using the disjunctive syllogism.

91. Label the statements.
    c: We are going to have a Caesar salad.
    e: We buy some eggs.
In symbolic form the argument is:

$c \rightarrow e$

$\sim e$
_____

$\therefore \sim c$

The argument is valid using modus tollens.

92. Label the statements
    l: We serve lasagna.
    d: Eva will not come to our dinner party.
In symbolic form the argument is:

$l \rightarrow d$

$\sim l$
_____

$\therefore \sim d$

The argument is invalid using the fallacy of the inverse.

93. The argument is valid.

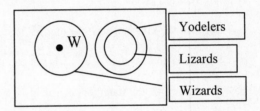

94. The argument is invalid.

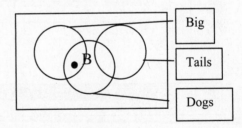

95. The argument is invalid.

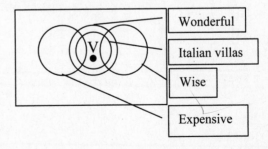

96. The argument is valid.

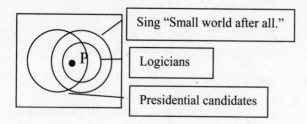

CHAPTER 3 TEST

1. a.   The sentence "Look for the cat" is a command. It is not a statement.

   b.   The sentence "Clark Kent is afraid of the dark" is either true or false, so the sentence is a statement.

2. a.   All trees are green.

   b.   Some of the kids had seen the movie.

3. a.   Since 5 is not less than 4, and 5 is not equal to 4, the statement is false.

   b.   Since $-2 = -2$, the statement is true.

4. a.   False.

p	q	r	(p	$\vee$	~q)	$\wedge$	(~r	$\wedge$	q)
T	F	T	T	T	T	F	F	F	F
			1	5	2	7	3	6	4

   b.   True.

p	q	r	(r	$\vee$	~p)	$\vee$	[(p	$\vee$	~q)	$\leftrightarrow$	(q	$\rightarrow$	r)]
T	F	T	T	T	F	T	T	T	T	T	F	T	T
			1	7	2	11	3	8	4	10	5	9	6

5.

p	q	~	(p	$\wedge$	~q)	$\vee$	(q	$\rightarrow$	p)
T	T	T	T	F	F	T	T	T	T
T	F	F	T	T	T	T	F	T	T
F	T	T	F	F	F	T	T	F	F
F	F	T	F	F	T	T	F	T	F
		6	1	5	2	8	3	7	4

6.

p	q	r	(r	↔	~q)	∧	(p	→	q)
T	T	T	T	F	F	F	T	T	T
T	T	F	F	T	F	T	T	T	T
T	F	T	T	T	T	F	T	F	F
T	F	F	F	F	T	F	T	F	F
F	T	T	T	F	F	F	F	T	T
F	T	F	F	T	F	T	F	T	T
F	F	T	T	T	T	T	F	T	F
F	F	F	F	F	T	F	F	T	F
			1	5	2	7	3	6	4

7. Let b represent the statement "Elle ate breakfast." Let l represent the statement "Elle took a lunch break." In symbolic form, the original sentence is $\sim b \wedge \sim l$. An equivalent form is $\sim(b \vee l)$. Thus an equivalent statement is "It is not the case that Elle ate breakfast or took a lunch break."

8. A tautology is a statement that is always true.

9. $\sim p \vee q$

10. a. The statement is of the form "q, if p," where p is $|x| = |y|$ and q is $x = y$. This statement is false when p is true and q is false. It is true in all other cases. When $x = 1$ and $y = -1$, then $|x| = |y|$ is true but $x = y$ is false. So the statement is false.

b. When $x = 2$, $y = 1$, and $z = -1$, the statement $x > y$ is true, but $xz > yz$ is false. So, p is true and q is false, and the statement is false.

11. a. *Converse:* If $x > 4$, then $x + 7 > 11$.

b. *Inverse:* If $x + 7 \leq 11$, then $x \leq 4$.

c. *Contrapositive:* If $x \leq 4$, then $x + 7 \leq 11$.

12. The standard form is:
$$p \to q$$
$$\underline{p}$$
$$\therefore q$$

13. The standard form is:
$$p \to q$$
$$\underline{q \to r}$$
$$\therefore p \to r$$

14.

p	q	First premise $(p \wedge \sim q) \wedge (\sim p \to q)$	Second premise p	Conclusion $\sim q$
T	T	F	T	F
T	F	T	T	T
F	T	F	F	F
F	F	F	F	T

The premises are true in row 2. Because the conclusion is true and the premises are both true, the argument is valid.

15.

p	q	r	First premise r	Second premise $p \rightarrow \sim r$	Third premise $\sim p \rightarrow q$	Conclusion $p \wedge q$
T	T	T	T	F	T	T
T	T	F	F	T	T	T
T	F	T	T	F	T	F
T	F	F	F	T	T	F
F	T	T	T	T	T	F
F	T	F	F	T	T	F
F	F	T	T	T	F	F
F	F	F	F	T	F	F

The premises are true in row 5. Because the conclusion is false and the premises are true, we know the argument is invalid.

16. Label the statements.
   t: We wish to win the talent contest.
   p: We must practice.
In symbolic form the argument is:

$t \rightarrow p$

$\sim t$
—————
$\therefore \sim p$

The argument is invalid using the fallacy of the inverse.

17. Label the statements.
   a: Gina will take a job in Atlanta.
   k: Gina will take a job in Kansas City.
In symbolic form the argument is:

$a \vee k$

$\sim a$
—————
$\therefore k$

The argument is valid using the disjunctive syllogism.

18. The argument is invalid.

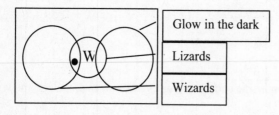

19. The argument is invalid.

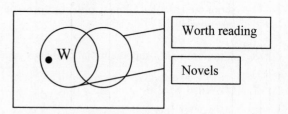

20. Label the statements.
   c: I cut my night class.
   p: I will go to the party.
In symbolic form the argument is:

$c \rightarrow p$

p
—————
$\therefore c$

The argument is invalid using the fallacy of the converse.

Chapter 4:
Numeration Systems and
Number Theory

EXERCISES 4.1

1. $46 = 40 + 6$

    ∩∩∩∩||||||

3. $103 = 100 + 3$

    ⌒|||

5. $2568 = 2000 + 500 + 60 + 8$

6.

7. $23,402 = 20,000 + 3,000 + 400 + 2$

9. $65,800 = 60,000 + 5,000 + 800$

11. $1,405,203 = 1,000,000 + 400,000 + 5,000 +$
    $200 + 3$

13. $2 \times 1,000 + 100 + 3 \times 10 + 4 \times 1 = 2,134$

15. $8 \times 100 + 4 \times 10 + 5 = 845$

17. $1 \times 1000 + 2 \times 100 + 3 \times 10 + 2 = 1232$

19. $2 \times 100,000 + 2 \times 10,000 + 1,000 + 10 + 1 =$
    $221,011$

21. $6 \times 10,000 + 5 \times 1000 + 7 \times 100 + 6 \times 10 + 9 =$
    $65,769$

23. $5 \times 1,000,000 + 1 \times 100,000 \times 2 \times 10,000 + 2 \times$
    $1000 + 4 \times 100 + 6 = 5,122,406$

25.

    $= 94$

27.

    $= 666$

29.

    $= 32$

31.

    Borrow 1 heel bone and change it to ten strokes.
    Borrow 1 scroll and change it to 10 heel bones.

    $= 56$

33. $DCL = 500 + 100 + 50 = 650$

35. $MCDIX = 1000 + (500 - 100) + (10 - 1)$
    $= 1409$

37. $MCCXL = 1,000 + 100 + 100 + (50 - 10) = 1,240$

39. $DCCCXL = 500 + 100 + 100 + 100 + (50 - 10)$
    $= 840$

41. $\overline{IX}XLIV = (10 - 1) \times 1000 + (50 - 10) +$
    $(5 - 1) = 9,044$

43. $\overline{XI}CDLXI = 11 \times 1000 + (500 - 100) + 50 + 10 +$
    $1 = 11,461$

45. $157 = 100 + 50 + 5 + 1 + 1 = CLVII$

47. $542 = 500 + (50 - 10) + 2 = DXLII$

49. $1197 = 1000 + 100 + (100 - 10) + 5 + 1 + 1 =$
MCXCVII

51. $787 = 500 + 100 + 100 + 50 + 10 + 10 + 10 + 5$
$+ 2 = DCCLXXXVII$

53. $683 = 500 + 100 + 50 + 10 + 10 + 10 + 3 =$
DCLXXXIII

55. $6898 = 6 \times 1000 + 500 + 100 + 100 + 100 +$
$(100 - 10) + 5 + 3 = \overline{VI}DCCCXCVIII$

57.

	1	63
	2	126
	4	252
√	8	504
	8	504

$8 \times 63 = 504$

59.

√	1	29
√	2	58
√	4	116
	7	203

$7 \times 29 = 203$

61.

√	1	35
	2	70
	4	140
	8	280
√	16	560
	17	595

$17 \times 35 = 595$

63.

√	1	108
√	2	216
√	4	432
	8	864
√	16	1728
	23	2484

$23 \times 108 = 2484$

65. a. Answers will vary. For example, it may be
easier to perform addition and subtraction
in the Egyptian numeration system.

b. Answers will vary. For example, less
symbols are needed to express some
numbers in the Roman numeration system.

EXERCISES 4.2

1. $48 = (4 \times 10^1) + (8 \times 10^0)$

3. $420 = (4 \times 10^2) + (2 \times 10^1) + (0 \times 10^0)$

5. $6803 = (6 \times 10^3) + (8 \times 10^2) + (0 \times 10^1) + (3 \times 10^0)$

7. $10,208 = (1 \times 10^4) + (0 \times 10^3) + (2 \times 10^2) +$
$(0 \times 10^1) + (8 \times 10^0)$

9. $400 + 50 + 6 = 456$

11. $5000 + 70 + 6 = 5076$

13. $30,000 + 5000 + 400 + 7 = 35,407$

15. $600,000 + 80,000 + 3000 + 40 = 683,040$

17. Expanding:
$$35 = (3 \times 10) + 5$$
$$+41 = (4 \times 10) + 1$$
$$= (7 \times 10) + 6 = 76$$

19. Expanding:
$$257 = (2 \times 100) + (5 \times 10) + 7$$
$$+138 = (1 \times 100) + (3 \times 10) + 8$$
$$= (3 \times 100) + (8 \times 10) + 15$$
$$= (3 \times 100) + (9 \times 10) + 5 = 395$$

21. Expanding:
$$1023 = (1 \times 1000) + (0 \times 100) + (2 \times 10) + 3$$
$$+1458 = (1 \times 1000) + (4 \times 100) + (5 \times 10) + 8$$
$$= (2 \times 1000) + (4 \times 100) + (7 \times 10) + 11$$
$$= (2 \times 1000) + (4 \times 100) + (8 \times 10) + 1$$
$$= 2481$$

23. Write 60 as $5 \times 10 + 10$.
$$62 = (5 \times 10) + 12$$
$$-35 = (3 \times 10) + 5$$
$$= (2 \times 10) + 7 = 27$$

25. Write 700 as $6 \times 100 + 10 \times 10$.

$$4725 = (4 \times 1000) + (6 \times 100) + (12 \times 10) + 5$$
$$\underline{-1362 = (1 \times 1000) + (3 \times 100) + (6 \times 10) + 2}$$
$$= (3 \times 1000) + (3 \times 100) + (6 \times 10) + 3$$
$$= 3363$$

27. Write 3000 as $2 \times 1000 + 10 \times 100$.

$$23,168 = 2 \cdot 10^4 + 2 \cdot 10^3 + 11 \cdot 10^2 + 6 \cdot 10^1 + 8$$
$$\underline{-12,857 = 1 \cdot 10^4 + 2 \cdot 10^3 + 8 \cdot 10^2 + 5 \cdot 10^1 + 7}$$
$$= 1 \cdot 10^4 + 0 \cdot 10^3 + 3 \cdot 10^2 + 1 \cdot 10^1 + 1$$
$$= 10,311$$

29. $20 + 3 = 23$

31. $(1 \times 60) + 30 + 7 = 97$

33. Expanding:

$$(20 \times 60^2) + (2 \times 60) + 10 + 3$$
$$= (20 \times 3600) + (2 \times 60) + 10 + 3$$
$$= 72,133$$

35. Expanding:

$$(10 \times 60^3) + (3 \times 60^2) + (11 \times 60) + 6$$
$$= (10 \times 216,000) + (3 \times 3600) + (660) + 6$$
$$= 2,171,466$$

37. $42 = 40 + 2$

<<<< ▼▼

39. $128 = 2 \times 60 + 8$

▼▼   ▼▼▼▼▼▼▼▼

41. $5678 = (1 \times 60^2) + (34 \times 60^1) + (38 \times 60^0)$

▼   <<< ▼▼▼▼
<<< ▼▼▼▼▼▼▼▼

43. $10,584 = (2 \times 60^2) + (56 \times 60^1) + (24 \times 60^0)$

▼▼   <<<<< ▼▼▼▼▼▼
<< ▼▼▼▼

45. $21,345 = (5 \times 60^2) + (55 \times 60^1) + (45 \times 60^0)$

▼▼▼▼▼   <<<<< ▼▼▼▼▼   <<<< ▼▼▼▼▼

47. Converting:

$$45$$
$$\underline{+23}$$
$$68 = 60 + 8$$

▼   ▼▼▼▼▼▼▼▼

49. Converting:

$$33 \cdot 60 + 42$$
$$\underline{+32 \cdot 60 + 21}$$
$$65 \cdot 60 + 63 = 66 \cdot 60 + 3 = 1 \cdot 60^2 + 6 \cdot 60 + 3$$

▼   ▼▼▼▼▼▼   ▼▼▼

51. Converting:

$$10 \cdot 60^2 + 31 \cdot 60 + 43$$
$$\underline{+ \ 1 \cdot 60^2 + 21 \cdot 60 + 32}$$
$$11 \cdot 60^2 + 52 \cdot 60 + 75 = 66 \cdot 60 + 3 =$$
$$11 \cdot 60^2 + 53 \cdot 60 + 15$$

<▼   <<<<< ▼▼▼   <▼▼▼▼▼

53. $9 \times 20 + 14 \times 1 = 194$

55. $5 \times 360 + 0 \times 20 + 3 \times 1 = 1803$

57. $2 \times 7200 + 0 \times 360 + 4 \times 20 + 12 \times 1 = 14,492$

59. $5 \times 7200 + 0 \times 360 + 5 \times 20 + 3 \times 1 = 36,103$

61. $137 = (6 \times 20^1) + (17 \times 20^0)$

63. $948 = (2 \times 360) + (11 \times 20^1) + (8 \times 20^0)$

65. $1693 = (4 \times 360) + (12 \times 20^1) + (13 \times 20^0)$

67. $7432 = (1 \times 7200) + (0 \times 360) + (11 \times 20^1) + (12 \times 20^0)$

69. a.   Anwers will vary.

   b.   Anwers will vary.

EXERCISE SET 4.3

1.   Converting:

$$243_{\text{five}} = 2 \times 5^2 + 4 \times 5^1 + 3 \times 5^0$$
$$= 2 \times 25 + 4 \times 5 + 3 \times 1$$
$$= 73$$

3.   Converting:

$$67_{\text{nine}} = 6 \times 9^1 + 7 \times 9^0$$
$$= 6 \times 9 + 7 \times 1$$
$$= 61$$

5.   Converting:

$$3154_{\text{six}} = 3 \times 6^3 + 1 \times 6^2 + 5 \times 6^1 + 4 \times 6^0$$
$$= 3 \times 216 + 1 \times 36 + 5 \times 6 + 4$$
$$= 718$$

7.   Converting:

$$13211_{\text{four}} = 1 \times 4^4 + 3 \times 4^3 + 2 \times 4^2 + 1 \times 4^1 + 1 \times 4^0$$
$$= 1 \times 256 + 3 \times 64 + 2 \times 16 + 1 \times 4 + 1$$
$$= 485$$

9.   Converting:

$$B5_{\text{sixteen}} = 11 \times 16^1 + 5 \times 16^0$$
$$= 11 \times 16 + 5$$
$$= 181$$

11.
```
5 |267
5 |53   2
5 |10   3
   2    0        267 = 2032_five
```

13.
```
6|1932
6|322   0
6|53    4
6|8     5
  1     2        1932 = 12540_six
```

15.
```
9|15306
9|1700   6
9|188    8
9|20     8
  2      2        15306 = 22886_nine
```

17.
```
2|4060
2|2030   0
2|1015   0
2|507    1
2|253    1
2|126    1
2|63     0
2|31     1
2|15     1
2|7      1
2|3      1
  1      1        4060 = 111111011100_two
```

19.
```
12|283
12|23      7
   1      11 (B in base twelve)
283 = 1B7_twelve
```

21. Expanding:

$$1101_{\text{two}} = 1 \times 2^3 + 1 \times 2^2 + 0 \times 2^1 + 1 \times 2^0$$
$$= 1 \times 8 + 1 \times 4 + 0 \times 2 + 1 \times 1$$
$$= 13$$

23. Expanding:

$$11011_{\text{two}} = 1 \times 2^4 + 1 \times 2^3 + 0 \times 2^2 + 1 \times 2^1 + 1 \times 2^0$$
$$= 1 \times 16 + 1 \times 8 + 0 \times 4 + 1 \times 2 + 1 \times 1$$
$$= 27$$

25. Expanding:

$$1100100_{two} = 1 \times 2^6 + 1 \times 2^5 + 0 \times 2^4 + 0 \times 2^3 + 1 \times 2^2$$
$$+ 0 \times 2^1 + 0 \times 2^0$$
$$= 1 \times 64 + 1 \times 32 + 0 \times 16 + 0 \times 8 + 1 \times 4$$
$$+ 0 \times 2 + 0 \times 1$$
$$= 100$$

27. Expanding:
$$10001011_{two} = 1 \times 2^7 + 0 \times 2^6 + 0 \times 2^5 + 0 \times 2^4 +$$
$$1 \times 2^3 + 0 \times 2^2 + 1 \times 2^1 + 1 \times 2^0$$
$$= 1 \times 128 + 0 \times 64 + 0 \times 32 + 0 \times 16$$
$$+ 1 \times 8 + 0 \times 4 + 1 \times 2 + 1 \times 1$$
$$= 139$$

29.
1	0	1	0	0	1
	double	dabble	double	double	dabble
	= 2	= 5	= 10	= 20	= 41

$$101001_{two} = 41$$

31.
1	0	1	1	0	1	0
	double	dabble	dabble	double	dabble	double
	= 2	= 5	= 11	= 22	= 45	= 90

$$1011010_{two} = 90$$

33.
1	0	1	0	0	1
	double	dabble	double	double	dabble
	= 2	= 5	= 10	= 20	= 41

1	1	0	1	0
dabble	dabble	double	dabble	double
= 83	= 167	= 334	= 669	= 1338

$$10100111010_{two} = 1338$$

35. First, convert the number to base ten.
$$34_{six} = 3 \times 6^1 + 4 \times 6^0$$
$$= 3 \times 6 + 4 \times 1$$
$$= 22$$

Now, change 22 into base eight.
```
8| 22
    2   6
```
$$34_{six} = 26_{eight}$$

37. First, convert the number to base ten.
$$878_{nine} = 8 \times 9^2 + 7 \times 9^1 + 8 \times 9^0$$
$$= 8 \times 81 + 7 \times 9 + 8 \times 1$$
$$= 719$$

Now, change 719 into base four.
```
4| 719
4| 179   3
4| 44    3
4| 11    0
    2    3
```
$$878_{nine} = 23033_{four}$$

39. First, convert the number to base ten.
$$1110_{two} = 1 \times 2^3 + 1 \times 2^2 + 1 \times 2^1 + 0 \times 2^0$$
$$= 14$$

Now, change 14 into base five.
```
5| 14
    2   4
```
$$1110_{two} = 24_{five}$$

41. First, convert the number to base ten.
$$3440_{eight} = 3 \times 8^3 + 4 \times 8^2 + 4 \times 8^1 + 0 \times 8^0$$
$$= 3 \times 512 + 4 \times 64 + 4 \times 8 + 0 \times 1$$
$$= 1824$$

Now, change 1824 into base nine.
```
9| 1824
9| 202   6
9| 22    4
    2    4
```
$$3440_{eight} = 2446_{nine}$$

43. First, convert the number to base ten.
$$56_{sixteen} = 5 \times 16^1 + 6 \times 16^0$$
$$= 5 \times 16 + 6 \times 1$$
$$= 86$$

Now change 86 into base eight.
```
8| 86
8| 10    6
    1    2
```
$$56_{sixteen} = 126_{eight}$$

45. First, convert the number to base ten.
$$A4_{twelve} = 10 \times 12^1 + 4 \times 12^0$$
$$= 10 \times 12 + 4 \times 1$$
$$= 124$$

Now, change 124 into base sixteen.
```
16| 124
    7    12 (C in base sixteen)
```
$$A4_{twelve} = 7C_{sixteen}$$

47. Use the tables given to write the 3-digit binary representation of each digit.

3	5	2	(base eight)
011	101	010	(base two)

$352_{eight} = 11101010_{two}$

49. Starting at the right, divide the numeral into groups of 3 digits, adding zeroes as needed. Replace each 3-digit group with its base eight symbol.

011	001	010	(base two)
3	1	2	(base eight)

$11001010_{two} = 312_{eight}$

51. Starting at the right, divide into groups of 4 digits, adding zeroes in front if needed. Replace each 4-digit group with its base sixteen symbol. You may write base ten first, if that is easier.

0001	0101	0001	(base two)
1	5	1	(base ten)
1	5	1	(base sixteen)

$101010001_{two} = 151_{sixteen}$

53. Use the tables given to write the 4-digit binary representation of each digit.

B	E	F	3	(base sixteen)
11	14	15	3	(base ten)
1011	1110	1111	0011	(base two)

$BEF3_{sixteen} = 1011111011110011_{two}$

55. Use the tables given to write the 4-digit binary representation of each digit or change to base ten digits and replace each one with a 4-digit binary equivalent.

B	A	5	C	F	(base 16)
11	10	5	12	15	(base 10)
1011	1010	0101	1100	1111	(base 2)

$BA5CF_{sixteen} = 10111010010111001111_{two}$

57. Answers will vary. Start at the left with the first number that is not a zero and move to the right. Every time you pass by a zero, triple your current number. Every time you pass by a 1, whipple. Whippling is accomplished by tripling your current number and adding 1. Every time you pass by a 2, zipple. Zippling is accomplished by tripling your current number and adding 2.

59. Converting:

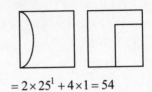

$= 2 \times 25^1 + 4 \times 1 = 54$

61. The numeral for 15 is the numeral for 3 rotated $90°$ counterclockwise.

63. To represent 8, overlay the symbol for 5 on the symbol for 3.

65. The first symbol is the symbol for 10, since it is a 2 rotated. It is on the left so it is to be multiplied by 25. The second symbol is the symbol for 6, since it is the overlay of 5 and 1. The number is 256.

67. Answers will vary.

EXERCISE SET 4.4

1. Adding:

$$\begin{array}{ccc} & 1 & \\ 2 & 0 & 4_{five} \\ + 1 & 2 & 3_{five} \\ \hline 3 & 3 & 2_{five} \end{array}$$

3. Adding:

$$\begin{array}{cccc} 1 & & 1 & \\ 5 & 6 & 2 & 5_{seven} \\ + & 6 & 3 & 4_{seven} \\ \hline 6 & 5 & 6 & 2_{seven} \end{array}$$

5. Adding:

$$\begin{array}{ccccccc} 1 & 1 & & 1 & 1 & 1 & \\ & 1 & 1 & 0 & 1 & 0 & 1_{\text{two}} \\ + & & 1 & 0 & 0 & 1 & 1_{\text{two}} \\ \hline 1 & 0 & 0 & 1 & 0 & 0 & 0_{\text{two}} \end{array}$$

7. Adding:

$$\begin{array}{ccccc} & 1 & 1 & 1 & \\ & & 8 & B & 5_{\text{twelve}} \\ + & & 5 & 7 & 8_{\text{twelve}} \\ \hline & 1 & 2 & 7 & 1_{\text{twelve}} \end{array}$$

9. Adding:

$$\begin{array}{ccccc} & 1 & 1 & 1 & \\ & C & 4 & 8 & 9_{\text{sixteen}} \\ + & & B & A & D_{\text{sixteen}} \\ \hline & D & 0 & 3 & 6_{\text{sixteen}} \end{array}$$

11. Adding:

$$\begin{array}{ccccc} 1 & 1 & 1 & & \\ & 4 & 3 & 5_{\text{six}} \\ + & 2 & 4 & 5_{\text{six}} \\ \hline 1 & 1 & 2 & 4_{\text{six}} \end{array}$$

13. Borrow 5 fives from the twenty-fives column.

$$\begin{array}{cccc} & & 3 & 13 \\ & 4 & 3 & 4_{\text{five}} \\ - & 1 & 4 & 3_{\text{five}} \\ \hline & 2 & 4 & 1_{\text{five}} \end{array}$$

15. Borrow 8 sixty-fours and 8 eights.

$$\begin{array}{cccc} & 12 & 12 & \\ 6 & 13 & & \\ 7 & 3 & 2 & 5_{\text{eight}} \\ - & 5 & 6 & 3_{\text{eight}} \\ \hline 6 & 5 & 4 & 2_{\text{eight}} \end{array}$$

17. Borrow 2 ones from the twos column. Borrow 2 twos by first borrowing 2 fours from the eights column. Then borrow 2 eights from the sixteens column.

$$\begin{array}{ccccc} & 10 & 1 & & \\ 0 & 0 & 10 & 10 & 10 \\ 1 & 1 & 0 & 1 & 0_{\text{two}} \\ - & 1 & 0 & 1 & 1_{\text{two}} \\ \hline & 1 & 1 & 1 & 1_{\text{two}} \end{array}$$

19. Borrow 2 ones from the twos column by first borrowing 2 twos from the fours column. Then borrow 2 eights from the sixteens column. Next borrow 2 sixteens from the thirty-twos column by first borrowing 2 thirty-twos from the sixty-fours column. Finally, borrow 2 sixty-fours from the one hundred twenty-eights column.

$$\begin{array}{cccccccc} & 10 & 1 & 10 & & & 1 & \\ 0 & 0 & 10 & 0 & 10 & 0 & 10 & 10 \\ 1 & 1 & 0 & 1 & 0 & 1 & 0 & 0_{\text{two}} \\ - & 1 & 0 & 1 & 1 & 0 & 1 & 1_{\text{two}} \\ \hline & 1 & 1 & 1 & 1 & 0 & 0 & 1_{\text{two}} \end{array}$$

21. Borrow 12 ones from the twelves column for a total of 19 or 17_{twelve} ones.

$$\begin{array}{cccc} & & 9 & 17 \\ 4 & 3 & A & 7_{\text{twelve}} \\ - & 2 & 8 & 9_{\text{twelve}} \\ \hline 4 & 1 & 1 & A_{\text{twelve}} \end{array}$$

So, $43A7_{\text{twelve}} - 289_{\text{twelve}} = 411A_{\text{twelve}}$

23. Borrow 9 ones from the nines column and 9 nines from the eighty-ones column.

$$\begin{array}{cccc} & & 15 & 12 \\ & 6 & 16 & \\ & 7 & 6 & 2_{\text{nine}} \\ - & 3 & 6 & 7_{\text{nine}} \\ \hline & 3 & 8 & 4_{\text{nine}} \end{array}$$

25. $3 \times 5 = 15 = 23_{\text{six}}$

$3 \times 4 + 2 = 14 = 22_{\text{six}}$

$$\begin{array}{cccc} & 2 & 2 & \\ & 1 & 4 & 5_{\text{six}} \\ \times & & & 3_{\text{six}} \\ \hline & 5 & 2 & 3_{\text{six}} \end{array}$$

27. $2 \times 2 = 4 = 11_{three}$

    $2 \times 1 + 1 = 3 = 10_{three}$

    $2 \times 2 + 1 = 5 = 12_{three}$

1	1	1	
	2	1	2_{three}
$\times$			2_{three}
1	2	0	1_{three}

29. $5 \times 4 = 20 = 24_{eight}$

    $5 \times 5 + 2 = 27 = 33_{eight}$

    $5 \times 7 + 2 = 37 = 45_{eight}$

	2	3	2	
	7	3	5	4_{eight}
$\times$				5_{eight}
4	5	2	3	4_{eight}

31. Multiplying:

1	0	1	0	1	0_{two}
			$\times$	1	0_{two}
0	0	0	0	0	0
1	0	1	0	1	0
1	0	1	0	1	0_{two}

33. $5 \times 3 = 15 = 17_{eight}$

    $5 \times 5 + 1 = 26 = 32_{eight}$

    $5 \times 4 + 3 = 23 = 27_{eight}$

    $2 \times 5 = 10 = 12_{eight}$

    $2 \times 4 + 1 = 9 = 11_{eight}$

	1	1		
		4	5	3_{eight}
$\times$			2	5_{eight}
	2	7	2	7
1	1	2	6	
1	4	2	0	7_{eight}

35. $2 \times 3 = 6 = 12_{four}$

    $2 \times 2 + 1 = 5 = 11_{four}$

    $2 \times 3 + 1 = 7 = 13_{four}$

    $3 \times 3 = 9 = 21_{four}$

    $3 \times 2 + 2 = 8 = 20_{four}$

    $3 \times 3 + 2 = 11 = 23_{four}$

    $3 \times 1 + 2 = 5 = 11_{four}$

		2	2	2	
		1	3	2	3_{four}
$\times$			1	3	2_{four}
		3	3	1	2
1	1	3	0	1	
1	3	2	3		
3	2	1	2	2	2_{four}

37. $5_{sixteen} \times D_{sixteen} = 5 \times 13 = 65 = 41_{sixteen}$

    $5_{sixteen} \times A_{sixteen} + 4_{sixteen} = 5 \times 10 + 4 = 54$
    $$= 36_{sixteen}$$

    $5_{sixteen} \times B_{sixteen} + 3_{sixteen} = 5 \times 11 + 3 = 58$
    $$= 3A_{sixteen}$$

	3	4	
B	A	$D_{sixteen}$	
$\times$		$5_{sixteen}$	
3	A	6	$1_{sixteen}$

39. Dividing:

    $$2_{four} \overline{)1 \ 3 \ 2_{four}} \quad 3 \ 3_{four}$$

	3	3	$_{four}$
2_{four})	1 3	2	$_{four}$
	1 2		
	1	2	
	1	2	
		0	

41. Dividing:

	3	3	$_{four}$
3_{four})	2 3	1	$_{four}$
	2 1		
	2	1	
	2	1	
		0	

43. Dividing:

$$
\begin{array}{r}
1\ 2\ 2\ 3_{\text{six}} \\
4_{\text{six}}\overline{)5\ 3\ 4\ 1_{\text{six}}} \\
\underline{4} \\
1\ 3 \\
\underline{1\ 2} \\
1\ 4 \\
\underline{1\ 2} \\
2\ 1 \\
\underline{2\ 0} \\
1 \quad \text{R }1_{\text{six}}
\end{array}
$$

45. Dividing:

$$
\begin{array}{r}
1\ 1\ 1\ 0_{\text{two}} \\
11_{\text{two}}\overline{)1\ 0\ 1\ 0\ 1\ 0_{\text{two}}} \\
\underline{1\ 1} \\
1\ 0\ 0 \\
\underline{1\ 1} \\
1\ 1 \\
\underline{1\ 1} \\
0
\end{array}
$$

47. Dividing:

$$
\begin{array}{r}
A\ 8_{\text{twelve}} \\
5_{\text{twelve}}\overline{)4\ 5\ 7_{\text{twelve}}} \\
\underline{4\ 2} \\
3\ 7 \\
\underline{3\ 4} \\
3 \quad \text{R }3_{\text{twelve}}
\end{array}
$$

49. Dividing:

$$
\begin{array}{r}
1\ 4_{\text{five}} \\
12_{\text{five}}\overline{)2\ 3\ 4_{\text{five}}} \\
\underline{1\ 2} \\
1\ 1\ 4 \\
\underline{1\ 0\ 3} \\
1\ 1 \quad \text{R }11_{\text{five}}
\end{array}
$$

51. Expand both numerals.

$$143_x = 1 \cdot x^2 + 4 \cdot x^1 + 3 \cdot x^0 = x^2 + 4x + 3$$

$$10200_{\text{three}} = 1 \cdot 3^4 + 0 \cdot 3^3 + 2 \cdot 3^2 + 0 \cdot 3^1 + 0 \cdot 3^0$$
$$= 99$$

Set them equal to each other and solve the quadratic equation.

$$x^2 + 4x + 3 = 99$$
$$x^2 + 4x - 96 = 0$$
$$(x+12)(x-8) = 0$$
$$x = -12,\ x = 8$$

Since the base must be positive, $x = 8$.

53. a. $384 + 245 = 629$

  b. Convert 384 and 245 to base two.

$$
\begin{array}{ll}
2\underline{|\ 384} & 2\underline{|\ 245} \\
2\underline{|\ 192}\ \ 0 & 2\underline{|\ 122}\ \ 1 \\
2\underline{|\ 96}\ \ \ 0 & 2\underline{|\ 61}\ \ \ 0 \\
2\underline{|\ 48}\ \ \ 0 & 2\underline{|\ 30}\ \ \ 1 \\
2\underline{|\ 24}\ \ \ 0 & 2\underline{|\ 15}\ \ \ 0 \\
2\underline{|\ 12}\ \ \ 0 & 2\underline{|\ 7}\ \ \ \ 1 \\
2\underline{|\ 6}\ \ \ \ 0 & 2\underline{|\ 3}\ \ \ \ 1 \\
2\underline{|\ 3}\ \ \ \ 0 & \quad\ \ 1\ \ \ \ 1 \\
\quad\ \ 1\ \ \ \ 1 &
\end{array}
$$

$384 = 110000000_{\text{two}}$

$245 = 11110101_{\text{two}}$

  c. Add the results in base two.

$$
\begin{array}{r}
1 \\
1\ 1\ 0\ 0\ 0\ 0\ 0\ 0\ 0_{\text{two}} \\
1\ 1\ 1\ 1\ 0\ 1\ 0\ 1_{\text{two}} \\
\hline
1\ 0\ 0\ 1\ 1\ 1\ 0\ 1\ 0\ 1_{\text{two}}
\end{array}
$$

  d. Convert the answer from part c. to base ten.

$$1001110101_{\text{two}} = 1 \cdot 2^9 + 0 \cdot 2^8 + 0 \cdot 2^7 + 1 \cdot 2^6 + 1 \cdot 2^5 +$$
$$1 \cdot 2^4 + 0 \cdot 2^3 + 1 \cdot 2^2 + 0 \cdot 2^1 + 1 \cdot 2^0$$
$$= 629$$

  e. The answers are the same.

55. a. $247 \times 26 = 6422$

    b.   Convert 247 and 26 to base two.

    ```
    2| 247
    2| 123    1        2| 26
    2| 61     1        2| 13    0
    2| 530    1        2| 6     1
    2| 15     0        2| 3     0
    2| 7      1           1     1
    2| 3      1
       1      1
    ```

    $247 = 11110111_{two}$

    $26 = 11010_{two}$

    c.   Multiply the results in base two.

    ```
            1  1  1  1  0  1  1  1 two
                        1  1  0  1  0 two
    ─────────────────────────────────────
            0  0  0  0  0  0  0  0  0
         1  1  1  1  0  1  1  1
      0  0  0  0  0  0  0  0
   1  1  1  1  0  1  1  1
1  1  1  1  0  1  1  1
    ─────────────────────────────────────
 1  1  0  0  1  0  0  0  1  0  1  1  0 two
    ```

    d.   Convert the answer from part c to base ten.

    $$1100100010110_{two} = 1 \cdot 2^{12} + 1 \cdot 2^{11} + 0 \cdot 2^{10}$$
    $$+ 0 \cdot 2^9 + 1 \cdot 2^8 + 0 \cdot 2^7$$
    $$+ 0 \cdot 2^6 + 0 \cdot 2^5 + 1 \cdot 2^4$$
    $$+ 0 \cdot 2^3 + 1 \cdot 2^2 + 1 \cdot 2^1$$
    $$+ 0 \cdot 2^0$$
    $$= 6422$$

    e.   The answers are the same.

57. The product of the ones digits is $4 \times 4 = 16$.
    Since the answer has a ones digit of 2, the base is
    either 14 or 7. Checking both shows that the
    base is 7.

59. In the base one numeration system, 0 would be
    the only numeral and the place values would be
    $1^0, 1^1, 1^2, 1^3, \ldots$, each of which equals 1. Thus 0
    is the only number you could write using a base
    one numeration system.

61. M = 1, A = 4, S = 3, and O = 0

EXERCISE SET 4.5

1. $20 = 1 \times 20 = 2 \times 10 = 4 \times 5$
       The divisors are 1, 2, 4, 5, 10, and 20.

3. $65 = 1 \times 65 = 5 \times 13$
       The divisors are 1, 5, 13, and 65.

5. $41 = 1 \times 41$. The divisors are 1 and 41.

7. $110 = 1 \times 110 = 2 \times 55 = 5 \times 22 = 10 \times 11$
       The divisors are 1, 2, 5, 10, 11, 22, 55, 110.

9. $385 = 1 \times 385 = 5 \times 77 = 7 \times 55 = 11 \times 35$
       The divisors are 1, 5, 7, 11, 35, 55, 77, 385.

11.   The divisors of 21 are 1, 3, 7, and 21. Thus 21 is a
       composite number.

13.   The only divisors of 37 are 1 and 37. Thus 37 is
       a prime number.

15.   The only divisors of 101 are 1 and 101. Thus
       101 is a prime number.

17.   The only divisors of 79 are 1 and 79. Thus 79 is
       a prime number.

19.   The divisors of 203 are 1, 7, 29, and 203. Thus
       203 is a composite number.

21.   2: Yes. 210 is even.
       3: Yes. $2 + 1 + 0 = 3$, which is divisible by 3.
       4: No. 10 is not divisible by 4.
       5: Yes. 210 ends in a 0.
       6: Yes. 210 is divisible by both 2 and 3.
       8: No. 210 is not divisible by 8.
       9: No. $2 + 1 + 0 = 3$ and 3, which is not
          divisible by 9.
       10: Yes. 210 ends in a 0.

23.   2: No. 51 is not even
       3: Yes. $5 + 1 = 6$ and 6 is divisible by 3.
       4: No. 51 is not divisible by 4.
       5: No. 51 does not end in a 0 or 5.
       6: No. 51 is not divisible by both 2 and 3.
       8: No. 51 is not divisible by 8
       9: No. $5 + 1 = 6$ and 6 is not divisible by 9.
       10: No, because it does not end in a 0.

25. 2: Yes. 2568 is even.
    3: Yes. $2 + 5 + 6 + 8 = 21$, which is divisible by 3.
    4 : Yes. 68 is divisible by 4.
    5: No. 2568 does not end in a 0 or 5.
    6: Yes. 2568 is divisible by both 2 and 3.
    8: Yes. 568 is divisible by 8.
    9: No. $2 + 5 + 6 + 8 = 21$, which is not divisible by 9.
    10: No. 2568 does not end in a 0.

27. 2: Yes. 4190 is even.
    3: No. $4 + 1 + 9 + 0 = 14$, which is not divisible by 3.
    4: No. 90 is not divisible by 4.
    5: Yes. 4190 ends in a 0.
    6: No. 4190 is not divisible by both 2 and 3.
    8: No. 190 is not divisible by 8.
    9: No. $4 + 1 + 9 + 0 = 14$ and 14 is not divisible by 9.
    10: Yes. 4190 ends in a 0.

29. Factoring:
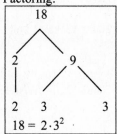
$18 = 2 \cdot 3^2$

31. Factoring:
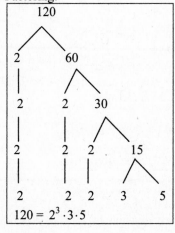
$120 = 2^3 \cdot 3 \cdot 5$

33. Factoring:
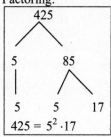
$425 = 5^2 \cdot 17$

35. Factoring:
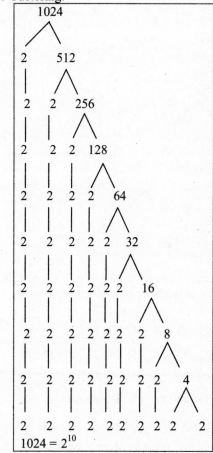
$1024 = 2^{10}$

37. Factoring:

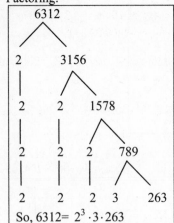

So, 6312 = $2^3 \cdot 3 \cdot 263$

39. Factoring:

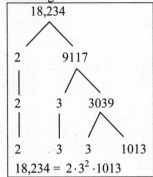

18,234 = $2 \cdot 3^2 \cdot 1013$

41. Cross out all multiples of 2, 3, 5, 7, 11, and 13. The rest of the numbers (bold) are prime.

2 **3** 4 **5** 6 **7** 8 9 10

11 12 **13** 14 15 16 **17** 18 **19** 20

21 22 **23** 24 25 26 27 28 **29** 30

31 32 33 34 35 36 **37** 38 39 40

41 42 **43** 44 45 46 **47** 48 49 50

51 52 **53** 54 55 56 57 58 **59** 60

61 62 63 64 65 66 **67** 68 69 70

71 72 **73** 74 75 76 77 78 **79** 80

81 82 **83** 84 85 86 87 88 **89** 90

91 92 93 94 95 96 **97** 98 99 100

101 102 **103** 104 105 106 **107** 108 **109** 110

111 112 **113** 114 115 116 117 118 119 120

121 122 123 124 125 126 **127** 128 129 130

131 132 133 134 135 136 **137** 138 **139** 140

141 142 143 144 145 146 147 148 **149** 150

151 152 153 154 155 156 **157** 158 159 160

161 162 **163** 164 165 166 **167** 168 169 170

171 172 **173** 174 175 176 177 178 **179** 180

181 182 183 184 185 186 187 188 189 190

191 192 **193** 194 195 196 **197** 198 **199** 200

43. 3 and 5, 5 and 7, 11 and 13, 17 and 19, 29 and 31, 41 and 43, 59 and 61, 71 and 73, 101 and 103, 107 and 109, 137 and 139, 149 and 151, 179 and 181, 191 and 193, 197 and 199.

45. 311 and 313, or 347 and 349.

47. a.   24 = 5 + 19 also 24 = 11 + 13

    b.   50 = 19 + 31 also 50 = 3 + 47

    c.   86 = 3 + 83 also 86 = 7 + 79 and others

    d.   144 = 5 + 139 also 144 = 71 + 73

    e.   210 = 11 + 199 also 210 = 103 + 107

    f.   264 = 7 + 257 also 264 = 13 + 251 and others

49. The double of the ones digit is 4. 18 − 4 = 14, which is divisible by 7. Thus 182 is divisible by 7.

51. The double of the ones digit is 2. 100 − 2 = 98, which is divisible by 7. Thus 1001 is divisible by 7.

53. The double of the ones digit is 2.
1156 − 2 = 1154
The double of the ones digit is 8.
115 − 8 = 107, which is not divisible by 7. Thus 11,561 is not divisible by 7.

55. The double of the ones digit is 12
20,431 − 12 = 20,419.
The double of the ones digit is 18.
2041 − 18 = 2023
The double of the ones digit is 6. 202 − 6 = 196.
The double of the ones digit is 12. 19 − 12 = 7, which is clearly divisible by 7. Thus 204,316 is divisible by 7.

57. Four times the ones digit is 4. 9 + 4 = 13, which is divisible by 13. Thus 91 is divisible by 13.

59. Four times the ones digit is 20. $20 + 188 = 208$.
Four times the ones digit is 32. $32 + 20 = 52$, which is divisible by 13. Thus 1885 is divisible by 13.

61. Four times the ones digit is 28.
$28 + 1450 = 1478$.
Four times the ones digit is 32. $32 + 147 = 179$.
Four times the ones digit is 36. $36 + 17 = 53$, which is not divisible by 13 Thus 14,507 is not divisible by 13.

63. Four times the ones digit is 4. $4 + 1335 = 1339$.
Four times the ones digit is 36. $36 + 133 = 169$, which is divisible by 13. Thus 13,351 is divisible by 13.

65. a. If $n = 1$, the two values are 2 and 0. If $n = 2$, the values are 3 and 1. If $n = 3$, the values are 7 and 5. Thus $n = 3$ is the smallest value.

    b. The first few values are given in part a). If $n = 4$, the values are 25 and 23. Thus $n = 4$ is the smallest value.

67. To determine whether a number is divisible by 17, multiply the ones digit of the given number by 5. Find the difference between this result and the number formed by omitting the ones digit from the given number. Keep repeating this procedure until you obtain a small final difference. If the final difference is divisible by 17, then the given number is divisible by 17. If the final difference is not divisible by 17, then the given number is not divisible by 17.

69. $60 = 6 \cdot 10 = 3 \cdot 2 \cdot 2 \cdot 5 = 2^2 \cdot 3^1 \cdot 5^1$
Adding one to each exponent gives the numbers 3, 2 and 2. Their product is $3 \cdot 2 \cdot 2 = 12$. There are 12 distinct divisors of 60.

71. $297 = 3^3 \cdot 11$
Adding one to each exponent gives 4 and 2. Their product is 8. There are 8 distinct divisors of 297.

73. $360 = 36 \cdot 10 = 2 \cdot 3 \cdot 2 \cdot 3 \cdot 2 \cdot 5 = 2^3 \cdot 3^2 \cdot 5^1$
Adding one to each exponent gives the numbers 4, 3 and 2. Their product is $4 \cdot 3 \cdot 2 = 24$. There are 24 distinct divisors of 360.

EXERCISE SET 4.6

1. The proper factors of 18 are 1,2,3,6,9. Their sum is $21 > 18$, so 18 is abundant.

3. The proper factors of 91 are 1,7,13. Their sum is $21 < 91$, so 91 is deficient.

5. The only proper factor of 19 is 1. Since $1 < 19$, 19 is deficient.

7. The proper factors of 204 are 1,2,3,4,6,12, 17,34,68,51,102. Their sum is $300 > 204$, so 204 is abundant.

9. The proper factors of 610 are 1,2,5,10,61,122,305. Their sum is $506 < 610$, so 610 is deficient.

11. The proper factors of 291 are 1,3,97. Their sum is $101 < 291$, so 291 is deficient.

13. The proper factors of 176 are 1, 2, 4, 8, 11, 16, 22, 44, 88. Their sum is $196 > 176$, so 176 is abundant.

15. The proper factors of 260 are 1, 2, 4, 5, 10, 13, 20, 26, 52, 65, 130. Their sum $328 > 260$, so 260 is abundant.

17. $2^3 - 1 = 8 - 1 = 7$ which is prime.

19. $2^7 - 1 = 128 - 1 = 127$ which is prime.

21. Euclid's theorem says that for n prime and a prime Mersenne number $2^n - 1$, $2^{n-1}(2^n - 1)$ is a perfect number. Letting $n = 127$ yields the perfect number $2^{127-1}(2^{127} - 1) = 2^{126}(2^{127} - 1)$.

23. The number of digits in the number, b^x is the greatest integer of $(x \log b) + 1$.
We have $b = 2$ and $x = 17$.
$(x \log b) + 1 = (17 \log 2) + 1 \approx 6.117$
The greatest integer is 6, so the number of digits in $2^{17} - 1$ is 6.

25. The number of digits in the number, b^x is the greatest integer of $(x \log b) + 1$.

We have $b = 2$ and $x = 1,398,269$.
$(x \log b) + 1 = (1,398,269 \log 2) + 1 \approx 420921.91$
The greatest integer is 420,921, so the number of digits in $2^{1398269} - 1$ is 420,921.

27. The number of digits in the number, b^x is the greatest integer of $(x \log b) + 1$.
We have $b = 2$ and $x = 6,972,593$.
$(x \log b) + 1 = (6,972,593 \log 2) + 1 \approx$
2,098,960.6
The greatest integer is 2,098,960, so the number of digits in $2^{6972593} - 1$ is 2,098,960.

29. Substituting, we have $9^5 + 15^5 = z^5$
Evaluating, we have
$59,049 + 759,375 = 818,424 = z^5$
However, the fifth root of $818,424$ is approximately 15.226, which is not a natural number.
So, $x = 9$, $y = 15$ and $n = 5$ do not yield a solution to the equation, $x^n + y^n = z^n$

31. a. False. For instance, if $n = 11$, then
$2^{11} - 1 = 2047 = 23 \cdot 89$

b. False. Fermat's Last Theorem was the last of Fermat's theories (conjectures) that other mathematicians were able to establish.

c. True.

d. Conjecture

33. a. Fermat's Little Theorem states $a^n - a$ is divisible by n if n is prime.
Substituting gives $12^7 - 12 = 35,831,796$, which is divisible by 7.

b. Fermat's Little Theorem states $a^n - a$ is divisible by n if n is prime.
Substituting gives $8^{11} - 8 = 8,589,934,584$, which is divisible by 11.

35. $8128 = 1^3 + 3^3 + 5^3 + \ldots + 13^3 + 15^3$

37. a. The divisors of 6 are 1, 2, 3, 6.
$\frac{1}{1} + \frac{1}{2} + \frac{1}{3} + \frac{1}{6} = \frac{6}{6} + \frac{3}{6} + \frac{2}{6} + \frac{1}{6} = \frac{12}{6} = 2$

b. The divisors of 28 are 1, 2, 4, 7, 14, 28.
$\frac{1}{1} + \frac{1}{2} + \frac{1}{4} + \frac{1}{7} + \frac{1}{14} + \frac{1}{28} =$
$\frac{28}{28} + \frac{14}{28} + \frac{7}{28} + \frac{4}{28} + \frac{2}{28} + \frac{1}{28} = \frac{56}{28} = 2$

39. The first 5 Fermat numbers corresponding formed using $m = 0, 1, 2, 3$, and 4 are all primes. In 1732, Euler discovered that the sixth Fermat number 4,294,967,297, formed using $m = 5$, is not a prime number because it is divisible by 641.

CHAPTER 4 REVIEW EXERCISES

1. $4,506,325 = 4,000,000 + 500,000 + 6000 + 300 + 20 + 5$

2. $3,124,043 = 3,000,000 + 100,000 + 20,000 + 4000 + 40 + 3$

3. $2 \times 100,000 + 3 \times 1000 + 2 \times 10,000 + 1 \times 10 + 3 = 223,013$

4. $2 \times 100,000 + 3 \times 100 + 1 \times 1000 + 4 + 5 \times 10 + 2 \times 10,000 = 221,354$

5. CCCXLIX $= 100 + 100 + 100 + (50 - 10) + (10 - 1) = 349$

6. DCCLXXIV $= 500 + 100 + 100 + 50 + 10 + 10 + (5 - 1) = 774$

7. Rewriting:
$\overline{\text{IX}}\text{DCXL} = (10,000 - 1000) + 500$
$\qquad + 100 + (50 - 10)$
$\qquad = 9640$

8. Rewriting:
$\overline{\text{XCII}}\text{CDXLIV} = (100,000 - 10,000)$
$\qquad + 2000 + (500 - 100)$
$\qquad + (50 - 10) + (5 - 1)$
$\qquad = 92,444$

9. $567 = 500 + 60 + 7 = \text{DLXVII}$

10. $823 = 800 + 20 + 3 = \text{DCCCXXIII}$

11. $2489 = 2000 + 400 + 80 + 9 = \text{MMCDLXXXIX}$

12. $1335 = 1000 + 300 + 30 + 5 = \text{MCCCXXXV}$

13. $432 = 400 + 30 + 2 = (4 \times 10^2) + (3 \times 10^1) + (2 \times 10^0)$

14. $456{,}327 = 400{,}000 + 50{,}000 + 6000 + 300 + 20 + 7 = (4 \times 10^5) + (5 \times 10^4) + (6 \times 10^3) + (3 \times 10^2) + (2 \times 10^1) + (7 \times 10^0)$

15. $5{,}000{,}000 + 30{,}000 + 8000 + 200 + 4 = 5{,}038{,}204$

16. $300{,}000 + 80{,}000 + 7000 + 900 + 60 = 387{,}960$

17. $(13 \times 60) + (21 \times 1) = 801$

18. $(26 \times 60) + (43 \times 1) = 1603$

19. $(21 \times 60^2) + (14 \times 60) + (1 \times 1) = 76{,}441$

20. $(24 \times 60^2) + (16 \times 60) + (33 \times 1) = 87{,}393$

21. $721 = (12 \times 60) + (1 \times 1)$

22. $1080 = (18 \times 60) + (0 \times 1)$

23. $12{,}543 = (3 \times 3600) + (29 \times 60) + (3 \times 1)$

24. $19{,}281 = (5 \times 3600) + (21 \times 60) + (21 \times 1)$

25. $(9 \times 20) + (14 \times 1) = 194$

26. $(13 \times 20) + (7 \times 1) = 267$

27. $(6 \times 360) + (0 \times 20) + (18 \times 1) = 2178$

28. $(18 \times 360) + (5 \times 20) + (0 \times 1) = 6580$

29. $522 = (1 \times 360) + (8 \times 20) + (2 \times 1)$

30. $346 = (17 \times 20) + (6 \times 1)$

31. $1862 = (5 \times 360) + (3 \times 20) + (2 \times 1)$

32. $1987 = (5 \times 360) + (9 \times 20) + (7 \times 1)$

33. $45_{\text{six}} = (4 \times 6^1) + (5 \times 6^0) = 29$

34. $172_{\text{nine}} = (1 \times 9^2) + (7 \times 9^1) + (2 \times 9^0) = 146$

35. $\text{E3}_{\text{sixteen}} = (14 \times 16^1) + (3 \times 16^0) = 227$

36. $1\text{BA}_{\text{twelve}} = (1 \times 12^2) + (11 \times 12^1) + (10 \times 12^0) = 286$

37. $346_{\text{nine}} = (3 \times 9^2) + (4 \times 9^1) + (6 \times 9^0) = 285 = (1 \times 6^3) + (1 \times 6^2) + (5 \times 6^1) + (3 \times 6^0) = 1153_{\text{six}}$

38. $1532_{\text{six}} = (1 \times 6^3) + (5 \times 6^2) + (3 \times 6^1) + (2 \times 6^0) = 416 = (6 \times 8^2) + (4 \times 8^1) + (0 \times 8^0) = 640_{\text{eight}}$

39. $275_{twelve} = (2 \times 12^2) + (7 \times 12^1) + (5 \times 12^0) =$
    $377 = (4 \times 9^2) + (5 \times 9^1) + (8 \times 9^0) = 458_{nine}$

40. $67A_{sixteen} = (6 \times 16^2) + (7 \times 16^1) + (10 \times 16^0) =$
    $1658 = (11 \times 12^2) + (6 \times 12^1) + (2 \times 12^0) =$
    $B62_{twelve}$

41. 011   100   (base 2)
     3   4   (base 8)
    $11100_{two} = 34_{eight}$

42. 001   010   100   (base 2)
     1   2   4   (base 8)
    $1010100_{two} = 124_{eight}$

43. 0011   1000   1101   (base 2)
     3   8   D   (base 16)
    $1110001101_{two} = 38D_{sixteen}$

44. 0111   0101   0100   (base 2)
     7   5   4   (base 16)
    $11101010100_{two} = 754_{sixteen}$

45. 2   5   (base 8)
    010   101   (base 2)
    $25_{eight} = 10101_{two}$

46. 1   4   7   2   (base 8)
    001   100   111   010   (base 2)
    $1472_{eight} = 1100111010_{two}$

47. 4   A   (base 16)
    0100   1010   (base 2)
    $4A_{sixteen} = 1001010_{two}$

48. C   7   2   (base 16)
    1100   0111   0010   (base 2)
    $C72_{sixteen} = 110001110010_{two}$

49. Adding:

$$
\begin{array}{r}
{\scriptstyle 1 \quad 1} \\
2 \quad 3 \quad 5_{six} \\
+ \quad\quad 1 \quad 4 \quad 4_{six} \\
\hline
4 \quad 2 \quad 3_{six}
\end{array}
$$

50. Adding:

$$
\begin{array}{r}
{\scriptstyle 1 \quad 1} \\
6 \quad 7 \quad 3_{eight} \\
+ \quad\quad 3 \quad 4 \quad 5_{eight} \\
\hline
1 \quad 2 \quad 4 \quad 0_{eight}
\end{array}
$$

51. Subtracting:

$$
\begin{array}{r}
{\scriptstyle 6 \quad 12} \\
6 \quad 7 \quad 2_{nine} \\
- \quad\quad 1 \quad 3 \quad 5_{nine} \\
\hline
5 \quad 3 \quad 6_{nine}
\end{array}
$$

52. Subtracting:

$$
\begin{array}{r}
{\scriptstyle 2 \quad 12} \\
1 \quad 3 \quad 3 \quad 2_{four} \\
- \quad\quad\quad 2 \quad 1 \quad 3_{four} \\
\hline
1 \quad 1 \quad 1 \quad 3_{four}
\end{array}
$$

53. Multiplying:

$$
\begin{array}{r}
{\scriptstyle 2 \quad 1} \\
5 \quad 4 \quad 2_{eight} \\
\times \quad\quad\quad 2 \quad 5_{eight} \\
\hline
3 \quad 3 \quad 5 \quad 2 \\
1 \quad 3 \quad 0 \quad 4 \\
\hline
1 \quad 6 \quad 4 \quad 1 \quad 2_{eight}
\end{array}
$$

54. Multiplying:

$$
\begin{array}{r}
{\scriptstyle 2 \quad 1} \\
3 \quad 4 \quad 2 \quad 1_{five} \\
\times \quad\quad\quad\quad 4 \quad 3_{five} \\
\hline
2 \quad 1 \quad 3 \quad 1 \quad 3 \\
3 \quad 0 \quad 2 \quad 3 \quad 4 \\
\hline
3 \quad 2 \quad 4 \quad 2 \quad 0 \quad 3_{five}
\end{array}
$$

55. Dividing:

$$
\begin{array}{r}
1\,1\,1\,0\,0_{two} \\
11_{two}\,\overline{)\,1\,0\,1\,0\,1\,0\,1_{two}} \\
\underline{1\,1} \\
1\,0\,0 \\
\underline{1\,1} \\
1\,1 \\
\underline{1\,1} \\
0\,0\,1_{two}
\end{array}
$$

Quotient: 11100_{two} Remainder: 1_{two}

56. Dividing:

$$
12_{four} \overline{)\begin{array}{c} 2 \quad 1_{four} \\ 3 \quad 2 \quad 1_{four} \end{array}}
$$

$$
\begin{array}{r} 3 \ 0 \\ \hline 2 \ 1 \\ 1 \ 2 \\ \hline 3_{four} \end{array}
$$

Quotient: 21_{four} Remainder: 3_{four}

57. $45 = 3 \cdot 3 \cdot 5 = 3^2 \cdot 5$

58. $54 = 2 \cdot 3 \cdot 3 \cdot 3 = 2 \cdot 3^3$

59. $153 = 3 \cdot 3 \cdot 17 = 3^2 \cdot 17$

60. $285 = 3 \cdot 5 \cdot 19$

61. 501 is divisible by 3, so it is a composite number.

62. 781 is divisible by 11, so it is a composite number.

63. 689 is divisible by 13, so it is a composite number.

64. 1003 is divisible by 17, so it is a composite number.

65. The proper factors of 28 are 1, 2, 4, 7, and 14. Their sum is $1 + 2 + 4 + 7 + 14 = 28$, so 28 is a perfect number.

66. The proper factors of 81 are 1, 3, 9, and 27. Their sum is $1 + 3 + 9 + 27 = 40 < 81$, so 81 is a deficient number.

67. The proper factors of 144 are 1, 2, 3, 4, 6, 8, 9, 12, 16, 18, 24, 36, 48, and 72. Their sum is $1 + 2 + 3 + 4 + 6 + 8 + 9 + 12 + 16 + 18 + 24 + 36 + 48 + 72 = 259 > 144$, so 144 is an abundant number.

68. The proper factors of 200 are 1, 2, 4, 5, 8, 10, 20, 25, 40, 50, and 100. Their sum is $1 + 2 + 4 + 5 + 8 + 10 + 20 + 25 + 40 + 50 + 100 = 265 > 200$, so 200 is an abundant number.

69. Let $n = 61$ and use Euclid's procedure.
$2^{60}(2^{61} - 1)$

70. Let $n = 1279$ and use Euclid's procedure.
$2^{1278}(2^{1279} - 1)$

71.

	1	46
	2	92
	4	184
x	8	368
	8	368

$8 \times 46 = 368$

72.

x	1	57
	2	114
	4	228
x	8	456
	9	513

$9 \times 57 = 513$

73.

	1	83
x	2	166
x	4	332
x	8	664
	14	1162

$14 \times 83 = 1162$

74.

x	1	143
	2	286
x	4	572
	8	1144
x	16	2288
	21	3003

$21 \times 143 = 3003$

75.

1	1	0	0	1	1
	dabble	double	double	dabble	dabble
	= 3	= 6	= 12	= 25	= 51

0	1	0
double	dabble	double
= 102	= 205	= 410

$110011010_{two} = 410$

76.

1	0	0	0	1	0
	double	double	double	dabble	double
	= 2	= 4	= 8	= 17	= 34

1	0	1
dabble	double	dabble
= 69	= 138	= 277

$100010101_{two} = 277$

77.
1	0	0	0	0	0
	double	double	double	double	double
	$=2$	$=4$	$=8$	$=16$	$=32$

1	0	0	0	1
dabble	double	double	double	dabble
$=65$	$=130$	$=260$	$=520$	$=1041$

$10000010001_{two} = 1041$

78.
1	1	0	0	1	0
	dabble	double	double	dabble	double
	$=3$	$=6$	$=12$	$=25$	$=50$

1	0	0	0	0
dabble	double	double	double	double
$=101$	$=202$	$=404$	$=808$	$=1616$

$11001010000_{two} = 1616$

79. A base ten number is divisible by 3 provided the sum of the digits of the number is divisible by 3.

80. A number is divisible by 6 provided it is divisible by 2 and by 3.

81. Every composite number can be written as a unique product of prime numbers (disregarding the order of the factors).

82. zero

83. $132,049 \log 2 + 1 \approx 39,751.7$ The number has 39,751 digits.

84. $2,976,221 \log 2 + 1 \approx 895,932.8$ The number has 895,932 digits.

CHAPTER 4 TEST

1. $3124 = 3000 + 100 + 20 + 4$

   ꜰꜰꜰꝯ∩∩∩||||

2. $4000 + 200 + 60 + 3 = 4263$

3. $1000 + (500 - 100) + (50 - 10) + 5 + 1 + 1 = 1447$

4. $2609 = 2000 + 600 + 9$

   MMDCIX

5. $67,485 = 60,000 + 7000 + 400 + 80 + 5 =$
   $(6 \times 10^4) + (7 \times 10^3) + (4 \times 10^2) + (8 \times 10^1) + (5 \times 10^0)$

6. $500,000 + 30,000 + 200 + 80 + 4 = 530,284$

7. $(10 \times 60^2) + (21 \times 60) + (14 \times 1) = 37,274$

8. $9675 = (2 \times 3600) + (41 \times 60) + (15 \times 1)$

   𒌋𒌋 𒌌𒌌𒌌𒌋 𒌌𒌋𒌋𒌋𒌋

9. $(3 \times 360) + (11 \times 20) + (5 \times 1) = 1305$

10. $502 = (1 \times 360) + (7 \times 20) + (2 \times 1)$

11. Converting:
    $3542_{six} = (3 \times 6^3) + (5 \times 6^2) + (4 \times 6^1) + (2 \times 6^0) = 854$

12. a. Converting:
    $2148 = (4 \times 8^3) + (1 \times 8^2) + (4 \times 8^1) + (4 \times 8^0) = 4144_{eight}$

    b. Converting:
    $2148 = (1 \times 12^3) + (2 \times 12^2) + (11 \times 12^1) + (0 \times 12^0) = 12BO_{twelve}$

13. $4_{eight} = 100_{two}$
    $5_{eight} = 101_{two}$
    $6_{eight} = 110_{two}$
    $7_{eight} = 111_{two}$
    $4567_{eight} = 100101110111_{two}$

14. 1010 1011 0111   (base 2)
    A   B   7   (base 16)
    $101010110111_{two} = AB7_{sixteen}$

15. Adding:
	1	1	
		3	4_{five}
+		2	3_{five}
	1	1	2_{five}

16. Subtracting:

$$\begin{array}{r} \overset{5}{\cancel{4}}\ \overset{10}{\cancel{6}}\ 2_{eight} \\ -\ \ 1\ 4\ 7_{eight} \\ \hline 3\ 1\ 3_{eight} \end{array}$$

24. Let $n = 17$. Then Euclid's Procedure gives
    $2^{16}(2^{17}-1)$.

17. Multiplying:

$$\begin{array}{r} 1\ 0\ 1\ 1\ 1\ 0_{two} \\ \times\ \quad\quad 1\ 0\ 1_{two} \\ \hline 1\ 0\ 1\ 1\ 1\ 0 \\ 0\ 0\ 0\ 0\ 0\ 0 \\ 1\ 0\ 1\ 1\ 1\ 0 \\ \hline 1\ 1\ 1\ 0\ 0\ 1\ 1\ 0_{two} \end{array}$$

18. Dividing:

$$\begin{array}{r} 6\ \ 1_{seven} \\ 5_{seven}\overline{)4\ 3\ 1_{sevem}} \\ 4\ 2 \\ \hline 1\ \ 1 \\ \quad\ 5 \\ \hline \quad\ 3_{seven} \end{array}$$

    Quotient: 61_{seven} Remainder: 3_{seven}

19. $230 = 2 \cdot 5 \cdot 23$

20. 1001 is divisible by 7, so it is a composite number.

21. a. The number is not even. No.

    b. The digits add up to 45, which is divisible by 3. Yes.

    c. The ones digit does not end in a 5 or 0. No.

22. a. The last two digits are divisible by 4. Yes.

    b. The number is even; however, the digits add up to 43, which is not divisible by 3, so the number is not divisible by 3. No.

    c. The digits add up to 18 and 25, and their difference is not divisible by 11. No.

23. The proper factors of 96 are 1, 2, 3, 4, 6, 8, 12, 16, 24, 32, and 48. Their sum is $1 + 2 + 3 + 4 + 6 + 8 + 12 + 16 + 24 + 32 + 48 = 156 > 96$, so 96 is an abundant number.

Chapter 5: Applications of Equations

EXERCISE SET 5.1

1. Possible answer: An expression contains one or more terms, the addends of the expression. $8x + 1$ is an expression. An equation expresses the equality of two expressions. $2y - 4 = 5$ is an equation.

3. By substituting the solution for the variable in the original equation. If the resulting equation is a true equation, then the solution is correct.

5. $b = 21 - 9 = 12$.

7. $b = 11 + 11 = 22$.

9. $z = \dfrac{-48}{6} = -8$.

11. $x = \dfrac{15(-4)}{3} = -20$.

13. $x = -2(-4) = 8$.

15. Solving:
$$4 - 2b = 2 - 4b$$
$$4 - 4 - 2b + 4b = 2 - 4 - 4b + 4b$$
$$2b = -2$$
$$b = -1$$

17. Solving:
$$5x - 3 = 9x - 7$$
$$5x - 5x - 3 + 7 = 9x - 5x - 7 + 7$$
$$4 = 4x$$
$$x = 1$$

19. Solving:
$$3m + 5 = 2 - 6m$$
$$3m + 6m + 5 - 5 = 2 - 5 - 6m + 6m$$
$$9m = -3$$
$$m = -\frac{1}{3}$$

21. Solving:
$$5x + 7 = 8x + 5$$
$$5x - 5x + 7 - 5 = 8x - 5x + 5 - 5$$
$$3x = 2$$
$$x = \frac{2}{3}$$

23. Solving:
$$4b + 15 = 3 - 2b$$
$$4b + 2b + 15 - 15 = 3 - 15 - 2b + 2b$$
$$6b = -12$$
$$b = -2$$

25. Solving:
$$9n - 15 = 3(2n - 1)$$
$$9n - 15 = 6n - 3$$
$$9n - 6n - 15 + 15 = 6n - 6n - 3 + 15$$
$$3n = 12$$
$$n = 4$$

27. Solving:
$$5(3 - 2y) = 3 - 4y$$
$$15 - 10y = 3 - 4y$$
$$15 - 3 - 10y + 10y = 3 - 3 - 4y + 10y$$
$$6y = 12$$
$$y = 2$$

29. Solving:
$$2(3b + 5) - 1 = 10b + 1$$
$$6b - 6b + 10 - 1 - 1 = 10b - 6b + 1 - 1$$
$$4b = 8$$
$$b = 2$$

31. Solving:
$$4a + 3 = 7 - (5 - 8a)$$
$$4a - 4a + 3 - 2 = 2 - 2 + 8a - 4a$$
$$4a = 1$$
$$a = \frac{1}{4}$$

33. Solving:
$$4(3y + 1) = 2(y - 8)$$
$$12y + 4 = 2y - 16$$
$$12y - 2y + 4 - 4 = 2y - 2y - 16 - 4$$
$$10y = -20$$
$$y = -2$$

35. Solving:
$$3(x-4)=1-(2x-7)$$
$$3x-12=1-2x+7$$
$$3x+2x-12+12=1+7+12-2x+2x$$
$$5x=20$$
$$x=4$$

37. Solving:
$$\frac{x}{8}+2=\frac{3x}{4}-3$$
$$8\left(\frac{x}{8}+2\right)=8\left(\frac{3x}{4}-3\right)$$
$$x-x+16+24=6x-x-24+24$$
$$5x=40$$
$$x=8$$

39. Solving:
$$\frac{2}{3}+\frac{3x+1}{4}=\frac{5}{3}$$
$$12\left(\frac{2}{3}+\frac{3x+1}{4}\right)=12\left(\frac{5}{3}\right)$$
$$8+9x+3=20$$
$$11-11+9x=20-11$$
$$9x=9$$
$$x=1$$

41. Solving:
$$\frac{3}{4}(x-8)=\frac{1}{2}(2x+4)$$
$$4\left(\frac{3}{4}(x-8)\right)=4\left(\frac{1}{2}(2x+4)\right)$$
$$3(x-8)=2(2x+4)$$
$$3x-24=4x+8$$
$$3x-3x-24-8=4x-3x+8-8$$
$$x=-32$$

43. $P=0.02076\,L$
Substitute 350 for P.
$$\frac{350}{0.02076}=L$$
$$16,859.34\approx L$$
The maximum amount is $16,859.34.

45. $T=\dfrac{L}{S}$

$3\text{ min}\left(\dfrac{60\sec}{1\min}\right)=180$ seconds.

Substitute 180 for T and $\dfrac{15}{2}$ for S.

$$180=\frac{L}{\frac{15}{2}}$$
$$180\left(\frac{15}{2}\right)=L$$
$$1350=L$$
The length of the tape is 1350 inches.

47. $P=15+\dfrac{1}{2}D$
Substitute 45 for P.
$$45=15+\frac{1}{2}D$$
$$2(45-15)=D$$
$$60=D$$
The depth is 60 feet.

49. $t=17.08-0.0067\,y$
Substitute 4 for t.
$$4=17.08-0.0067\,y$$
$$\frac{17.08-4}{0.0067}=y$$
$$1952\approx y$$
The year 1952.

51. $C=\dfrac{1}{4}D-45$
Substitute -3 for C.
$$-3=\frac{1}{4}D-45$$
$$4(45-3)=D$$
$$168=D$$
The car will slide 168 feet.

53. $N=7C-30$
Substitute 100 for N.
$$100=7C-30$$
$$130=7C$$
$$18.6\approx C$$
The temperature is approximately $18.6°C$.

55. $H=0.8(200-A)$
Substitute 20 for H.
$$20=0.8(200-H)$$
$$20-160=-0.8H$$
$$\frac{140}{0.8}=H$$
$$175=H$$
The bowler's average score is 175.

57. Let O be the original value of the car, and V the current value.

$$V = \frac{3}{5}O$$

$$13,200 = \frac{3}{5}O$$

$$\frac{5}{3}(13,200) = \frac{5}{3}\left(\frac{3}{5}O\right)$$

$$22,000 = O$$

The original value is $22,000.

59. a. Let T be the average tuition and fees at public colleges, and P the average tuition and fees at private colleges.

$$P = 5T + 171$$

$$12,146 = 5T + 171$$

$$11,975 = 5T$$

$$2,395 = T$$

The average tuition and fees at public colleges is $2,395.

b. Let T be the average tuition and fees at public colleges, and P the average tuition and fees at private colleges.

$$P = 4T + 934$$

$$19,710 = 4T + 934$$

$$18,776 = 4T$$

$$4,694 = T$$

The average tuition and fees at public colleges is $4,694.

61. Let M be the amount of the monthly payment, and P the purchase price.

$$P = 450 + 24M$$

$$3276 = 450 + 24M$$

$$\frac{3276 - 450}{24} = M$$

$$117.75 = M$$

The monthly payment is $117.75.

63. a. Let F be the lift-off thrust, in kilograms, of the Friendship 7, and D the lift-off thrust, in kilograms, of the Discovery.

$$D = -85,000 + 20F$$

$$3,175,000 = -85,000 + 20F$$

$$\frac{3,175,000 + 85,000}{20} = F$$

$$163,000 = F$$

Friendship 7's lift-off thrust was 163,000 kg.

b. Let F be the weight, in kilograms, of the Friendship 7, and D the weight, in kilograms, of the Discovery.

$$D = 290 + 36F$$

$$69,770 = 290 + 36F$$

$$\frac{69,770 - 290}{36} = F$$

$$1930 = F$$

Friendship 7's weight was 1930 kg.

65. Let n be the regular hourly wage. Then the overtime wage is $1.5n$.

$$40n + 7(1.5n) = 631.25$$

$$40n + 10.5n = 631.25$$

$$50.5n = 631.25$$

$$n = 12.50$$

The regular hourly wage is $12.50.

67. Let h be the number of horizontal pixels, and v the number of vertical pixels.

$$h = 400 + \frac{1}{2}v$$

$$1040 = 400 + \frac{1}{2}v$$

$$2(1040 - 400) = v$$

$$1280 = v$$

There are 200 vertical pixels.

69. Let t be the length of the call, and c the cost of the call.

$$c = 1.42 + .65(t - 3) \quad [t \geq 3]$$

$$10.52 = 1.42 + .65(t - 3)$$

$$\frac{10.52 - 1.42 + 1.95}{.65} = t$$

$$17 = t$$

The length of the call is more than 16 minutes but not over 17 minutes.

71. $b = P - a - c$

73. $R = \dfrac{E}{I}$

75. $r = \dfrac{I}{Pt}$

77. $C = \dfrac{5}{9}(F - 32)$

79. $t = \dfrac{A-P}{Pr}$

81. $f = \dfrac{T+gm}{m}$

83. $S = C - Rt$

85. $b_2 = \dfrac{2A}{h} - b_1$

87. $h = \dfrac{S}{2\pi r} - r$

89. $y = 2 - \dfrac{4}{3}x$

91. $x = \dfrac{y - y_1}{m} + x_1$

93. Solving:
$$9 - (6 - 4x) = 3(1 - 2x)$$
$$9 - 6 + 4x = 3 - 6x$$
$$3 - 3 + 4x + 6x = 3 - 3 - 6x + 6x$$
$$10x = 0$$
$$x = 0$$

95. Solving:
$$7(x - 20) - 5(x - 22) = 2x - 30$$
$$7x - 140 - 5x + 110 = 2x - 30$$
$$2x - 2x - 30 + 30 = 2x - 2x - 30 + 30$$
$$0 = 0$$
Every real number is a solution to this equation.

97. Solving:
$$ax + b = cx + d$$
$$ax + b - b = cx + d - b$$
$$ax - cx = cx - cx + d - b$$
$$x(a - c) = d - b$$
$$x = \dfrac{d - b}{a - c}$$
No, the solution is not valid when $a = c$ and $b \neq d$, as the denominator would equal zero, and x would then be undefined.

EXERCISE SET 5.2

1. Examples will vary.

3. The purpose is to allow the conversion of one country's currency to another country's currency.

5. Explanations will vary.

7. The cross-products method is a shortcut for multiplying each side of the proportion by the least common multiple of the denominators.

9. $\dfrac{138 \text{ miles}}{6 \text{ gallons}} = \dfrac{23 \text{ miles}}{1 \text{ gallon}} = 23$ miles per gallon of gasoline

11. $\dfrac{100 \text{ meters}}{8 \text{ seconds}} = \dfrac{12.5 \text{ meters}}{1 \text{ second}} = 12.5$ meters per second

13. $\dfrac{1000 \text{ square feet}}{2.5 \text{ gallons}} = \dfrac{400 \text{ square feet}}{1 \text{ gallon}} = 400$ square feet of wall per gallon of paint

15. $\dfrac{680,400 \text{ gallons}}{2.5 \text{ minutes}} = \dfrac{272,160 \text{ gallons}}{1 \text{ minute}} = 272,160$ gallons of fuel per minute

17. $\dfrac{\$2.79}{32 \text{ ounces}} \approx \dfrac{\$.0872}{1 \text{ ounce}}$

$\dfrac{\$2.09}{24 \text{ ounces}} \approx \dfrac{\$.0871}{1 \text{ ounce}}$

The 24-ounce jar is more economical.

19. $\left(\dfrac{\$16.50}{1 \text{ hour}}\right)\left(\dfrac{40 \text{ hours}}{1 \text{ week}}\right)\left(\dfrac{52 \text{ weeks}}{1 \text{ year}}\right) = \$34,320$

$\left(\dfrac{\$650}{1 \text{ week}}\right)\left(\dfrac{52 \text{ weeks}}{1 \text{ year}}\right) = \$33,800$

$\left(\dfrac{\$2840}{1 \text{ month}}\right)\left(\dfrac{12 \text{ months}}{1 \text{ year}}\right) = \$34,080$

A wage of \$16.50 per hour is the best choice.

21. a. Australia: $\dfrac{19,626,000}{2,938,000} \approx 7$

India: $\dfrac{1,067,018,000}{1,146,000} \approx 931$

U.S.: $\dfrac{287,998,000}{3,535,000} \approx 81$

Australia

b. India has 850 more people per square mile than the U.S.

23. Using ratios:
$$\frac{6.0374 \text{ krona}}{\$1} = \frac{x \text{ krona}}{\$10,000}$$
$$x = 60,374$$
60,374 krona

25. Using ratios:
$$\frac{11.443 \text{ pesos}}{\$1} = \frac{x \text{ pesos}}{\$38,000}$$
$$x = 434,834$$
434,834 pesos

27. Alabama: $\frac{260,639}{83,404} \approx 3.125$

Alaska: $\frac{56,246}{17,999} \approx 3.125$

Arizona: $\frac{307,864}{98,516} \approx 3.125$

Arkansas: $\frac{155,690}{49,821} \approx 3.125$

California: $\frac{1,446,550}{462,896} = 3.125$

Colorado: $\frac{271,694}{86,942} \approx 3.125$

Connecticut: $\frac{175,501}{56,160} \approx 3.125$

Delaware: $\frac{51,806}{16,578} \approx 3.125$

For each state, the ratio is 3.125 to 1.

29. $\dfrac{\text{Number of students}}{\text{Number of faculty}} = \dfrac{4722+6024}{815} \approx 13$

13:1, 13 to 1. There are 13 students per faculty member at Syracuse University.

31. University of Connecticut

33. Dividing:
$$\frac{\text{Monthly debt}}{\text{Monthly income}} =$$
$$\frac{1800+104+27+354+199}{3400+83+640+34} \approx \frac{3}{5}$$
No.

35. Solving:
$$\frac{3}{y} = \frac{7}{40}$$
$$7y = 120$$
$$y \approx 17.14$$

37. Solving:
$$\frac{16}{d} = \frac{25}{40}$$
$$25d = 640$$
$$d = 25.6$$

39. Solving:
$$\frac{120}{c} = \frac{144}{25}$$
$$144c = 3000$$
$$c \approx 20.83$$

41. Solving:
$$\frac{4}{a} = \frac{9}{5}$$
$$9a = 20$$
$$a \approx 2.22$$

43. Solving:
$$\frac{1.2}{2.8} = \frac{b}{32}$$
$$2.8b = 38.4$$
$$b \approx 13.71$$

45. Solving:
$$\frac{2.5}{0.6} = \frac{165}{x}$$
$$2.5x = 99$$
$$x = 39.6$$

47. Solving:
$$\frac{y}{2.54} = \frac{132}{640}$$
$$640y = 335.28$$
$$y \approx 0.52$$

49. Solving:
$$\frac{12.5}{m} = \frac{102}{55}$$
$$102m = 687.5$$
$$m \approx 6.74$$

51. Solving:
$$\frac{5}{4} = \frac{x}{36,000}$$
$$4x = 180,000$$
$$x = 45,000$$
$45,000

53. Solving:

$$\frac{2}{80} = \frac{x}{220}$$

$$80x = 440$$

$$x = 5.5$$

5.5 mg

55. Solving:

$$\frac{1}{3} = \frac{5}{x}$$

$$x = 15$$

$$\frac{1}{3} = \frac{8}{y}$$

$$y = 24$$

15 feet by 24 feet

57. Solving:

$$\frac{2}{3} = \frac{x}{240,000}$$

$$3x = 480,000$$

$$x = 160,000$$

160,000 people

59. Solving:

$$\frac{7,000 \text{ miles}}{4 \text{ months}} = \frac{x \text{ miles}}{3 \text{ years}}$$

$$\frac{7,000 \text{ miles}}{4 \text{ months}} = \frac{x \text{ miles}}{3 \text{ years}}\left(\frac{1 \text{ year}}{12 \text{ months}}\right)$$

$$4x = 252,000$$

$$x = 63,000$$

63,000 miles

61. The amount of fat in a pancake is proportional to the area of the pancake, which is proportional to the square of the pancake's diameter.

$$\frac{4^2}{5} = \frac{6^2}{x}$$

$$16x = 180$$

$$x = 11.25$$

11.25 grams

63. a. True.

   b. Comparing:

   If $\frac{a}{b} = \frac{c}{d}$, then $ad = bc$.

   If $\frac{b}{a} = \frac{d}{c}$, then $ad = bc$.

   The statement is true.

c. Comparing:

   If $\frac{a}{b} = \frac{c}{d}$, then $ad = bc$.

   If $\frac{a}{c} = \frac{b}{d}$, then $ad = bc$.

   The statement is true.

d. Comparing:

   If $\frac{a}{b} = \frac{c}{d}$, then $ad = bc$.

   If $\frac{a}{d} = \frac{c}{b}$, then $ab = cd$.

   The statement is false.

65. a. Different time slots and different shows reach audiences of different sizes, and require different rates.

   b. Advertisers have target gender and age groups for their products, and look for shows that best reach their target consumers.

   c. Nielsen ratings quantify the number of viewers watching a particular program.

EXERCISE SET 5.3

1. Answers will vary.

3. $300\% = 3$. Multiplying by 300% is the same as multiplying by 3.

5. Employee B's salary is now the highest because Employee B had the highest initial salary, and the percent raises were the same for all three employees.

7. $\frac{1}{2} = 0.5 = 50\%$

9. Converting:

$$40\% = 0.4$$

$$40\% = 40\left(\frac{1}{100}\right) = \frac{40}{100} = \frac{2}{5}$$

$$\frac{2}{5} = 0.4 = 40\%$$

11. Converting:

    $.0.7 = 70\%$

    $$70\% = 70\left(\frac{1}{100}\right) = \frac{70}{100} = \frac{7}{10}$$

    $$\frac{7}{10} = 0.7 = 70\%$$

13. $\dfrac{11}{20} = 0.55 = 55\%$

15. Converting:

    $15.625\% = 0.15625$

    $$15.625\% = 15.625\left(\frac{1}{100}\right) = \frac{15.625}{100}$$

    $$= \frac{15,625}{100,000} = \frac{5}{32}$$

    $$\frac{5}{32} = 0.15625 = 15.625\%$$

17. a.   73

    b.   More fans approved.

    c.   7%. The total percents should add up to 100%. $100\% - 73\% - 20\% = 7\%$.

19. Solving:

    $0.14(190) = A$

    $\quad 26.6 = A$

    $\$26.6$ billion

21. Solving:

    $$\frac{\text{Percent}}{100} = \frac{\text{amount}}{\text{base}}$$

    $$\frac{p}{100} = \frac{293}{1236}$$

    $1236p = 29,300$

    $\quad p \approx 23.7$

    23.7% were irked most by tailgaters.

23. a.   Solving:

    $$\frac{\text{Percent}}{100} = \frac{\text{amount}}{\text{base}}$$

    $$\frac{21.0}{100} = \frac{290}{B}$$

    $21.0B = 29,000$

    $\quad B \approx 1381$

    The total amount is $1381.

b.   Solving:

$$\frac{\text{Percent}}{100} = \frac{\text{amount}}{\text{base}}$$

$$\frac{26.9}{100} = \frac{438}{B}$$

$26.9B = 43,800$

$\quad B \approx 1628$

The total amount is $1628.

25. a.   Solving:

    $0.43(112) = A$

    $\quad 48.2 \approx A$

    48.2 hours

    b.   Solving:

    $0.3(112) = A$

    $\quad 33.6 = A$

    33.6 hours

    c.   Prefer:

    $0.23(112) = A$

    $\quad 25.8 \approx A$

    Actual:

    $0.2(112) = A$

    $\quad 22.4 = A$

    $25.8 - 22.4 = 3.4$

    3.4 hours

27. a.   1998-2000:

    $$\frac{p}{100} = \frac{0.8}{9.6}$$

    $9.6p = 80$

    $\quad p \approx 8.3$

    Growth was 8.3%.

    2000-2002:

    $$\frac{p}{100} = \frac{0.6}{10.4}$$

    $10.4p = 60$

    $\quad p \approx 5.8$

    Growth was 5.8%.

    2002-2004:

    $$\frac{p}{100} = \frac{0.2}{11}$$

    $11p = 20$

    $\quad p \approx 1.8$

    Growth was 1.8%.

2004-2006:

$$\frac{p}{100} = \frac{0.2}{11.2}$$

$$11.2p = 20$$

$$p \approx 1.8$$

Growth was 1.8%.
The percent increase is the highest in 1998-2000.

b.   The percent increase is the lowest in 2004-2006.

c.   Growth increases more slowly.

29. a.   Arizona:

$$\frac{53}{1000} = \frac{x}{2,229,300}$$

$$1000x = 118,152,900$$

$$x \approx 118,153$$

118,153 high-tech employees

California:

$$\frac{62}{1000} = \frac{x}{10,081,300}$$

$$1000x = 625,040,600$$

$$x \approx 625,041$$

625,041 high-tech employees

Colorado:

$$\frac{75}{1000} = \frac{x}{2,200,500}$$

$$1000x = 165,037,500$$

$$x \approx 165,038$$

165,038 high-tech employees

Maryland:

$$\frac{51}{1000} = \frac{x}{2,798,400}$$

$$1000x = 142,718,400$$

$$x \approx 142,718$$

142,718 high-tech employees

Massachusetts:

$$\frac{75}{1000} = \frac{x}{3,291,100}$$

$$1000x = 246,832,500$$

$$x \approx 246,833$$

246,833 high-tech employees

Minnesota:

$$\frac{55}{1000} = \frac{x}{4,173,500}$$

$$1000x = 229,542,500$$

$$x \approx 229,543$$

229,543 high-tech employees

New Hampshire:

$$\frac{78}{1000} = \frac{x}{663,000}$$

$$1000x = 51,714,000$$

$$x = 51,714$$

51,714 high-tech employees

New Jersey:

$$\frac{55}{1000} = \frac{x}{4,173,500}$$

$$1000x = 229,542,500$$

$$x \approx 229,543$$

229,543 high-tech employees

Vermont:

$$\frac{52}{1000} = \frac{x}{332,200}$$

$$1000x = 17,274,400$$

$$x \approx 17,274$$

17,274 high-tech employees

Virginia:

$$\frac{52}{1000} = \frac{x}{3,575,000}$$

$$1000x = 185,900,000$$

$$x = 185,900$$

185,900 high-tech employees

b.   New Hampshire has the highest rate, and California has the largest number.

c.   118,153 + 625,041 + 165,038 + 142,718 + 246,833 + 229,543 + 51,714 + 229,543 + 17,274 + 185,900 = 2,011,757, which is more than half of 4 million. So, more than half of the high-tech employees work in the 10 states listed.

d.   Massachusetts:

$$\frac{p}{100} = \frac{75}{1000}$$
$$1000p = 7500$$
$$p = 7.5$$

7.5%

New Jersey:

$$\frac{p}{100} = \frac{55}{1000}$$
$$1000p = 5500$$
$$p = 5.5$$

5.5%

Virginia:

$$\frac{p}{100} = \frac{52}{1000}$$
$$1000p = 5200$$
$$p = 5.2$$

5.2%

e.   5.3%

f.   The rate is ten times the percent.

31. a.   Using ratios:

$$\frac{x}{100} = \frac{3,500,000 - 350,000}{350,000}$$
$$350,000x = 315,000,000$$
$$x = 900$$

900% increase

b.   Using ratios:

$$\frac{x}{100} = \frac{5,600,000 - 3,500,000}{3,500,000}$$
$$3,500,000x = 210,000,000$$
$$x = 60$$

60% increase

c.   Using ratios:

$$\frac{x}{100} = \frac{5,600,000 - 350,000}{350,000}$$
$$350,000x = 525,000,000$$
$$x = 1500$$

1500% increase

d.   Explanations may vary.

33. a.   Home health aides

b.   Software engineers

c.   Solving:
$$0.40(158,000) = A$$
$$63,200 = A$$
63,200 people

d.   Solving:
$$0.37(234,000) = A$$
$$86,580 = A$$
$$234,000 + 86,580 = 320,580$$
320,580 people

e.   Home health aides:
$$0.47(615,000) = A$$
$$289,050 = A$$
$$615,000 + 289,050 = 904,050$$
Computer support specialists:
$$0.97(506,000) = A$$
$$490,820 = A$$
$$506,000 + 490,820 = 996,820$$
It is expected that more people will be
employed as computer support specialists.

f.   The percent increases are based on different
original employment figures.

g.   Answers will vary.

35. $(1.05)(1.06)(1.07) < (1.06)^3$ Your salary is less.

37. Calculate the percent increase in fuel efficiency.
$$\frac{x}{100} = \frac{80 - 28}{28}$$
$$28x = 5200$$
$$x \approx 185.7$$
Determine the amount of money spent when
$1500 is saved.
$$1.857(B) = 1500$$
$$B = \frac{1500}{1.857}$$
$$B \approx 807.8$$
Solve for the time required.

$\$807.8\left(\dfrac{1\text{ gal}}{\$2.00}\right)\left(\dfrac{80\text{ mi}}{1\text{ gal}}\right)\approx 32{,}312\text{ mi}$

$32{,}312\text{ mi}\left(\dfrac{1\text{ year}}{10{,}000\text{ mi}}\right)=3.23\text{ years}$

$3.23\times 12\approx 38.8$

39 months

39. a. Solving:
$(0.101)(105{,}500{,}000)=A$
$10{,}655{,}500=A$
10.7 million TV households
$(0.17)B=10{,}655{,}500$
$B\approx 62{,}679{,}412$
62.7 million TV households

  b. Solving:
$(0.056)(105{,}500{,}000)=A$
$5{,}908{,}000=A$
5.9 million TV households
$(0.11)B=5{,}908{,}000$
$B\approx 53{,}709{,}091$
53.7 million TV households

  c. Solving:
$PB=A$
$(0.075)(105{,}500{,}000)=A$
$7{,}912{,}500=A$
$\dfrac{19{,}781{,}000}{7{,}912{,}500}=2.5$
2.5 people per TV household

  d. Answers will vary.

EXERCISE SET 5.4

1. The principle of Zero Products states that if the product of two factors is zero, then one of the two factors equals zero. A second-degree equation must be written in standard form so that the variable expression is equal to zero. Then, when the variable expression is factored, the Principle of Zero Products can be used to set each factor equal to zero.

3. Answers will vary.

5. Solving:
$r^2-3r=10$
$r^2-3r-10=10-10$
$(r-5)(r+2)=0$
$r-5=0 \quad r+2=0$
$r=5 \quad\quad r=-2$

7. Solving:
$t^2=t+1$
$t^2-t-1=t-t+1-1$
$t^2-t-1=0$
$t=\dfrac{1\pm\sqrt{(-1)^2-4(1)(-1)}}{2}$
$t=\dfrac{1-\sqrt5}{2},\dfrac{1+\sqrt5}{2}$
$t=-0.618,1.618$

9. Solving:
$y^2-6y=4$
$y^2-6y-4=4-4$
$y^2-6y-4=0$
$y=\dfrac{6\pm\sqrt{(-6)^2-4(1)(-4)}}{2}$
$y=3-\sqrt{13},\,3+\sqrt{13}$
$y=-0.606,6.606$

11. Solving:
$9z^2-18z=0$
$9z(z-2)=0$
$9z=0 \quad z-2=0$
$z=0 \quad\quad z=2$

13. Solving:
$z^2=z+4$
$z^2-z-4=z-z+4-4$
$z^2-z-4=0$
$z=\dfrac{1\pm\sqrt{(-1)^2-4(1)(-4)}}{2}$
$z=\dfrac{1-\sqrt{17}}{2},\dfrac{1+\sqrt{17}}{2}$
$z=-1.562,2.562$

15. Solving:
$$2s^2 = 4s + 5$$
$$2s^2 - 4s - 5 = 4s - 4s + 5 - 5$$
$$2s^2 - 4s - 5 = 0$$
$$s = \frac{4 \pm \sqrt{(-4)^2 - 4(2)(-5)}}{4}$$
$$s = \frac{2 - \sqrt{14}}{2}, \frac{2 + \sqrt{14}}{2}$$
$$s = -0.871, 2.871$$

17. Solving:
$$r^2 = 4r + 7$$
$$r^2 - 4r - 7 = 4r - 4r + 7 - 7$$
$$r^2 - 4r - 7 = 0$$
$$r = \frac{4 \pm \sqrt{(-4)^2 - 4(1)(-7)}}{2}$$
$$r = 2 - \sqrt{11}, \ 2 + \sqrt{11}$$
$$r = -1.317, 5.317$$

19. Solving:
$$2x^2 = 9x + 18$$
$$2x^2 - 9x - 18 = 9x - 9x + 18 - 18$$
$$(2x + 3)(x - 6) = 0$$
$$2x + 3 = 0 \qquad x - 6 = 0$$
$$x = -\frac{3}{2} \qquad x = 6$$

21. Solving:
$$6x - 11 = x^2$$
$$6x - 6x - 11 + 11 = x^2 - 6x + 11$$
$$0 = x^2 - 6x + 11$$
$$x = \frac{6 \pm \sqrt{(-6)^2 - 4(1)(11)}}{2}$$
$$x = \frac{6 \pm \sqrt{-8}}{2}$$

$\sqrt{-8}$ is not a real number. The equation has no real number solutions.

23. Solving:
$$4 - 15u = 4u^2$$
$$4 - 4 - 15u + 15u = 4u^2 + 15u - 4$$
$$0 = (4u - 1)(u + 4)$$
$$4u - 1 = 0 \qquad u + 4 = 0$$
$$u = \frac{1}{4} \qquad u = -4$$

25. Solving:
$$6y^2 - 4 = 5y$$
$$6y^2 - 5y - 4 = 5y - 5y$$
$$(3y - 4)(2y + 1) = 0$$
$$3y - 4 = 0 \qquad 2y + 1 = 0$$
$$y = \frac{4}{3} \qquad y = -\frac{1}{2}$$

27. Solving:
$$y - 2 = y^2 - y - 6$$
$$y - y - 2 + 2 = y^2 - y - y - 6 + 2$$
$$0 = y^2 - 2y - 4$$
$$y = \frac{2 \pm \sqrt{(-2)^2 - 4(1)(-4)}}{2}$$
$$y = 1 - \sqrt{5}, \ 1 + \sqrt{5}$$
$$y = -1.236, 3.236$$

29. Solving:
$$36 = -16t^2 + 60t$$
$$36 - 36 = -16t^2 + 60t - 36$$
$$0 = 4t^2 - 15t + 9$$
$$0 = (4t - 3)(t - 3)$$
$$4t - 3 = 0 \qquad t - 3 = 0$$
$$t = \frac{3}{4} \qquad t = 3$$

The height of the ball will be 36 feet at 0.75 second after the ball is hit and 3 seconds after the ball is hit.

31. Verifying:
$$T(1) = 0.5(1)^2 + 0.5(1) = 1$$
$$T(2) = 0.5(2)^2 + 0.5(2) = 3$$
$$T(3) = 0.5(3)^2 + 0.5(3) = 6$$
$$T(4) = 0.5(4)^2 + 0.5(4) = 10$$

$$55 = 0.5n^2 + 0.5n$$
$$55 - 55 = 0.5n^2 + 0.5n - 55$$
$$0 = 0.5n^2 + 0.5n - 55$$
$$0 = n^2 + n - 110$$
$$0 = (n+11)(n-10)$$
$$n+11 = 0 \qquad n-10 = 0$$
$$n = -11 \qquad n = 10$$
The number of rows is 10.

33. a. $d = 0.05(60)^2 + (60) = 240$ feet

   b. Solving:
   $$75 = 0.05r^2 + r$$
   $$75 - 75 = 0.05r^2 + r - 75$$
   $$0 = 0.05r^2 + r - 75$$
   $$0 = r^2 + 20r - 1500$$
   $$0 = (r+50)(r-30)$$
   $$r+50 = 0 \qquad r-30 = 0$$
   $$r = -50 \qquad r = 30$$
   The car was going at 30 miles per hour.

35. Solving:
   $$0 = -16t^2 + 88t + 1$$
   $$t = \frac{-88 \pm \sqrt{(88)^2 - 4(-16)(1)}}{-32}$$
   $$t = \frac{11 - \sqrt{122}}{4}, \frac{11 + \sqrt{122}}{4}$$
   The hangtime of the football is 5.51 seconds.

37. Solving:
   $$5 = -0.005x^2 + 1.2x + 10$$
   $$5 - 5 = -0.005x^2 + 1.2x + 10 - 5$$
   $$0 = -0.005x^2 + 1.2x + 5$$
   $$0 = x^2 - 240x - 1000$$
   $$x = \frac{240 \pm \sqrt{(-240)^2 - 4(1)(-1000)}}{2}$$
   $$x = 120 - 10\sqrt{154}, \ 120 + 10\sqrt{154}$$
   The water is 244.10 feet away from the tugboat.

39. $-0.002(36)^2 + 0.36(36) \approx 10.37 > 8$

   When the ball reaches the goal, it will be too high off the ground to go into the net.

41. Solving:
   $$300 = -16t^2 + 200t$$
   $$300 - 300 = -16t^2 + 200t - 300$$
   $$0 = 4t^2 - 50t + 75$$
   $$t = \frac{50 \pm \sqrt{(-50)^2 - 4(4)(75)}}{8}$$
   $$t = \frac{25 - 5\sqrt{13}}{4}, \frac{25 + 5\sqrt{13}}{4}$$
   The rocket will be at 300 feet off the ground at 1.74 seconds after the launch and 10.76 seconds after the launch.

43. Substituting $x = 125$:
   $$y = 0.0099(125)^2 - 0.8084(125) + 16.8750 \approx 70.5$$
   A first-class stamp will be 71 cents.

45. If the discriminant is not a perfect square, the radical in the equation will not simplify to a whole number.

47. Solving:
   $$x^2 + 16ax + 48a^2 = 0$$
   $$(x+4a)(x+12a) = 0$$
   $$x+4a = 0 \qquad x+12a = 0$$
   $$x = -4a \qquad x = -12a$$

49. Solving:
   $$2x^2 + 3bx + b^2 = 0$$
   $$(2x+b)(x+b) = 0$$
   $$2x+b = 0 \qquad x+b = 0$$
   $$x = -\frac{b}{2} \qquad x = -b$$

51. Solving:
   $$2x^2 - xy - 3y^2 = 0$$
   $$(2x-3y)(x+y) = 0$$
   $$2x-3y = 0 \qquad x+y = 0$$
   $$x = \frac{3y}{2} \qquad x = -y$$

53. Calculate the discriminant:
   $$b^2 - 4ac = b^2 - 4(1)(-1) = b^2 + 4$$
   When the discriminant of a quadratic equation is greater than or equal to zero, the equation has real solutions. As $b^2 + 4$ is always greater than 0, $x^2 + bx - 1 = 0$ always has real solutions.

CHAPTER 5 REVIEW EXERCISES

1. Solving:
$$5x + 3 = 10x - 17$$
$$5x - 5x + 3 + 17 = 10x - 5x - 17 + 17$$
$$5x = 20$$
$$x = 4$$

2. Solving:
$$3x + \frac{1}{8} = \frac{1}{2}$$
$$3x + \frac{1}{8} - \frac{1}{8} = \frac{1}{2} - \frac{1}{8}$$
$$3x = \frac{3}{8}$$
$$x = \frac{1}{8}$$

3. Solving:
$$6x + 3(2x - 1) = -27$$
$$6x + 6x - 3 + 3 = -27 + 3$$
$$12x = -24$$
$$x = -2$$

4. Solving:
$$\frac{5}{12} = \frac{n}{8}$$
$$12n = (5)(8) = 40$$
$$n = \frac{40}{12} = \frac{10}{3}$$

5. Solving:
$$4y^2 + 9 = 0$$
$$4y^2 + 9 - 9 = -9$$
$$4y^2 = -9$$
$$y^2 = \frac{-9}{4}$$
$$y = \sqrt{\frac{-9}{4}}$$

$\sqrt{\frac{-9}{4}}$ is not a real number. The equation has no real solutions.

6. Solving:
$$x^2 - x = 30$$
$$x^2 - x - 30 = 30 - 30$$
$$(x + 5)(x - 6) = 0$$
$$x + 5 = 0 \quad x - 6 = 0$$
$$x = -5 \qquad x = 6$$

7. Solving:
$$x^2 = 4x - 1$$
$$x^2 - 4x + 1 = 4x - 4x - 1 + 1$$
$$x^2 - 4x + 1 = 0$$
$$x = \frac{-(-4) \pm \sqrt{(-4)^2 - (4)(1)(1)}}{(2)(1)}$$
$$x = 2 + \sqrt{3},\, 2 - \sqrt{3}$$

8. Solving:
$$x + 3 = x^2$$
$$x - x + 3 - 3 = x^2 - x - 3$$
$$0 = x^2 - x - 3$$
$$x = \frac{1 \pm \sqrt{(-1)^2 - (4)(1)(-3)}}{(2)(1)}$$
$$x = \frac{1 + \sqrt{13}}{2}, \frac{1 - \sqrt{13}}{2}$$

9. Solving:
$$4x + 3y = 12$$
$$4x - 4x + 3y = 12 - 4x$$
$$\frac{1}{3} \cdot 3y = \frac{1}{3} \cdot (12 - 4x)$$
$$y = 4 - \frac{4x}{3}$$

10. Solving:
$$f = v + at$$
$$f - v = v - v + at$$
$$\frac{f - v}{a} = \frac{1}{a}(at)$$
$$t = \frac{f - v}{a}$$

11. Solving:
$$101 = -0.005x + 113.25$$
$$101 - 113.25 = -0.005x + 113.25 - 113.25$$
$$0.005x = 12.25$$
$$x = 2450$$
The elevation is 2450 feet.

12. Solving:
$$100 = 4 + 32t$$
$$100 - 4 = 4 - 4 + 32t$$
$$32t = 96$$
$$t = 3$$
It takes 3 seconds for the velocity to increase from 4 feet per second to 100 feet per second.

13. Solving:
$$100(80 - T) = 50(T - 20)$$
$$8000 - 100T = 50T - 1000$$
$$8000 + 1000 - 100T = 50T + 1000 - 1000$$
$$9000 + 100T - 100T = 50T + 100T$$
$$9000 = 150T$$
$$T = 60$$
The final temperature is 60°C.

14. Solving:
$$4.97 = 4.25 + 0.08m$$
$$4.97 - 4.25 = 4.25 - 4.25 + 0.08m$$
$$0.72 = 0.08m$$
$$m = 9$$
The service was used for 39 minutes.

15. Solving:
$$\frac{326.6}{11.5} = \frac{x}{1}$$
$$(326.6)(1) = 11.5x$$
$$x = 28.4 \text{ miles per gallon}$$

16. $\dfrac{350,000 - 280,000}{280,000} = \dfrac{70,000}{280,000} = \dfrac{1}{4}$

17. Let c be the price of cropland in 1997:
$$1720 = 2c - 800$$
$$2520 = 2c$$
$$1260 = c$$
Cropland had an average value of $1,260 per acre in 1997.

18. a. New York:
$$\frac{8,008,000}{321.8} \approx 24,885$$
Los Angeles:
$$\frac{3,695,000}{467.4} \approx 7905$$
Chicago:
$$\frac{2,896,000}{228.469} \approx 12,676$$
Houston:
$$\frac{1,954,000}{594.03} \approx 3289$$
Philadelphia:
$$\frac{1,519,000}{136} \approx 11,169$$
New York, Chicago, Philadelphia, Los Angeles, Houston

   b. 24,885 − 3289 = 21,596 more people per square mile

19. a. $\dfrac{305 + 435}{49} \approx 15$

15 : 1, or 15 to 1. There are 15 students per faculty member at the university.

   b. Arizona State:
$$\frac{14,875 + 16,054}{1722} \approx 18$$
Embry-Riddle:
$$\frac{1181 + 244}{87} \approx 16$$
Grand Canyon:
$$\frac{450 + 880}{96} \approx 14$$
NAU:
$$\frac{4468 + 6566}{711} \approx 16$$
Arizona:
$$\frac{11,283 + 12,822}{1495} \approx 16$$
Grand Canyon University has the lowest student-faculty ratio; Arizona State has the highest.

   c. Embry-Riddle Aeronautical University, Northern Arizona University and the University of Arizona have the same student-faculty ratio.

20. Using ratios:

$$\frac{3}{3+7} = \frac{a}{350,000}$$

$$10a = (3)(350,000) = 1,050,000$$

$$a = 105,000$$

Department A has $105,000 allocated, and Department B has 350,000 − 105,000 = $245,000 allocated.

21. Using ratios:

$$\frac{3}{4} = \frac{t}{10}$$

$$4t = (3)(10) = 30$$

$$t = 7.5$$

7.5 tablespoons of fertilizer are required.

22. a.   No.

b. $$\frac{23+22+15+14+8}{18} = \frac{82}{18} = \frac{41}{9}$$

$41 : 9$

c.   Using ratios:

$$\frac{41}{41+9} = \frac{x}{2230}$$

$$50x = (41)(2230) = 91,430$$

$$x = 1828.6$$

$1828.6 billion will be spent on fixed expenditures.

d.   Using ratios:

$$\frac{22}{100} = \frac{x}{2230}$$

$$100x = (22)(2230) = 49060$$

$$x = 490.6$$

$490.6 billion will be spent on fixed expenditures.

23. a. $$\frac{170,931}{170,931 + 164,119} \approx 0.51 = 51.0\%$$

51.0% of the projected population in 2025 is female.

b.   2050:

$$\frac{200,696}{200,696 + 193,234} \approx 0.509 = 50.9\%$$

Less than one percent.

24. Using ratios:

$$\frac{38.6}{100} = \frac{283,700}{x}$$

$$38.6x = (100)(283,700) = 28,370,000$$

$$x \approx 735,000$$

The population of San Francisco is 735,000.

25. Using ratios:

$$\frac{p}{100} = \frac{34.2}{34.2 + 4.4 + 3.7 + 7.3}$$

$$49.6p = (34.2)(100) = 3420$$

$$p \approx 69$$

69% of all people waiting for organ transplants are waiting for kidney transplants.

26. a.   Using ratios:

$$\frac{p}{100} = \frac{24-4}{24}$$

$$24p = (20)(100) = 2000$$

$$p = 83.\overline{3}$$

The fat content is decreased by $83.\overline{3}\%$.

b.   The cholesterol content is decreased by 100%.

c.   Using ratios:

$$\frac{p}{100} = \frac{280-140}{280}$$

$$280p = (140)(100) = 14,000$$

$$p = 50$$

The calorie content is decreased by 50%.

27. Using ratios:

$$\frac{p}{100} = \frac{785.1 - 656.3}{785.1}$$

$$785.1p = (128.8)(100) = 12,880$$

$$p \approx 16.4$$

The reduction represented a 16.4% decrease.

28. a.   Ages 9-10

b.   Ages 15-16

c.   Total number of girls playing youth soccer:
23,805 + 45,181 + 46,758 + 39,939 + 26,157 + 11,518 + 4430 = 197,788

$$\frac{45,181 + 46,758}{197,788} \approx 0.465$$

46.5% of girls playing youth soccer are ages 7-10, which is less than half of all the girls playing.

d. Using ratios:
$$\frac{38}{62} = \frac{4430}{x}$$
$$38x = (4430)(62) = 274{,}660$$
$$x \approx 7230$$
About 7230 boys ages 17 to 18 play youth soccer.

e. Using ratios:
$$\frac{36}{100} = \frac{197{,}788}{x}$$
$$36x = (197{,}788)(100) = 19{,}778{,}800$$
$$x \approx 549{,}400$$
549,400 young people play youth soccer.

29. Solving:
$$25 = 30t - 5t^2$$
$$25 - 25 = -25 + 30t - 5t^2$$
$$0 = t^2 - 6t + 5$$
$$0 = (t-1)(t-5)$$
$$t - 1 = 0 \quad t - 5 = 0$$
$$t = 1 \qquad t = 5$$
The rocket is 25 meters above the cliff 1 second and at 5 seconds after it is shot.

30. Solving:
$$18 = -16t^2 + 32t + 6$$
$$18 - 18 = -16t^2 + 32t + 6 - 18$$
$$0 = 4t^2 - 8t + 3$$
$$0 = (2t-3)(2t-1)$$
$$t = 0.5, 1.5$$
The ball is 18 feet off the ground 0.5 second and 1.5 seconds after it is thrown.

CHAPTER 5 TEST

1. Solving:
$$\frac{x}{4} - 3 = \frac{1}{2}$$
$$\frac{x}{4} - 3 + 3 = \frac{1}{2} + 3$$
$$4 \cdot \frac{x}{4} = 4 \cdot \frac{7}{2}$$
$$x = 14$$

2. Solving:
$$x + 5(3x - 20) = 10(x - 4)$$
$$16x - 100 = 10x - 40$$
$$16x - 10x - 100 = 10x - 10x - 40$$
$$6x - 100 + 100 = 100 - 40$$
$$6x = 60$$
$$x = 10$$

3. Solving:
$$\frac{7}{16} = \frac{x}{12}$$
$$16x = 84$$
$$x = \frac{84}{16} = \frac{21}{4}$$

4. Solving:
$$x^2 = 12x - 27$$
$$x^2 - 12x + 27 = 12x - 12x - 27 + 27$$
$$(x-3)(x-9) = 0$$
$$x - 3 = 0 \quad x - 9 = 0$$
$$x = 3 \qquad x = 9$$

5. Solving:
$$3x^2 - 4x = 1$$
$$3x^2 - 4x - 1 = 1 - 1$$
$$3x^2 - 4x - 1 = 0$$
$$x = \frac{-(-4) \pm \sqrt{(-4)^2 - (4)(3)(-1)}}{(2)(3)}$$
$$x = \frac{2 + \sqrt{7}}{3}, \frac{2 - \sqrt{7}}{3}$$
$$x = -0.215, 1.549$$

6. Solving:
$$x - 2y = 15$$
$$x - x - 2y = 15 - x$$
$$\frac{1}{-2} \cdot -2y = \frac{1}{-2} \cdot (15 - x)$$
$$y = \frac{x - 15}{2}$$

7. Solving:
$$C = \frac{5}{9}(F - 32)$$
$$\frac{9}{5}C = F - 32$$
$$F = \frac{9}{5}C + 32$$

8. Solving:
$$63 = 12.4L + 32$$
$$12.4L = 63 - 32 = 31$$
$$L = 2.5$$
The duration of the last eruption was 2.5 minutes.

9. Solving:
$$78 = 15 + 7(d - 1)$$
$$78 = 15 + 7d - 7$$
$$70 = 7d$$
$$d = 10$$
The book was ten days overdue.

10. Using ratios:
$$\frac{246.6}{4.5} = \frac{x}{1}$$
$$(246.6)(1) = 4.5x$$
$$x = 54.8 \text{ miles per hour}$$

11. Solving:
$$4218 = 3 + 5c$$
$$5c = 4215$$
$$c = 843$$
Central Park has 843 acres.

12. a. Ty Cobb:
$$\frac{11,429}{4191} \approx 2.727$$
Billy Hamilton:
$$\frac{6284}{2163} \approx 2.905$$

Roger Hornsby:
$$\frac{8137}{2930} \approx 2.777$$
Joe Jackson:
$$\frac{4981}{1774} \approx 2.808$$
Tris Speaker:
$$\frac{10,195}{3514} \approx 2.901$$
Ted Williams:
$$\frac{7706}{2654} \approx 2.904$$

b. Ty Cobb, Roger Hornsby, Joe Jackson, Tris Speaker, Ted Williams, Billy Hamilton

13. $\frac{20}{35} = \frac{4}{7}$

14. Using ratios:
$$\frac{5}{5 + 3} = \frac{x}{180,000}$$
$$8x = (5)(180,000) = 900,000$$
$$x = 112,500$$
The amounts are \$112,500 and 180,000 − 112,500 = \$67,500.

15. Using ratios:
$$\frac{0.5}{50} = \frac{x}{275}$$
$$50x = 137.5$$
$$x = 2.75$$
2.75 pounds of plant food should be used.

16. a. Miami-Dade
$$\frac{2.5 - 2.0}{2.5} = 0.2$$
The number of working farms decreased 20% between 1977 and 1997.

b. Using ratios:
$$\frac{13.4}{1000} = \frac{x}{703,000}$$
$$1000x = 9,420,200$$
$$x \approx 9420$$
There are 9420 violent crimes committed in Baltimore.

17. $\dfrac{172,481}{1,200,000} \approx 0.144$

14.4% of registrations are Lab retrievers.

18. a. $\dfrac{59.3}{59.3 + 69.2 + 77.2 + 82.5} \approx 0.206$

    20.6% of the total number is expected to be sold in 2004.

    b. The percent increase is greatest between 2004 and 2005.

    $\dfrac{69.2 - 59.3}{59.3} \approx 0.167$

    $\dfrac{77.2 - 69.2}{69.2} \approx 0.116$

    $\dfrac{82.5 - 77.2}{77.2} \approx 0.069$

    c. $\dfrac{82.5 - 59.3}{59.3} \approx 0.391$

    The number of digital cameras sold increased by 39.1% between 2004 and 2007.

19. a. $\dfrac{2.5 - 2.0}{2.5} = 0.2$

    The number of working farms decreased 20% between 1977 and 1997.

    b. 0.20 (2.0 million) = 0.4 million
    The number of working farms will decrease by 0.4 million. There will be 2.0 − 0.4 = 1.6 million working farms in 2017.

    c. Answers will vary.

20. Solving:
$$10 = -16t^2 + 28t + 6$$
$$10 - 10 = -16t^2 + 28t + 6 - 10$$
$$0 = -16t^2 + 28t - 4$$
$$0 = 4t^2 - 7t + 1$$
$$t = \frac{-(-7) \pm \sqrt{(-7)^2 - (4)(4)(1)}}{(2)(4)}$$
$$t = \frac{7 + \sqrt{33}}{8}, \frac{7 - \sqrt{33}}{8}$$

The shot put is 10 feet above the ground at 0.2 second and 1.6 seconds.

Chapter 6: Applications of Functions

EXERCISE SET 6.1

1.

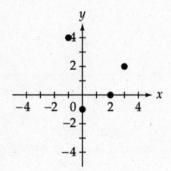

3.

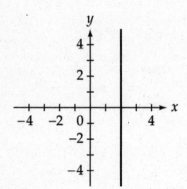

5.

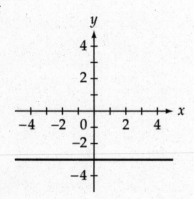

7.

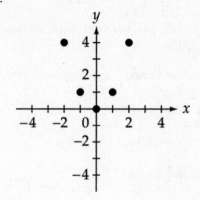

9.

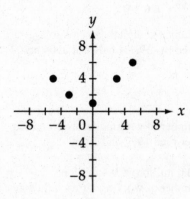

11.

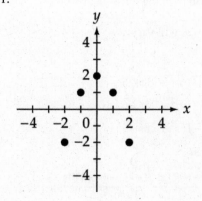

13.

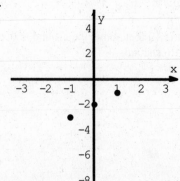

15.

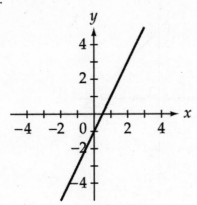

17.

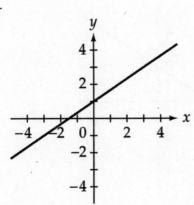

19.

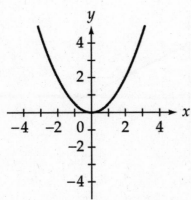

21.

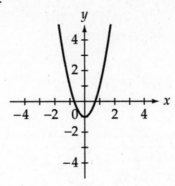

23.

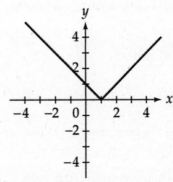

25. $f(x) = 2x + 7$
$f(-2) = 2(-2) + 7$
$= 3$

27. $f(t) = t^2 - t - 3$
$f(3) = 3^2 - 3 - 3$
$= 3$

29. $v(s) = s^3 + 3s^2 - 4s - 2$
$v(-2) = (-2)^3 + 3(-2)^2 - 4(-2) - 2$
$= 10$

31. $T(p) = \dfrac{p^2}{p-2}$

$T(0) = \dfrac{(0)^2}{0-2} = 0$

33. a. $P(s) = 4s$
$P(4) = 4(4) = 16$
The perimeter of the square is 16 meters.

b. $P(s) = 4s$
$P(5) = 4(5) = 20$
The perimeter of the square is 20 feet.

35. a. $h(t) = -16t^2 + 80t + 4$
         $h(2) = -16(2)^2 + 80(2) + 4 = 100$
         The height of the ball is 100 feet.

    b. $h(t) = -16t^2 + 80t + 4$
         $h(4) = -16(4)^2 + 80(4) + 4 = 68$
         The height of the ball is 68 feet.

37. a. $s(t) = \dfrac{1087\sqrt{t+273}}{16.52}$

         $s(0) = \dfrac{1087\sqrt{0+273}}{16.52} \approx 1087$

         The speed is approximately 1087 feet per second.

    b. $s(t) = \dfrac{1087\sqrt{t+273}}{16.52}$

         $s(25) = \dfrac{1087\sqrt{25+273}}{16.52} \approx 1136$

         The speed is approximately 1136 feet per second.

39. a.   The original concentration is the value of P when $x = 0$ grams are added.

         $P(x) = \dfrac{100x+100}{x+10}$

         $P(0) = \dfrac{100(0)+100}{0+10} = 10$

         The original concentration is 10%.

    b. $P(x) = \dfrac{100x+100}{x+10}$

         $P(5) = \dfrac{100(5)+100}{5+10} = 40$

         The concentration is 40%.

41.

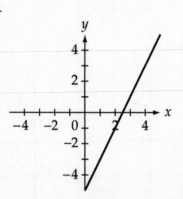

43.

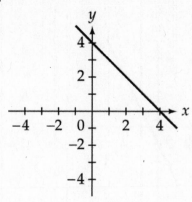

45.

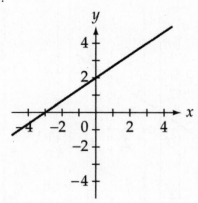

47.

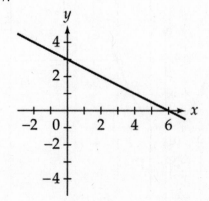

49.

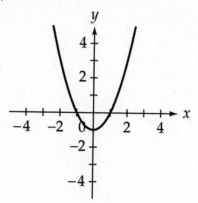

51.

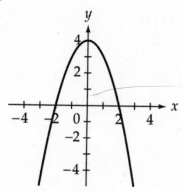

53.

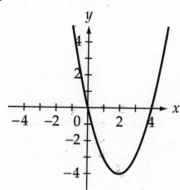

55.

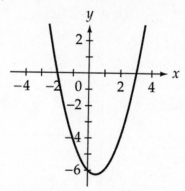

57. The width of the rectangle is 6 units and the
    length is 8 units.
    $A = lw$
    $A = (6)(8) = 48$
    The area of the rectangle is 48 square units.

59. No. A function cannot have different elements
    in the range corresponding to one element in the
    domain.

61. $f(x) = 2x + 5$
    $f(a) = 2a + 5$
    Since $f(a) = 9$, $2a + 5 = 9$, so $a = 2$.

63. $f(a, b) = a + b$
    $f(2, 5) = 2 + 5 = 7$
    $g(a, b) = (a)(b)$
    $g(2, 5) = (2)(5) = 10$
    $f(2, 5) + g(2, 5) = 7 + 10 = 17$

65. A graph of $f(x) = x^2 + 3$ shows that the lowest
    point occurs at $(0, 3)$, so the value of x for which
    $f(x)$ is smallest is $x = 0$.

EXERCISE SET 6.2

1. Solving:
    $f(x) = 3x - 6$
    $0 = 3x - 6$
    $3x = 6$
    $x = 2$
    x-intercept: $(2, 0)$

    $f(x) = 3x - 6$
    $f(0) = 3(0) - 6$
    $= -6$
    y-intercept: $(0, -6)$

3. Solving:

$$y = \frac{2}{3}x - 4$$

$$0 = \frac{2}{3}x - 4$$

$$-\frac{2}{3}x = -4$$

$$x = 6$$

x-intercept: $(6, 0)$

$$y = \frac{2}{3}x - 4$$

$$y = \frac{2}{3}(0) - 4$$

$$= -4$$

y-intercept: $(0, -4)$

5. Solving:

$$y = -x - 4$$

$$0 = -x - 4$$

$$x = -4$$

x-intercept: $(-4, 0)$

$$y = -x - 4$$

$$y = -(0) - 4$$

$$= -4$$

y-intercept: $(0, -4)$

7. Solving:

$$3x + 4y = 12$$

$$3x + 4(0) = 12$$

$$3x = 12$$

$$x = 4$$

x-intercept: $(4, 0)$

$$3x + 4y = 12$$

$$3(0) + 4y = 12$$

$$4y = 12$$

$$y = 3$$

y-intercept: $(0, 3)$

9. Solving:

$$2x - 3y = 9$$

$$2x - 3(0) = 9$$

$$2x = 9$$

$$x = \frac{9}{2}$$

x-intercept: $\left(\frac{9}{2}, 0\right)$

$$2x - 3y = 9$$

$$2(0) - 3y = 9$$

$$-3y = 9$$

$$y = -3$$

y-intercept: $(0, -3)$

11. Solving:

$$\frac{x}{2} + \frac{y}{3} = 1$$

$$\frac{x}{2} + \frac{0}{3} = 1$$

$$\frac{x}{2} = 1$$

$$x = 2$$

x-intercept: $(2, 0)$

$$\frac{x}{2} + \frac{y}{3} = 1$$

$$\frac{0}{2} + \frac{y}{3} = 1$$

$$\frac{y}{3} = 1$$

$$y = 3$$

y-intercept: $(0, 3)$

13. Solving:

$$x - \frac{y}{2} = 1$$

$$x - \frac{0}{2} = 1$$

$$x = 1$$

x-intercept: $(1, 0)$

$$x - \frac{y}{2} = 1$$

$$0 - \frac{y}{2} = 1$$

$-\dfrac{y}{2} = 1$

$y = -2$

y-intercept: $(0, -2)$

15. Finding the intercept:

$f(x) = 7x - 30$

$0 = 7x - 30$

$-7x = -30$

$x = \dfrac{30}{7}$

x-intercept: $\left(\dfrac{30}{7}, 0\right)$; At $\dfrac{30}{7}°$ C the cricket stops chirping.

17. Finding the intercepts:

$T(x) = 3x - 15$

$0 = 3x - 15$

$-3x = -15$

$x = 5$

The intercept on the horizontal axis is $(5, 0)$. This means that it takes 5 minutes for the temperature of the object to reach $0°$ F.

$T(x) = 3x - 15$

$T(0) = 3(0) - 15$

$T(0) = -15$

The intercept on the vertical axis is $(0, -15)$. This means that the temperature of the object is $-15°$ F before it is removed from the freezer.

19. $m = \dfrac{1-3}{3-1} = \dfrac{-2}{2} = -1$

21. $m = \dfrac{5-4}{2-(-1)} = \dfrac{1}{3}$

23. $m = \dfrac{5-3}{(-4)-(-1)} = \dfrac{2}{-3} = -\dfrac{2}{3}$

25. $m = \dfrac{0-3}{4-0} = \dfrac{-3}{4} = -\dfrac{3}{4}$

27. $m = \dfrac{(-2)-4}{2-2} = \dfrac{-6}{0}$; the slope is undefined.

29. $m = \dfrac{(-2)-5}{(-3)-2} = \dfrac{-7}{-5} = \dfrac{7}{5}$

31. $m = \dfrac{3-3}{(-1)-2} = \dfrac{0}{-3} = 0$

33. $m = \dfrac{5-4}{(-2)-0} = \dfrac{1}{-2} = -\dfrac{1}{2}$

35. $m = \dfrac{4-(-1)}{(-3)-(-3)} = \dfrac{5}{0}$; the slope is undefined.

37. $m = \dfrac{240-80}{6-2} = \dfrac{160}{4} = 40$; The slope is 40, which means the motorist was traveling at 40 miles per hour.

39. $m = \dfrac{14,325-4,000}{70,350-29,050} = \dfrac{10,325}{41,300} = 0.25$; The slope is 0.25, which means the tax rate for an income range of \$29,050 to \$70,350 is 25%.

41. $m = \dfrac{5000-0}{14.54-0} = \dfrac{5000}{14.54} \approx 343.9$; The slope is approximately 343.9, which means the runner traveled at a rate of 343.9 meters per minute.

43.

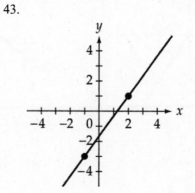

45.

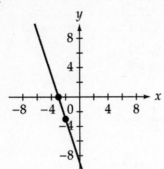

47.

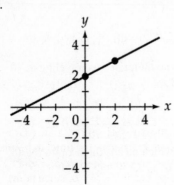

49.

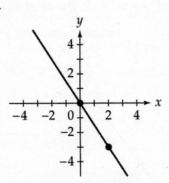

51.

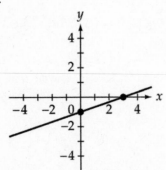

53. Line A represents the distance traveled by Lois in x hours, since it has the greatest slope. Line B represents the distance traveled by Tanya in x

hours, since its slope is smaller than Lois' and greater than the difference between them. Line C represents the distance between them in x hours, since it has the smallest slope, equal to the difference between the slopes of the other two lines, or $9 - 6 = 3$ km/hr.

55. No.

57. Graph the point (2, 3) and then find another point on the line with slope 2 by moving 2 units up and 1 unit to the right. Then move 2 units up and 1 unit right again. The value when $x = 4$ is $y = 7$.

59. Graph the point (1, 4) and then find another point on the line with slope $\frac{2}{3}$, which is equivalent to $\frac{-2}{-3}$, by moving 2 units down and 3 units to the left. The value when $x = -2$ is $y = 2$.

61. It rotates the line counterclockwise.

63. It raises the line on the rectangular coordinate system.

65. No. A vertical line through (4, 0) does not have a y-intercept.

EXERCISE SET 6.3

1. $y - y_1 = m(x - x_1)$
    $y - 5 = 2(x - 0)$
    $y - 5 = 2x$
    $y = 2x + 5$

3. $y - y_1 = m(x - x_1)$
    $y - 7 = -3(x - (-1))$
    $y - 7 = -3(x + 1)$
    $y - 7 = -3x - 3$
    $y = -3x + 4$

5. $y - y_1 = m(x - x_1)$
    $y - 5 = -\frac{2}{3}(x - 3)$
    $y - 5 = -\frac{2}{3}x + 2$
    $y = -\frac{2}{3}x + 7$

7. $y - y_1 = m(x - x_1)$

   $y - (-3) = 0(x - (-2))$

   $y + 3 = 0$

   $y = -3$

9. Find the slope of the line.

   $m = \dfrac{y_2 - y_1}{x_2 - x_1} = \dfrac{5 - 2}{3 - 0} = 1$

   Use the point-slope formula to find the equation of the line.

   $y - y_1 = m(x - x_1)$

   $y - 2 = 1(x - 0)$

   $y = x + 2$

11. Find the slope of the line.

    $m = \dfrac{y_2 - y_1}{x_2 - x_1} = \dfrac{0 - 3}{2 - 0} = -\dfrac{3}{2}$

    Use the point-slope formula to find the equation of the line.

    $y - y_1 = m(x - x_1)$

    $y - 3 = -\dfrac{3}{2}(x - 0)$

    $y = -\dfrac{3}{2}x + 3$

13. Find the slope of the line.

    $m = \dfrac{y_2 - y_1}{x_2 - x_1} = \dfrac{-1 - 0}{0 - 2} = \dfrac{1}{2}$

    Use the point-slope formula to find the equation of the line.

    $y - y_1 = m(x - x_1)$

    $y - 0 = \dfrac{1}{2}(x - 2)$

    $y = \dfrac{1}{2}x - 1$

15. Find the slope of the line.

    $m = \dfrac{y_2 - y_1}{x_2 - x_1} = \dfrac{(-5) - 5}{2 - (-2)} = -\dfrac{10}{4} = -\dfrac{5}{2}$

    Use the point-slope formula to find the equation of the line.

    $y - y_1 = m(x - x_1)$

    $y - 5 = -\dfrac{5}{2}(x - (-2))$

    $y = -\dfrac{5}{2}x$

17. Let $R(x)$ represent the number of rooms rented and let x represent the price for each room. Then $R(75) = 500$. The number of rooms rented decreases by 6 for each \$10 increase in price.

    Therefore, the slope is $\dfrac{-6}{10} = -\dfrac{3}{5}$.

    $y - y_1 = m(x - x_1)$

    $R - 500 = -\dfrac{3}{5}(x - 75)$

    $R = -\dfrac{3}{5}x + 545$

    Evaluate the function when $x = 100$.

    $R(x) = -\dfrac{3}{5}x + 545$

    $R(100) = -\dfrac{3}{5}(100) + 545$

    $R(100) = 485$

    When the room rate is \$100 per night, 485 rooms will be rented.

19. Let D represent the number of miles the plane will travel and let t represent the time, in hours. Then when $t = 2$, $D = 830$, so an ordered pair is (2, 830). Also, when $t = 0$, $D = 0$, since at the time of take-off, the plane has traveled 0 miles. So, a second ordered pair is (0, 0).

    Find the slope.

    $m = \dfrac{D_2 - D_1}{t_2 - t_1} = \dfrac{830 - 0}{2 - 0} = 415$

    Create the linear function, using the slope of 415 and the y-intercept of 0.

    $D(t) = mt + b$

    $D(t) = 415t$

    Evaluate the function when $t = 4.5$ hours.

    $D(t) = 415t$

    $D(4.5) = 415(4.5) = 1867.5$

    In $4\dfrac{1}{2}$ hours, the plane has traveled 1867.5 miles.

21. Let N represent the number of economy cars that would be sold per month and let x be the price of the truck. Then when $x = 9{,}000$, $N = 50{,}000$, and when $x = 8{,}750$, $N = 55{,}000$. So the ordered pairs are (9,000, 50,000) and (8,750, 55,000).

    Find the slope.

    $m = \dfrac{N_2 - N_1}{x_2 - x_1} = \dfrac{55{,}000 - 50{,}000}{8{,}750 - 9{,}000} = -20$

Use the point-slope formula to find the equation of the line.

$N - N_1 = m(x - x_1)$

$N - 50,000 = -20(x - 9,000)$

$N = -20x + 230,000$

Evaluate the function when $x = 8,500$ dollars.

$N(x) = -20x + 230,000$

$N(8,500) = -20(8,500) + 230,000$

$N(8,500) = 60,000$

At a price of \$8,500, there should be 60,000 economy cars sold.

23. a.   Using the calculator, the regression equation is $y = 0.56x + 41.71$.

     b.   Evaluate the regression equation when $x = 85$.
          $y = 0.56x + 41.71$
          $= 0.56(85) + 41.71$
          $= 89.31$
          ≈ 89
          A person whose stress test score was 85 has a diastolic blood pressure of approximately 89.

25. a.   Using the calculator, the regression equation is $y = 42.50x + 2613.76$.

     b.   Evaluate the regression equation when $x = 14$ (since 2010 is 14 years after 1996).
          $y = 42.50(14) + 2613.76$
          $= 3208.76$
          ≈ 3209
          In 2010, about 3209 thousand students will be expected to graduate from high school.

27. a.   Using the calculator, the regression equation is $y = -1.35x + 106.98$.

     b.   Evaluate the regression equation when $x = 45$.
          $y = -1.35x + 106.98$
          $= -1.35(45) + 106.98$
          $= 46.23 \approx 46$
          At $45°$ N latitude, the expected maximum temperature in January is $46°$ F.

29. Answers will vary. Any points on the line $y = -x + 3$.

31. Solutions to the same linear equation lie on the same line. The equation of the line through the first two points, $(0, 1)$, and $(4, 9)$, is $y = 2x + 1$. When $x = 3$, $y = 2(3) + 1 = 7$, so $n = 7$.

33. No. The three points do not lie on a straight line.

35. Let y represent the steepness of the hill and let x represent the speed of the car. Then when $x = 77$, $y = 5$, and when $x = 154$, $y = -2$.

     Find the slope.
     $$m = \frac{5 - (-2)}{77 - 154} = -\frac{7}{77} = -\frac{1}{11}$$

     Use the point-slope formula to find the equation of the line.
     $y - y_1 = m(x - x_1)$

     $y - 5 = -\frac{1}{11}(x - 77)$

     $y = -\frac{1}{11}x + 12$

     Evaluate the function when x = 99.

     $y = -\frac{1}{11}x + 12$

     $y = -\frac{1}{11}(99) + 12$

     $y = 3$

     The car is climbing $3°$.

EXERCISE SET 6.4

1. $x = -\dfrac{b}{2a} = -\dfrac{0}{2(1)} = 0$

     Find y by replacing x by 0 in the equation.
     $y = x^2 - 2$
     $y = 0^2 - 2$
     $y = -2$
     Vertex: $(0, -2)$

3. $x = -\dfrac{b}{2a} = -\dfrac{0}{2(-1)} = 0$

     Find y by replacing x by 0 in the equation.
     $y = -x^2 - 1$
     $y = -0^2 - 1$
     $y = -1$
     Vertex: $(0, -1)$

5. $x = -\dfrac{b}{2a} = -\dfrac{0}{2(-\frac{1}{2})} = 0$

    Find y by replacing x by 0 in the equation.

    $y = -\dfrac{1}{2}x^2 + 2$

    $y = -\dfrac{1}{2}(0)^2 + 2$

    $y = 2$

    Vertex: $(0, 2)$

7. $x = -\dfrac{b}{2a} = -\dfrac{0}{2(2)} = 0$

    Find y by replacing x by 0 in the equation.

    $y = 2x^2 - 1$

    $y = 2(0)^2 - 1$

    $y = -1$

    Vertex: $(0, -1)$

9. $x = -\dfrac{b}{2a} = -\dfrac{(-1)}{2(1)} = \dfrac{1}{2}$

    Find y by replacing x by $\dfrac{1}{2}$ in the equation.

    $y = x^2 - 3x + 2$

    $y = \left(\dfrac{1}{2}\right)^2 - \left(\dfrac{1}{2}\right) - 2$

    $y = -\dfrac{9}{4}$

    Vertex: $\left(\dfrac{1}{2}, -\dfrac{9}{4}\right)$

11. $x = -\dfrac{b}{2a} = -\dfrac{(-1)}{2(2)} = \dfrac{1}{4}$

    Find y by replacing x by $\dfrac{1}{4}$ in the equation.

    $y = 2x^2 - x - 5$

    $y = 2\left(\dfrac{1}{4}\right)^2 - \left(\dfrac{1}{4}\right) - 5$

    $y = -\dfrac{41}{8}$

    Vertex: $\left(\dfrac{1}{4}, -\dfrac{41}{8}\right)$

13. Solving:

    $y = 2x^2 - 4x$

    $0 = 2x^2 - 4x$

    $0 = 2x(x - 2)$

    $2x = 0 \qquad\qquad x - 2 = 0$

    $x = 0 \qquad\qquad\ x = 2$

    x-intercepts: $(0, 0)$ and $(2, 0)$

15. Solving:

    $y = 4x^2 + 11x + 6$

    $0 = 4x^2 + 11x + 6$

    $0 = (4x + 3)(x + 2)$

    $4x + 3 = 0 \qquad\qquad x + 2 = 0$

    $x = -\dfrac{3}{4} \qquad\qquad x = -2$

    x-intercepts: $\left(-\dfrac{3}{4}, 0\right)$ and $(-2, 0)$

17. Solving:

    $y = x^2 + 2x - 1$

    $0 = x^2 + 2x - 1$

    $x = \dfrac{-b \pm \sqrt{b^2 - 4ac}}{2a}$

    $x = \dfrac{-(2) \pm \sqrt{(2)^2 - 4(1)(-1)}}{2(1)}$

    $x = \dfrac{-2 \pm \sqrt{8}}{2} = \dfrac{-2 \pm 2\sqrt{2}}{2} = -1 \pm \sqrt{2}$

    x-intercepts: $\left(-1 + \sqrt{2}, 0\right)$ and $\left(-1 - \sqrt{2}, 0\right)$

19. Solving:

    $y = -x^2 - 4x - 5$

    $0 = -x^2 - 4x - 5$

    $x = \dfrac{-b \pm \sqrt{b^2 - 4ac}}{2a}$

    $x = \dfrac{-(-4) \pm \sqrt{(-4)^2 - 4(-1)(-5)}}{2(-1)}$

    $x = \dfrac{4 \pm \sqrt{16 - 20}}{-2} = \dfrac{4 \pm \sqrt{-4}}{-2}$

    Since $\sqrt{-4}$ is not a real number, this graph has no x-intercepts.

21. Solving:

$$y = -x^2 + 4x + 1$$
$$0 = -x^2 + 4x + 1$$

$$x = \frac{-b \pm \sqrt{b^2 - 4ac}}{2a}$$

$$x = \frac{-(4) \pm \sqrt{(4)^2 - 4(-1)(1)}}{2(-1)}$$

$$x = \frac{-4 \pm \sqrt{20}}{-2} = \frac{-4 \pm 2\sqrt{5}}{-2} = 2 \pm \sqrt{5}$$

x-intercepts: $\left(2 + \sqrt{5}, 0\right)$ and $\left(2 - \sqrt{5}, 0\right)$

23. Solving:

$$y = 2x^2 - 5x - 3$$
$$0 = 2x^2 - 5x - 3$$
$$0 = (2x + 1)(x - 3)$$
$$2x + 1 = 0 \qquad x - 3 = 0$$
$$x = -\frac{1}{2} \qquad\qquad x = 3$$

x-intercepts: $\left(-\frac{1}{2}, 0\right)$ and $(3, 0)$

25. $x = -\dfrac{b}{2a} = -\dfrac{(-2)}{2(1)} = 1$

Find y by replacing x by 1 in the function.

$$f(x) = x^2 - 2x + 3$$
$$f(1) = (1)^2 - 2(1) + 3$$
$$= 2$$

The vertex is (1, 2). The minimum value of the function is 2.

27. $x = -\dfrac{b}{2a} = -\dfrac{(4)}{2(-2)} = 1$

Find y by replacing x by 1 in the function.

$$f(x) = -2x^2 + 4x - 5$$
$$f(1) = -2(1)^2 + 4(1) - 5$$
$$f(1) = -3$$

The vertex is $(1, -3)$. The maximum value of the function is -3.

29. $x = -\dfrac{b}{2a} = -\dfrac{(-5)}{2(1)} = \dfrac{5}{2}$

Find y by replacing x by $\dfrac{5}{2}$ in the function.

$$f(x) = x^2 - 5x + 3$$
$$f\left(\frac{5}{2}\right) = \left(\frac{5}{2}\right)^2 - 5\left(\frac{5}{2}\right) + 3$$
$$f\left(\frac{5}{2}\right) = -\frac{13}{4}$$

The vertex is $\left(\frac{5}{2}, -\frac{13}{4}\right)$. The minimum value of the function is $-\frac{13}{4}$.

31. $x = -\dfrac{b}{2a} = -\dfrac{(-1)}{2(-1)} = -\dfrac{1}{2}$

Find y by replacing x by $-\dfrac{1}{2}$ in the function.

$$f(x) = -x^2 - x + 2$$
$$f\left(-\frac{1}{2}\right) = -\left(-\frac{1}{2}\right)^2 - \left(-\frac{1}{2}\right) + 2$$
$$f\left(-\frac{1}{2}\right) = \frac{9}{4}$$

The vertex is $\left(-\frac{1}{2}, \frac{9}{4}\right)$. The maximum value of the function is $\frac{9}{4}$.

33. The vertex of $y = y = x^2 - 2x - 3$ is $(1, -4)$, so its minimum value is -4.

The vertex of $y = x^2 - 10x + 20$ is $(5, -5)$ so its minimum value is -5.

The vertex of $y = 3x^2 - 1$ is $(0, -1)$ so its minimum value is -1.
The largest minimum value is -1, so the answer is **c**.

35. Find the coordinates of the vertex.

$$t = -\frac{b}{2a} = -\frac{(80)}{2(-16)} = 2.5$$

$$s(t) = -16t^2 + 80t + 50$$
$$s(2.5) = -16(2.5)^2 + 80(2.5) + 50$$
$$s(2.5) = 150$$

The maximum height above the ground that the ball will attain is 150 feet.

37. Find the x-coordinate of the vertex.

$$x = -\frac{b}{2a} = -\frac{(-20)}{2(0.1)} = 100$$

The company should produce 100 lenses to minimize the average cost.

39. Find the coordinates of the vertex.

$$x = -\frac{b}{2a} = -\frac{(-0.8)}{2(0.25)} = 1.6$$

$h(x) = 0.25x^2 - 0.8x + 25$
$h(1.6) = 0.25(1.6)^2 - 0.8(1.6) + 25$
$h(1.6) = 24.36$

The minimum height of the cable above the bridge is 24.36 feet.

41. Find the maximum height of the orange.

$$t = -\frac{b}{2a} = -\frac{(32)}{2(-16)} = 1$$

$h(t) = -16t^2 + 32t + 4$

$h(1) = -16(1)^2 + 32(1) + 4$

$h(1) = 20$

The maximum height the orange will attain is 20 feet. The orange will be at 18 feet on its way up to the maximum height and on its way back down to the ground. So, the answer is yes.

43. Find the coordinates of the vertex.

$$t = -\frac{b}{2a} = -\frac{(90)}{2(-16)} \approx 2.81$$

$h(t) = -16t^2 + 90t + 15$
$h(2.81) = -16(2.81)^2 + 90(2.81) + 15$
$h(2.81) \approx 141.6$

The maximum height of the waterspout is about 141.6 feet.

45. a. Find the x-coordinate of the vertex.

$$x = -\frac{b}{2a} = -\frac{(1.476)}{2(-0.018)} = 41$$

The speed that will yield the maximum fuel efficiency is 41 miles per hour.

b. Find the y-coordinate of the vertex.

$E(v) = -0.018v^2 + 1.476v + 3.4$

$E(41) = -0.018(41)^2 + 1.476(41) + 3.4$

$E(41) = 33.658$

The maximum fuel efficiency is 33.658 miles per gallon.

47. Replace x with 2 and y with 5 in the equation.

$f(x) = x^2 - 3x + k$

$5 = (2)^2 - 3(2) + k$

$5 = -2 + k$

$7 = k$

49. First, find the value of k. Replace x with $-\frac{3}{2}$ and y with 0.

$f(x) = 2x^2 - 5x + k$

$0 = 2\left(-\frac{3}{2}\right)^2 - 5\left(-\frac{3}{2}\right) + k$

$0 = \frac{9}{2} + \frac{15}{2} + k$

$0 = 12 + k$

$k = -12$

Now, replace $f(x)$ with 0 and k with -12 and solve.

$0 = 2x^2 - 5x - 12$

$0 = (2x + 3)(x - 4)$

$2x + 3 = 0 \qquad x - 4 = 0$

$x = -\frac{3}{2} \qquad\qquad x = 4$

The other value of x for which $f(x) = 0$ is 4.

EXERCISE SET 6.5

1. a. $f(2) = 3^2 = 9$

   b. $f(0) = 3^0 = 1$

   c. $f(-2) = 3^{-2} = \frac{1}{9}$

3. a. $g(3) = 2^{3+1} = 2^4 = 16$

   b. $g(1) = 2^{1+1} = 2^2 = 4$

   c. $g(-3) = 2^{-3+1} = 2^{-2} = \frac{1}{4}$

5. a. $G(0) = \left(\frac{1}{2}\right)^{2(0)} = \left(\frac{1}{2}\right)^0 = 1$

   b. $G\left(\frac{3}{2}\right) = \left(\frac{1}{2}\right)^{2(3/2)} = \left(\frac{1}{2}\right)^3 = \frac{1}{8}$

c. $G(-2) = \left(\frac{1}{2}\right)^{2(-2)} = \left(\frac{1}{2}\right)^{-4} = 16$

7.   a. $h(4) = e^{4/2} = e^2 \approx 7.3891$

b. $h(-2) = e^{-2/2} = e^{-1} \approx 0.3679$

c. $h\left(\frac{1}{2}\right) = e^{(1/2)/2} = e^{1/4} \approx 1.2840$

9.   a. $H(-1) = e^{-(-1)+3} = e^4 \approx 54.5982$

b. $H(3) = e^{-(3)+3} = e^0 = 1$

c. $H(5) = e^{-(5)+3} = e^{-2} \approx 0.1353$

11.   a. $F(2) = 2^{2^2} = 2^4 = 16$

b. $F(-2) = 2^{(-2)^2} = 2^4 = 16$

c. $F\left(\frac{3}{4}\right) = 2^{(3/4)^2} = 2^{9/16} \approx 1.4768$

13.   a. $f(-2) = e^{-(-2)^2/2} = e^{-2} \approx 0.1353$

b. $f(2) = e^{-(2)^2/2} = e^{-2} \approx 0.1353$

c. $f(-3) = e^{-(-3)^2/2} = e^{-9/2} \approx 0.0111$

15.

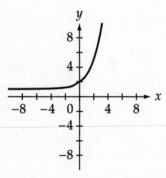

17.

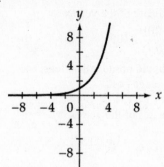

19.

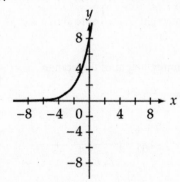

21.

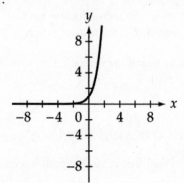

23.

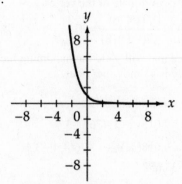

25. $A = P\left(1 + \dfrac{r}{n}\right)^{nt}$

$= 2500\left(1 + \dfrac{0.075}{365}\right)^{365(20)}$

$\approx 2500(4.480999) \approx 11{,}202.50$
After 20 years, there is \$11,202.50 in the account.

27. $A = Pe^{rt}$

$A = 2500\,e^{0.05(5)} \approx 3210.06$
The value of the investment is \$3210.06.

29. $A = 8\left(\dfrac{1}{2}\right)^{t/8}$

$A = 8\left(\dfrac{1}{2}\right)^{5/8} \approx 5.2$

The amount of iodine-131 present is 5.2 micrograms.

31. Make a table to find the pattern.

Note	Frequency
1 (concert A)	$440 = 440 \times 2^0 =$ $440 \times 2^{0/12}$
12 (next A)	$880 = 440 \times 2^1 =$ $440 \times 2^{12/12}$

The function to represent the frequency F of note n above concert A is
$F(n) = 440(2^{n/12})$.

33. Use the exponential equation $y = a \cdot b^t$ and the values of t and y to find the equation.

Substitute $t = 0$ (year 1900), $y = 8000$ into the equation:
$y = a \cdot b^t$

$8000 = a \cdot b^0$
$8000 = a$

Substitute $t = 100$ (year 2000), $y = 200{,}000{,}000$ and $a = 8000$ into the equation:
$y = a \cdot b^t$

$200{,}000{,}000 = 8000 \cdot b^{100}$

$\dfrac{200{,}000{,}000}{8000} = b^{100}$

$25{,}000 = b^{100}$

$(25{,}000)^{1/100} = b$
$1.1066 \approx b$

The exponential model is $y = 8000(1.1066)^t$.

Evaluate this function for $t = 110$, since the year 2010 is 110 years after 1900.
$y = 8000(1.1066)^t$

$y = 8000(1.1066)^{110} \approx 552{,}200{,}000$
The number of automobiles in the United States in 2010 will be about 552,200,000.

35. Use the exponential equation $y = a \cdot b^t$ and the values of t and y to find the equation.

Substitute $t = 0$ (time 0 days), $y = 100$ into the equation:
$y = a \cdot b^t$

$100 = a \cdot b^0$
$100 = a$
Substitute $t = 34.5$, $y = 75$ and $a = 100$ into the equation:
$y = a \cdot b^t$

$75 = 100 \cdot b^{34.5}$

$\dfrac{75}{100} = b^{34.5}$

$0.75^{1/34.5} = b$

$0.99 \approx b$

The exponential model is $y = 100(0.99)^t$.

37. a. Use the exponential regression function on your graphing calculator.
$y = 4.959(1.063)^x$

 b. To predict the amount of water, substitute 15 for x and evaluate.
$y = 4.959(1.063)^x$

$y = 4.959(1.063)^{15}$

$y \approx 12.4$
At $15\,^\circ C$, there will be 12.4 milliliters of water per cubic meter of air.

39. a.   The year 1980 corresponds to $t = 20$.
Substitute 20 for t and evaluate.

$$P(t) = \frac{280}{4 + 66e^{-0.021t}}$$

$$P(20) = \frac{280}{4 + 66e^{-0.021(20)}}$$

$$P(20) \approx 5.9$$

The population in 2000 was about 5.9 billion people.

b.   The year 2010 corresponds to $t = 30$.
Substitute 30 for t and evaluate.

$$P(t) = \frac{280}{4 + 66e^{-0.021t}}$$

$$P(30) = \frac{280}{4 + 66e^{-0.021(30)}}$$

$$P(30) \approx 7.2$$

The Earth's population in 2010 will be about 7.2 billion people.

c.   If $e^{-0.021t} \approx 0$, then

$$P(t) \approx \frac{280}{4 + 66(0)} = \frac{280}{4} = 70.$$

The maximum population the Earth can support is 70.0 billion people.

EXERCISE SET 6.6

1. $\log_7 49 = 2$

3. $\log_5 625 = 4$

5. $\log_{10} 0.0001 = -4$

7. $\log_{10} x = y$

9. $3^4 = 81$

11. $5^3 = 125$

13. $4^{-2} = \dfrac{1}{16}$

15. $e^y = x$

17. $\log_3 81 = x$
$3^x = 81$
$3^x = 3^4$
$x = 4$
$\log_3 81 = 4$

19. $\log 100 = x$
$10^x = 100$
$10^x = 10^2$
$x = 2$
$\log 100 = 2$

21. $\log_3 \dfrac{1}{9} = x$
$3^x = \dfrac{1}{9}$
$3^x = 3^{-2}$
$x = -2$
$\log_3 \dfrac{1}{9} = -2$

23. $\log_2 64 = x$
$2^x = 64$
$2^x = 2^6$
$x = 6$
$\log_2 64 = 6$

25. $\log_3 x = 2$
$3^2 = x$
$9 = x$

27. $\log_7 x = -1$
$7^{-1} = x$
$\dfrac{1}{7} = x$

29. $\log_3 x = -2$
$3^{-2} = x$
$\dfrac{1}{9} = x$

31. $\log_4 x = 0$
$4^0 = x$
$1 = x$

33. $\log x = 2.5$
$10^{2.5} = x$
$316.23 \approx x$

35. $\ln x = 2$
$e^2 = x$
$7.39 \approx x$

37. $\log x = 0.35$
$10^{0.35} = x$
$2.24 \approx x$

39. $\ln x = \dfrac{8}{3}$
$e^{8/3} = x$
$14.39 \approx x$

41.

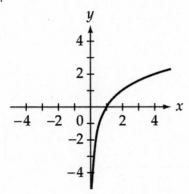

43.

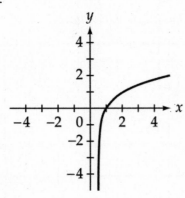

45.

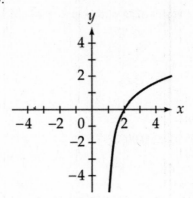

47. $\log P = -kd$
$\log P = -(0.2)(0.5)$
$\log P = -0.1$
$10^{-0.1} = P$
$0.79 \approx P$
The amount of light passing through the material is about 79%.

49. $D = 10(\log I + 16)$
$D = 10[\log(3.2 \times 10^{-10}) + 16] \approx 65$
Normal conversation is about 65 decibels.

51. $\text{pH} = -\log(\text{H}^+)$
$\text{pH} = -\log(0.045) \approx 1.35$
The digestive solution of the stomach has a pH of about 1.35

53. $M = \log \dfrac{I}{I_0}$
$M = \log\left(\dfrac{6,309,573 I_0}{I_0}\right)$
$M = \log 6,309,573 \approx 6.8$
The earthquake had a Richter scale magnitude of 6.8.

55. $M = \log \dfrac{I}{I_0}$
$8.9 = \log\left(\dfrac{I}{I_0}\right)$
$\dfrac{I}{I_0} = 10^{8.9} \approx 794,328,235$
$I = 794,328,235 I_0$
The earthquake had an intensity of $794,328,235 I_0$.

57. Find the intensity of an earthquake of magnitude 8.

$$M = \log \frac{I}{I_0}$$

$$8 = \log\left(\frac{I}{I_0}\right)$$

$$\frac{I}{I_0} = 10^8 \approx 100{,}000{,}000$$

$$I = 100{,}000{,}000 I_0$$

Find the intensity of an earthquake of magnitude 6.

$$M = \log \frac{I}{I_0}$$

$$6 = \log\left(\frac{I}{I_0}\right)$$

$$\frac{I}{I_0} = 10^6 \approx 1{,}000{,}000$$

$$I = 1{,}000{,}000 I_0$$

Since 100,000,000 is 100 times more than 1,000,000, the earthquake of magnitude 8 is 100 times stronger than an earthquake of magnitude 6.

59. $M = 5\log r - 5$

$-1.11 = 5\log r - 5$

$3.89 = 5\log r$

$\dfrac{3.89}{5} = \log r$

$0.778 = \log r$

$r = 10^{0.778} \approx 6.0$

The Earth is approximately 6.0 parsecs from Alpha Centauri.

61. $T = 14.29\ln(0.00411r + 1)$

$50 = 14.29\ln(0.00411r + 1)$

$\dfrac{50}{14.29} = \ln(0.00411r + 1)$

$0.00411r + 1 = e^{50/14.29}$

$0.00411r = e^{50/14.29} - 1$

$r = \dfrac{e^{50/14.29} - 1}{0.00411} \approx 7805.5$

Approximately 7805.5 billion barrels of oil are necessary to last 50 years.

63. $x = \dfrac{3}{2}$

$= \dfrac{10^{0.47712}}{10^{0.30103}} = 10^{0.47712 - 0.30103} = 10^{0.17609}$

In logarithmic form, the equation is $\log x = 0.17609$.

$10^{0.17609} \approx 1.5$

65. Answers will vary.

CHAPTER 6 REVIEW EXERCISES

1.

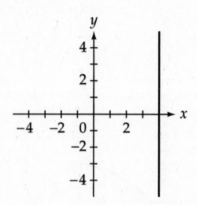

2.

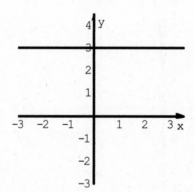

3.

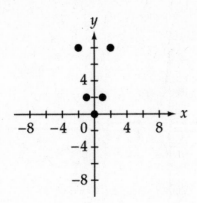

4.

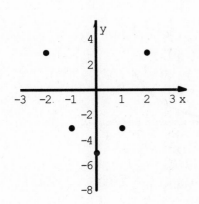

5.

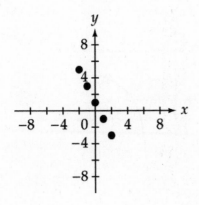

6.

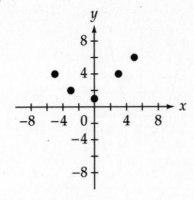

7.

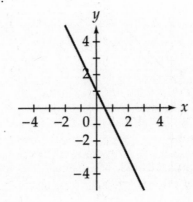

8.

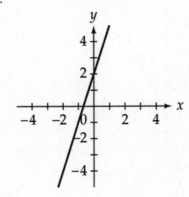

9.

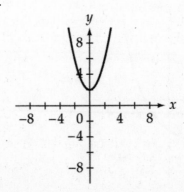

10.

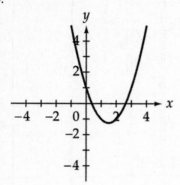

14.

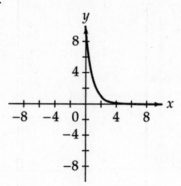

11.

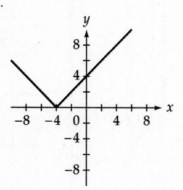

15.

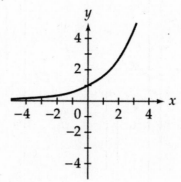

12.

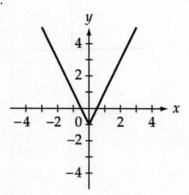

16.

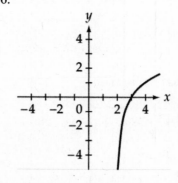

17. $f(x) = 4x - 5$

$f(-2) = 4(-2) - 5 = -13$

18. $g(x) = 2x^2 - x - 2$

$g(3) = 2(3)^2 - 3 - 2 = 13$

19. $s(t) = \dfrac{4}{3t - 5}$

$s(-1) = \dfrac{4}{3(-1) - 5} = \dfrac{4}{-8} = -\dfrac{1}{2}$

13.

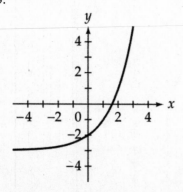

20. $R(s) = s^3 - 2s^2 + s - 3$

    $R(-2) = (-2)^3 - 2(-2)^2 + (-2) - 3 = -21$

21. $f(x) = 2^{x-3}$

    $f(5) = 2^{5-3} = 2^2 = 4$

22. $g(x) = \left(\dfrac{2}{3}\right)^x$

    $g(2) = \left(\dfrac{2}{3}\right)^2 = \dfrac{4}{9}$

23. $T(r) = 2e^r + 1$

    $T(2) = 2e^2 + 1 \approx 15.78$

24. a. $V(r) = \dfrac{4\pi r^3}{3}$

       $V(3) = \dfrac{4\pi(3)^3}{3} \approx 113.1$

       The volume is about 113.1 cubic inches.

    b. $V(r) = \dfrac{4\pi r^3}{3}$

       $V(12) = \dfrac{4\pi(12)^3}{3} \approx 7238.2$

       The volume of the sphere with radius 12 inches is about 7238.2 cubic centimeters.

25. a. $h(t) = -16t^2 + 96t + 5$

       $h(2) = -16(2)^2 + 96(2) + 5 = 133$

       The height of the ball is 133 feet.

    b. $h(t) = -16t^2 + 96t + 5$

       $h(4) = -16(4)^2 + 96(4) + 5 = 133$

       The height of the ball is 133 feet.

26. a. $P(x) = \dfrac{100x + 100}{x + 20}$

       $P(0) = \dfrac{100(0) + 100}{0 + 20} = 5$

       The original concentration of sugar is 5%.

    b. $P(x) = \dfrac{100x + 100}{x + 20}$

       $P(10) = \dfrac{100(10) + 100}{10 + 20} \approx 36.7$

       The concentration of sugar after 10 grams of sugar have been added is 36.7%.

27. $f(x) = 2x + 10$

    $0 = 2x + 10$

    $-2x = 10$

    $x = -5$

    x-intercept: $(-5, 0)$

    $f(x) = 2x + 10$

    $f(0) = 2(0) + 10 = 10$

    y-intercept: $(0, 10)$

28. $f(x) = \dfrac{3}{4}x - 9$

    $0 = \dfrac{3}{4}x - 9$

    $9 = \dfrac{3}{4}x$

    $12 = x$

    x-intercept: $(12, 0)$

    $f(x) = \dfrac{3}{4}x - 9$

    $f(0) = \dfrac{3}{4}(0) - 9$

    $f(0) = -9$

    y-intercept: $(0, -9)$

29. $3x - 5y = 15$

    $3x - 5(0) = 15$

    $3x = 15$

    $x = 5$

    x-intercept: $(5, 0)$

    $3x - 5y = 15$

    $3(0) - 5y = 15$

    $-5y = 15$

    $y = -3$

    y-intercept: $(0, -3)$

30. $4x + 3y = 24$

    $4x + 3(0) = 24$

    $4x = 24$

    $x = 6$

    x-intercept: $(6, 0)$

    $4x + 3y = 24$

    $4(0) + 3y = 24$

    $3y = 24$

    $y = 8$

    y-intercept: $(0, 8)$

31. $V(t) = 25,000 - 5000t$

    $V(0) = 25,000 - 5000(0)$

    $V(0) = 25,000$

    The intercept on the vertical axis is
    $(0, 25,000)$. This means that the value of the
    truck was \$25,000 when it was new.

    $V(t) = 25,000 - 5000t$

    $0 = 25,000 - 5000t$

    $5000t = 25,000$

    $t = 5$

    The intercept on the horizontal axis is
    $(5, 0)$. This means that after 5 years the truck
    will be worth \$0.

32. $m = \dfrac{-3 - 2}{2 - 3} = \dfrac{-5}{-1} = 5$

33. $m = \dfrac{-1 - 4}{-3 - (-1)} = \dfrac{-5}{-2} = \dfrac{5}{2}$

34. $m = \dfrac{-5 - (-5)}{-4 - 2} = \dfrac{0}{-6} = 0$

35. $m = \dfrac{7 - 2}{5 - 5} = \dfrac{5}{0}$; the slope is undefined.

36. $m = \dfrac{8.1 - 9.9}{2007 - 2004} = \dfrac{-1.8}{3} = -0.6$

    The slope is -0.6, which means that the revenue
    from home video rentals is decreasing by
    \$600,000 annually.

37.

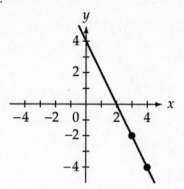

38.

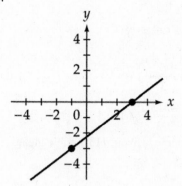

39. $y - y_1 = m(x - x_1)$

    $y - 3 = 2(x - (-2))$

    $y - 3 = 2x + 4$

    $y = 2x + 7$

40. $y - y_1 = m(x - x_1)$

    $y - (-4) = 1(x - 1)$

    $y + 4 = x - 1$

    $y = x - 5$

41. $y - y_1 = m(x - x_1)$

    $y - 1 = \dfrac{2}{3}(x - (-3))$

    $y - 1 = \dfrac{2}{3}x + 2$

    $y = \dfrac{2}{3}x + 3$

42. $y - y_1 = m(x - x_1)$

    $y - 1 = \dfrac{1}{4}(x - 4)$

    $y - 1 = \dfrac{1}{4}x - 1$

    $y = \dfrac{1}{4}x$

43.

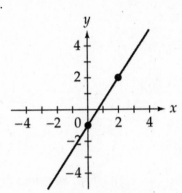

44. a.   Let x represent the number of square yards of carpet, and let $f(x)$ represent the cost. The cost of the carpet increases by \$25 for each additional square yard. So, the slope is 25. The initial cost of the carpet is \$100, for installation, so the y-intercept is 100.
$$y = mx + b$$
$$f(x) = 25x + 100$$

    b. $f(x) = 25x + 100$
$$f(32) = 25(32) + 100 = 900$$
The cost to carpet 32 square yards of floor space is \$900.

45. a.   Let A represent the number of gallons sold and let p represent the price relative to Western QuickMart. When the price increases by \$.02, the number of gallons sold decreases by 500 gallons.
$$m = \frac{\text{change in } A}{\text{change in } p} = \frac{-500}{0.02} = -25{,}000$$
When the prices are the same, the price relative to Western QuickMart is 0, so an ordered pair is
(0, 10,000). The y-intercept is 10,000.
$$y = mx + b$$
$$A = mp + b$$
$$A(p) = -25{,}000p + 10{,}000$$

    b.   The price is \$0.03 below Western QuickMart, so $p = -0.03$.
$$A(p) = -25{,}000p + 10{,}000$$
$$A(-0.03) = -25{,}000(-0.03) + 10{,}000$$
$$= 10{,}750$$
Valley Gas Mart will sell 10,750 gallons of gasoline.

46. a.   Using the calculator, the regression equation is
$$y = 0.16204x - 0.12674.$$

    b. $y = 0.16204x - 0.12674$
$$y = 0.16204(158) - 0.12674 \approx 25$$
The BMI is about 25.

47. $x = -\dfrac{b}{2a} = -\dfrac{2}{2(1)} = -1$
Find y by replacing x by -1 in the equation.
$$y = x^2 + 2x + 4$$
$$y = (-1)^2 + 2(-1) + 4$$
$$y = 3$$
Vertex: $(-1, 3)$

48. $x = -\dfrac{b}{2a} = -\dfrac{(-6)}{2(-2)} = -\dfrac{3}{2}$
Find y by replacing x by $-\dfrac{3}{2}$ in the equation.
$$y = -2x^2 - 6x + 1$$
$$y = -2\left(-\frac{3}{2}\right)^2 - 6\left(-\frac{3}{2}\right) + 1$$
$$y = \frac{11}{2}$$
Vertex: $\left(-\dfrac{3}{2}, \dfrac{11}{2}\right)$

49. $x = -\dfrac{b}{2a} = -\dfrac{6}{2(-3)} = 1$
Find y by replacing x by 1 in the equation.
$$y = -3x^2 + 6x - 1$$
$$y = -3(1)^2 + 6(1) - 1$$
$$y = 2$$
Vertex: $(1, 2)$

50. $x = -\dfrac{b}{2a} = -\dfrac{5}{2(1)} = -\dfrac{5}{2}$
Find y by replacing x by $-\dfrac{5}{2}$ in the equation.
$$y = x^2 + 5x - 1$$
$$y = \left(-\frac{5}{2}\right)^2 + 5\left(-\frac{5}{2}\right) - 1$$
$$y = -\frac{29}{4}$$
Vertex: $\left(-\dfrac{5}{2}, -\dfrac{29}{4}\right)$

51. $y = x^2 + x - 20$
    $0 = (x - 4)(x + 5)$

    $x - 4 = 0 \qquad x + 5 = 0$
    $x = 4 \qquad\qquad x = -5$
    x-intercepts: $(4, 0)$ and $(-5, 0)$

52. $y = x^2 + 2x - 1$
    $0 = x^2 + 2x - 1$

    $x = \dfrac{-b \pm \sqrt{b^2 - 4ac}}{2a}$

    $x = \dfrac{-(2) \pm \sqrt{(2)^2 - 4(1)(-1)}}{2(1)}$

    $x = \dfrac{-2 \pm \sqrt{8}}{2} = \dfrac{-2 \pm 2\sqrt{2}}{2} = -1 \pm \sqrt{2}$

    x-intercepts: $\left(-1 + \sqrt{2}, 0\right)$ and $\left(-1 - \sqrt{2}, 0\right)$

53. $f(x) = 2x^2 + 9x + 4$
    $0 = (2x + 1)(x + 4)$

    $2x + 1 = 0 \qquad x + 4 = 0$
    $x = -\dfrac{1}{2} \qquad\qquad x = -4$

    x-intercepts: $\left(-\dfrac{1}{2}, 0\right)$ and $(-4, 0)$

54. $y = x^2 + 4x + 6$
    $0 = x^2 + 4x + 6$

    $x = \dfrac{-b \pm \sqrt{b^2 - 4ac}}{2a}$

    $x = \dfrac{-(4) \pm \sqrt{(4)^2 - 4(1)(6)}}{2(1)}$

    $x = \dfrac{-4 \pm \sqrt{16 - 24}}{2} = \dfrac{-4 \pm \sqrt{-8}}{2}$

    Since $\sqrt{-8}$ is not a real number, the graph has no x-intercepts.

55. $x = -\dfrac{b}{2a} = -\dfrac{4}{2(-1)} = 2$

    Find y by replacing x by 2 in the function.
    $y = -x^2 + 4x + 1$
    $y = -(2)^2 + 4(2) + 1$
    $y = 5$
    The vertex is $(2, 5)$. The maximum value of the function is 5.

56. $x = -\dfrac{b}{2a} = -\dfrac{6}{2(2)} = -\dfrac{3}{2}$

    Find y by replacing x by $-\dfrac{3}{2}$ in the function.

    $y = 2x^2 + 6x - 3$

    $y = 2\left(-\dfrac{3}{2}\right)^2 + 6\left(-\dfrac{3}{2}\right) - 3$

    $y = -\dfrac{15}{2}$

    The vertex is $\left(-\dfrac{3}{2}, -\dfrac{15}{2}\right)$. The minimum value of the function is $-\dfrac{15}{2}$.

57. $x = -\dfrac{b}{2a} = -\dfrac{(-4)}{2(1)} = 2$

    Find y by replacing x by 2 in the function.
    $f(x) = x^2 - 4x - 1$
    $f(2) = 2^2 - 4(2) - 1$
    $f(2) = -5$
    The vertex is $(2, -5)$. The minimum value of the function is -5.

58. $x = -\dfrac{b}{2a} = -\dfrac{3}{2(-2)} = \dfrac{3}{4}$

    Find y by replacing x by $\dfrac{3}{4}$ in the function.

    $f(x) = -2x^2 + 3x - 1$

    $f\left(\dfrac{3}{4}\right) = -2\left(\dfrac{3}{4}\right)^2 + 3\left(\dfrac{3}{4}\right) - 1$

    $f\left(\dfrac{3}{4}\right) = \dfrac{1}{8}$

    The vertex is $\left(\dfrac{3}{4}, \dfrac{1}{8}\right)$. The maximum value of the function is $\dfrac{1}{8}$.

59. Find the coordinates of the vertex.

    $t = -\dfrac{b}{2a} = -\dfrac{80}{2(-16)} = \dfrac{5}{2}$

    $s(t) = -16t^2 + 80t + 25$

    $s(2.5) = -16\left(\dfrac{5}{2}\right)^2 + 80\left(\dfrac{5}{2}\right) + 25$

    $s(2.5) = 125$
    The maximum height that the rock will attain is 125 feet above the ground.

60. Find the x-coordinate of the vertex.

$$x = -\frac{b}{2a} = -\frac{(-40)}{2(0.01)} = 2000$$

The company should produce 2000 CD-RWs.

61. $A = P\left(1 + \dfrac{r}{n}\right)^{nt}$

$$= 5000\left(1 + \frac{0.06}{365}\right)^{365(15)}$$

$\approx 5000(2.459421) \approx 12{,}297.11$

After 15 years, there is \$12,297.11 in the account.

62. $A = 10\left(\dfrac{1}{2}\right)^{t/6}$

$$A(2) = 10\left(\frac{1}{2}\right)^{2/6} \approx 7.94$$

There will be 7.94 milligrams of the isotope present after 2 hours.

63. $A = 8\left(\dfrac{1}{2}\right)^{t/8}$

$$A(10) = 8\left(\frac{1}{2}\right)^{10/8} \approx 3.36$$

There will be 3.36 micrograms present after 10 days.

64. a. Make a table to find the pattern.

n = Number of bounces	H = Height of ball
0	6
1	$6(2/3) = 4$
2	$4(2/3)$ $= 6(2/3)^2$
3	$6(2/3)^2 (2/3)$ $= 6(2/3)^3$

$$H(n) = 6\left(\frac{2}{3}\right)^n$$

b. $H(n) = 6\left(\dfrac{2}{3}\right)^n$

$$H(5) = 6\left(\frac{2}{3}\right)^5 \approx 0.79$$

The height of the ball after the fifth bounce is about 0.79 feet.

65. a. $N = 250.2056(0.6506)^t$

b. Evaluate the function for $t = 8$.

$N = 250.2056(0.6506)^t$

$N(8) = 250.2056(0.6506)^8 \approx 8$

About 8 thousand people will attend the movie 8 weeks after it has been released.

66. $\log_3 243 = x$

$3^x = 243$

$3^x = 3^5$

$x = 5$

$\log_3 243 = 5$

67. $\log_2 \dfrac{1}{16} = x$

$2^x = \dfrac{1}{16}$

$2^x = 2^{-4}$

$x = -4$

$\log_2 \dfrac{1}{16} = -4$

68. $\log_4 \dfrac{1}{4} = x$

$4^x = \dfrac{1}{4}$

$4^x = 4^{-1}$

$x = -1$

$\log_4 \dfrac{1}{4} = -1$

69. $\log_2 64 = x$

$2^x = 64$

$2^x = 2^6$

$x = 6$

$\log_2 64 = 6$

70. $\log_4 x = 3$

$4^3 = x$

$64 = x$

71. $\log_3 x = \dfrac{1}{3}$

$3^{1/3} = x$

$1.4422 \approx x$

72. $\ln x = 2.5$

$e^{2.5} = x$

$12.1825 \approx x$

73. $\log x = 2.4$

$10^{2.4} = x$

$251.1886 \approx x$

74. $M = 5 \log r - 5$

$3.2 = 5 \log r - 5$

$8.2 = 5 \log r$

$\dfrac{8.2}{5} = \log r$

$10^{8.2/5} = r$

$43.7 \approx r$

The distance to the star is 43.7 parsecs.

75. $D = 10(\log I + 16)$

$D = 10[\log(0.01) + 16]$

$D = 140$

The pain threshold for most humans is 140 decibels.

CHAPTER 6 TEST

1. $s(t) = -3t^2 + 4t - 1$

$s(-2) = -3(-2)^2 + 4(-2) - 1 = -21$

2. $f(x) = 3^{x-4}$

$f(2) = 3^{2-4} = 3^{-2} = \dfrac{1}{9}$

3. $\log_5 125 = x$

$5^x = 125$

$5^x = 5^3$

$x = 3$

$\log_5 125 = 3$

4. $\log_6 x = 2$

$6^2 = x$

$36 = x$

5.

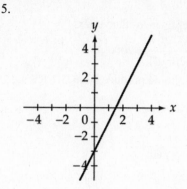

6.

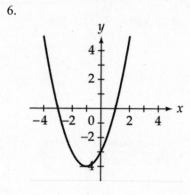

7.

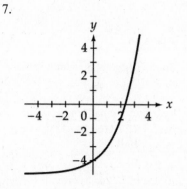

8.

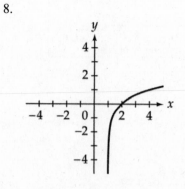

9. $m = \dfrac{(-4) - (-1)}{(-2) - 3} = \dfrac{-3}{-5} = \dfrac{3}{5}$

10. $y - y_1 = m(x - x_1)$

$y - 5 = \frac{2}{3}(x - 3)$

$y - 5 = \frac{2}{3}x - 2$

$y = \frac{2}{3}x + 3$

11. $x = -\frac{b}{2a} = -\frac{6}{2(1)} = -3$

Find y by replacing x with -3 in the equation.

$f(x) = x^2 + 6x - 1$

$f(-3) = (-3)^2 + 6(-3) - 1 = -10$

Vertex: $(-3, -10)$

12. $y = x^2 + 2x - 8$

$0 = (x - 2)(x + 4)$

$x - 2 = 0 \qquad x + 4 = 0$

$x = 2 \qquad\qquad x = -4$

x-intercepts: $(2, 0)$ and $(-4, 0)$

13. $x = -\frac{b}{2a} = -\frac{(-3)}{2(-1)} = -\frac{3}{2}$

Find y by replacing x with $-\frac{3}{2}$ in the function.

$y = -x^2 - 3x + 10$

$y = -\left(-\frac{3}{2}\right)^2 - 3\left(-\frac{3}{2}\right) + 10$

$y = \frac{49}{4}$

The vertex is $\left(-\frac{3}{2}, \frac{49}{4}\right)$. The maximum value of

the function is $\frac{49}{4}$.

14. $d = 250 - 100t$

$d = 250 - 100(0)$

$d = 250$

The vertical intercept is $(0, 250)$. This means that the plane starts 250 miles from its destination.

$d = 250 - 100t$

$0 = 250 - 100t$

$100t = 250$

$t = 2.5$

The horizontal intercept is $(2.5, 0)$. This means that it takes the plane 2.5 hours to reach its destination.

15. Find the coordinates of the vertex.

$t = -\frac{b}{2a} = -\frac{96}{2(-16)} = 3$

$s(t) = -16t^2 + 96t + 4$

$s(3) = -16(3)^2 + 96(3) + 4$

$s(3) = 148$

The maximum height above the ground that the ball will attain is 148 feet.

16. $A = 10\left(\frac{1}{2}\right)^{t/5}$

$A(8) = 10\left(\frac{1}{2}\right)^{8/5} \approx 3.30$

About 3.30 grams of the isotope will remain after 8 hours.

17. Find the intensity of an earthquake of magnitude 7.3.

$M = \log\frac{I}{I_0}$

$7.3 = \log\left(\frac{I}{I_0}\right)$

$\frac{I}{I_0} = 10^{7.3} \approx 19,952,623$

$I = 19,952,623 I_0$

Find the intensity of an earthquake of magnitude 7.0.

$M = \log\frac{I}{I_0}$

$7.0 = \log\left(\frac{I}{I_0}\right)$

$\frac{I}{I_0} = 10^7 = 10,000,000$

$I = 10,000,000 I_0$

Since 19,952,623 is about 1.995 times more than 10,000,000, the earthquake of magnitude 7.3 is about 2.0 times stronger than an earthquake of magnitude 7.0.

18. Let x represent the number of trees per acre and let y represent the average yield per tree. Then, when $x = 320$, $y = 260$, and when $x = 330$, $y = 245$. Find the slope.

$m = \frac{245 - 260}{330 - 320} = -1.5$

Find the equation of the linear model.

$$y - y_1 = m(x - x_1)$$
$$y - 260 = -1.5(x - 320)$$
$$y = -1.5x + 740$$

19. a. Using the calculator, the regression equation is $y = 40.5x + 659$.

    b. Evaluate the regression equation for $x = 8$, since the year 2005 is 8 years after 1997.
        $y = 40.5(8) + 659 = 983$
        The winning weight in 2005 is estimated to be 983 pounds.

20. $\log P = -kd$

     $\log(0.75) = -0.05d$

     $\dfrac{\log(0.75)}{-0.05} = d$

     $2.5 \approx d$

     At a depth of about 2.5 meters, 75% of the light will pass through the surface.

Chapter 7:
Mathematical Systems

EXERCISE SET 7.1

1. 8

3. 12

5. 2

7. 4

9. 4

11. 7

13. 11

15. 7

17. 0300

19. 0400

21. 2000

23. 2100

25. 3

27. 6

29. True. $5 \div 3 = 1$ remainder 2, $8 \div 3 = 2$ remainder 2.

31. False. $5 \div 4 = 1$ remainder 1, $20 \div 4 = 5$ remainder 0.

33. True. $21 \div 6 = 3$ remainder 3, $45 \div 6 = 7$ remainder 3.

35. False. $88 \div 9 = 9$ remainder 7, $5 \div 9 = 0$ remainder 5.

37. True. $100 \div 8 = 12$ remainder 4, $20 \div 8 = 2$ remainder 4.

39. Answers will vary. Adding a multiple of the modulus maintains the congruency. So, possible answers include
$8 + -1(6) = 2$
$8 + 0(6) = 8$
$8 + 1(6) = 14$
$8 + 2(6) = 20$
$8 + 3(6) = 26$
$8 + 4(6) = 32$
$8 + 5(6) = 38$
and so on.

41. $(9 + 15) \bmod 7 = 24 \bmod 7$
$24 \div 7 = 3$ remainder 3
So $(9 + 15) \bmod 7 \equiv 3$

43. $(5 + 22) \bmod 8 = 27 \bmod 8$
$27 \div 8 = 3$ remainder 3
So $(5 + 22) \bmod 8 \equiv 3$

45. $(42 + 35) \bmod 3 = 77 \bmod 3$
$77 \div 3 = 25$ remainder 2
So $(42 + 35) \bmod 3 \equiv 2$

47. $(37 + 45) \bmod 12 = 82 \bmod 12$
$82 \div 12 = 6$ remainder 10
So $(37 + 45) \bmod 12 \equiv 10$

49. $(19 - 6) \bmod 5 = 13 \bmod 5$
$13 \div 5 = 2$ remainder 3
So $(19 - 6) \bmod 5 \equiv 3$

51. $(48 - 21) \bmod 6 = 27 \bmod 6$
$27 \div 6 = 4$ remainder 3
So $(48 - 21) \bmod 6 \equiv 3$

53. $(8 - 15) \bmod 12 = -7 \bmod 12$
$-7 + 12 = 5$
So $(8 - 15) \bmod 12 \equiv 5$

55. $(15 - 32) \bmod 7 = -17 \bmod 7$
$-17 + 7 = -10$
$-10 + 7 = -3$
$-3 + 7 = 4$
So $(15 - 32) \bmod 7 \equiv 4$

57. $(6 \cdot 8) \bmod 9 = 48 \bmod 9$
$48 \div 9 = 5$ remainder 3
So $(6 \cdot 8) \bmod 9 \equiv 3$

59. $(9 \cdot 15) \bmod 8 = 135 \bmod 8$
$135 \div 8 = 16$ remainder 7
So $(9 \cdot 15) \bmod 8 \equiv 7$

61. $(14 \cdot 18) \bmod 5 = 252 \bmod 5$
$252 \div 5 = 50$ remainder 2
So $(14 \cdot 18) \bmod 5 \equiv 2$

63. a. $7 + 59 = 66$, which is positive. $66 \div 12 = 5$ remainder 6
It will be 6:00.

  b. $7 - 62 = -55$, which is negative.
$-55 + 12 = -43$
$-43 + 12 = -31$
$-31 + 12 = -19$
$-19 + 12 = -7$
$-7 + 12 = 5$
It was 5:00.

65. a. The number 5 is associated with Friday, and
        5 + 25 = 30, which is positive.
        30 ÷ 7 = 4 remainder 2
        It will be Tuesday 25 days from today.

    b. The number 5 is associated with Friday, and
        5 − 32 = − 27, which is negative.
        − 27 + 7 = − 20
        − 20 + 7 = − 13
        − 13 + 7 = − 6
        − 6 + 7 = 1
        It was Monday 32 days ago.

67. There are 10 years between the two dates, two of
    which (2008 and 2012) are leap years. So the
    total number of days between the two dates is
        10(365) + 2 = 3652 days
        3653 ÷ 7 = 521 remainder 5
    So, in 2015, Halloween will occur 5 days after
    Monday, which is Saturday.

69. There are 14 years between the two dates, three
    of which (2008, 2012 and 2016) are leap years.
    (Note: 2020 is itself a leap year, but the date in
    question occurs before the extra day, so does not
    affect the calculations.) So the total number of
    days between the two dates is
        14(365) + 3 = 5113 days
        5113 ÷ 7 = 730 remainder 3
    So, in 2010, Valentine's Day will occur 3 days
    after Tuesday, which is Friday.

71. Beginning with zero, substitute each whole
    number less than 3 into the congruence equation.

$x = 0$	$0 \not\equiv 10 \bmod 3$	NO
$x = 1$	$1 \equiv 10 \bmod 3$	YES
$x = 2$	$2 \not\equiv 10 \bmod 3$	NO

    The solutions are 1, 4, 7, 10, 13, 16, . . .

73. Beginning with zero, substitute each whole
    number less than 5 into the congruence equation.

$x = 0$	$2(0) \not\equiv 12 \bmod 5$	NO
$x = 1$	$2(1) \equiv 12 \bmod 5$	YES
$x = 2$	$2(2) \not\equiv 12 \bmod 5$	NO
$x = 3$	$2(3) \not\equiv 12 \bmod 5$	NO
$x = 4$	$2(4) \not\equiv 12 \bmod 5$	NO

    The solutions are 1, 6, 11, 16, 21, 26, . . .

75. Beginning with zero, substitute each whole
    number less than 4 into the congruence.

$x = 0$	$2(0) + 1 \equiv 5 \bmod 4$	YES
$x = 1$	$2(1) + 1 \not\equiv 5 \bmod 4$	NO
$x = 2$	$2(2) + 1 \equiv 5 \bmod 4$	YES
$x = 3$	$2(3) + 1 \not\equiv 5 \bmod 4$	NO

The solutions are 0, 2, 4, 6, 8, 10, 12, . . .

77. Beginning with zero, substitute each whole
    number less than 12 into the congruence.

$x = 0$	$2(0) + 3 \not\equiv 8 \bmod 12$	NO
$x = 1$	$2(1) + 3 \not\equiv 8 \bmod 12$	NO
$x = 2$	$2(2) + 3 \not\equiv 8 \bmod 12$	NO
$x = 3$	$2(3) + 3 \not\equiv 8 \bmod 12$	NO
$x = 4$	$2(4) + 3 \not\equiv 8 \bmod 12$	NO
$x = 5$	$2(5) + 3 \not\equiv 8 \bmod 12$	NO
$x = 6$	$2(6) + 3 \not\equiv 8 \bmod 12$	NO
$x = 7$	$2(7) + 3 \not\equiv 8 \bmod 12$	NO
$x = 8$	$2(8) + 3 \not\equiv 8 \bmod 12$	NO
$x = 9$	$2(9) + 3 \not\equiv 8 \bmod 12$	NO
$x = 10$	$2(10) + 3 \not\equiv 8 \bmod 12$	NO
$x = 11$	$2(11) + 3 \not\equiv 8 \bmod 12$	NO

    The congruent equation has no solutions.

79. Beginning with zero, substitute each whole
    number less than 4 into the congruence.

$x = 0$	$2(0) + 2 \equiv 6 \bmod 4$	YES
$x = 1$	$2(1) + 2 \not\equiv 6 \bmod 4$	NO
$x = 2$	$2(2) + 2 \equiv 6 \bmod 4$	YES
$x = 3$	$2(3) + 2 \not\equiv 6 \bmod 4$	NO

    The solutions are 0, 2, 4, 6, 8, 10, 12, . . .

81. Beginning with zero, substitute each whole
    number less than 8 into the congruence.

$x = 0$	$4(0) + 6 \not\equiv 5 \bmod 8$	NO
$x = 1$	$4(1) + 6 \not\equiv 5 \bmod 8$	NO
$x = 2$	$4(2) + 6 \not\equiv 5 \bmod 8$	NO
$x = 3$	$4(3) + 6 \not\equiv 5 \bmod 8$	NO
$x = 4$	$4(4) + 6 \not\equiv 5 \bmod 8$	NO
$x = 5$	$4(5) + 6 \not\equiv 5 \bmod 8$	NO
$x = 6$	$4(6) + 6 \not\equiv 5 \bmod 8$	NO
$x = 7$	$4(7) + 6 \not\equiv 5 \bmod 8$	NO

    The congruence equation has no solutions.

83. 4 + 5 = 9, so the additive inverse is 5.
    Substitute whole numbers less than 9 into the
    congruence $4x \equiv 1 \bmod 9$.

$x = 0$	$4(0) \not\equiv 1 \bmod 9$	NO
$x = 1$	$4(1) \not\equiv 1 \bmod 9$	NO
$x = 2$	$4(2) \not\equiv 1 \bmod 9$	NO
$x = 3$	$4(3) \not\equiv 1 \bmod 9$	NO
$x = 4$	$4(4) \not\equiv 1 \bmod 9$	NO
$x = 5$	$4(5) \not\equiv 1 \bmod 9$	NO
$x = 6$	$4(6) \not\equiv 1 \bmod 9$	NO
$x = 7$	$4(7) \equiv 1 \bmod 9$	YES

    The multiplicative inverse is 7.

85. 7 + 3 = 10, so the additive inverse is 3.
    Substitute whole numbers less than 10 into the
    congruence $7x \equiv 1 \bmod 10$

$x = 0$ $7(0) \not\equiv 1 \bmod 10$ NO
$x = 1$ $7(1) \not\equiv 1 \bmod 10$ NO
$x = 2$ $7(2) \not\equiv 1 \bmod 10$ NO
$x = 3$ $7(3) \equiv 1 \bmod 10$ YES
The multiplicative inverse is 3.

87. $3 + 5 = 8$, so the additive inverse is 5.
Substitute whole numbers less than 8 into the congruence $3x \equiv 1 \bmod 8$
$x = 0$ $3(0) \not\equiv 1 \bmod 8$ NO
$x = 1$ $3(1) \not\equiv 1 \bmod 8$ NO
$x = 2$ $3(2) \not\equiv 1 \bmod 8$ NO
$x = 3$ $3(3) \equiv 1 \bmod 8$ YES
The multiplicative inverse is 3.

89. Assume $x \equiv (2 \div 7) \bmod 8$; then
$7x \equiv 2 \bmod 8$.
$x = 0$ $7(0) \not\equiv 2 \bmod 8$ NO
$x = 1$ $7(1) \not\equiv 2 \bmod 8$ NO
$x = 2$ $7(2) \not\equiv 2 \bmod 8$ NO
$x = 3$ $7(3) \not\equiv 2 \bmod 8$ NO
$x = 4$ $7(4) \not\equiv 2 \bmod 8$ NO
$x = 5$ $7(5) \not\equiv 2 \bmod 8$ NO
$x = 6$ $7(6) \equiv 2 \bmod 8$ YES
The quotient is 6.

91. Assume $x \equiv (6 \div 4) \bmod 9$; then
$4x \equiv 6 \bmod 9$.
$x = 0$ $4(0) \not\equiv 6 \bmod 9$ NO
$x = 1$ $4(1) \not\equiv 6 \bmod 9$ NO
$x = 2$ $4(2) \not\equiv 6 \bmod 9$ NO
$x = 3$ $4(3) \not\equiv 6 \bmod 9$ NO
$x = 4$ $4(4) \not\equiv 6 \bmod 9$ NO

$x = 5$ $4(5) \not\equiv 6 \bmod 9$ NO
$x = 6$ $4(6) \equiv 6 \bmod 9$ YES
The quotient is 6.

93. Assume $x \equiv (5 \div 6) \bmod 7$; then
$6x \equiv 5 \bmod 7$.
$x = 0$ $6(0) \not\equiv 5 \bmod 7$ NO
$x = 1$ $6(1) \not\equiv 5 \bmod 7$ NO
$x = 2$ $6(2) \equiv 5 \bmod 7$ YES
The quotient is 2.

95. Assume $x \equiv (5 \div 8) \bmod 8$; then
$8x \equiv 5 \bmod 8$.
This congruence has no solution for x because $8x \equiv 0 \bmod 8$ for all whole number values of x.
Thus $5 \div 8$ has no solution in modulo 8 arithmetic.

97. In 3500 hours, the time will be
$(3500 + 3) \bmod 12$.
$3503 \div 12 = 291$ remainder 11
It will be 11:00.

99. Beginning with 0, substitute each whole number less than 11 into the congruence.
$x = 0$ $0^2 + 3(0) + 7 \not\equiv 2 \bmod 11$ NO
$x = 1$ $1^2 + 3(1) + 7 \not\equiv 2 \bmod 11$ NO
$x = 2$ $2^2 + 3(2) + 7 \not\equiv 2 \bmod 11$ NO
$x = 3$ $3^2 + 3(3) + 7 \not\equiv 2 \bmod 11$ NO
$x = 4$ $4^2 + 3(4) + 7 \equiv 2 \bmod 11$ YES
The solution is 4.

EXERCISE SET 7.2

1. $0(10) + 2(9) + 8(8) + 1(7) + 4(6) + 4(5) + 2(4) + 6(3) + 8(2) + 5(1) = 180 \equiv 4 \bmod 11 \not\equiv 0 \bmod 11$
$0 - 281 - 44268 - 5$ is not a valid ISBN.

3. $0(10) + 4(9) + 4(8) + 6(7) + 3(6) + 1(5) + 4(4) + 1(3) + 8(2) + 2(1) = 170 \equiv 5 \bmod 11 \not\equiv 0 \bmod 11$
$0 - 446 - 31418 - 2$ is not a valid ISBN.

5. $0(10) + 6(9) + 8(8) + 4(7) + 8(6) + 5(5) + 9(4) + 7(3) + 2(2) + 6(1) = 286 \equiv 0 \bmod 11$
$0 - 684 - 85972 - 6$ is a valid ISBN.

7. $0(10) + 3(9) + 8(8) + 5(7) + 5(6) + 0(5) + 4(4) + 2(3) + 0(2) + x \equiv 0 \bmod 11$
$178 + x \equiv 0 \bmod 11$
The next multiple of 11 is 187. $x = 187 - 178 = 9$

9. $0(10) + 8(9) + 0(8) + 2(7) + 7(6) + 1(5) + 3(4) + 3(3) + 2(2) + x \equiv 0 \bmod 11$
$158 + x \equiv 0 \bmod 11$
The next multiple of 11 is 165. $x = 165 - 158 = 7$

11. $0(10) + 1(9) + 4(8) + 0(7) + 4(6) + 4(5) + 4(4) + 1(3) + 7(2) + x \equiv 0 \bmod 11$
    $118 + x \equiv 0 \bmod 11$
    The next multiple of 11 is 121. $x = 121 - 118 = 3$

13. $0(10) + 5(9) + 1(8) + 7(7) + 8(6) + 8(5) + 4(4) + 4(3) + 1(2) + x \equiv 0 \bmod 11$
    $220 + x \equiv 0 \bmod 11$
    The next multiple of 11 is 220. $x = 220 - 220 = 0$

15. $0(3) + 5(1) + 1(3) + 0(1) + 0(3) + 0(1) + 0(3) + 3(1) + 1(3) + 3(1) + 9(3) + x \equiv 0 \bmod 10$
    $44 + x \equiv 0 \bmod 10$
    The next multiple of 10 is 50. $x = 50 - 44 = 6$

17. $0(3) + 3(1) + 7(3) + 0(1) + 0(3) + 0(1) + 3(3) + 7(1) + 7(3) + 1(1) + 7(3) + x \equiv 0 \bmod 10$
    $83 + x \equiv 0 \bmod 10$
    The next multiple of 10 is 90. $x = 90 - 83 = 7$

19. $0(3) + 2(1) + 5(3) + 5(1) + 0(3) + 0(1) + 8(3) + 1(1) + 1(3) + 2(1) + 1(3) + x \equiv 0 \bmod 10$
    $55 + x \equiv 0 \bmod 10$
    The next multiple of 10 is 60. $x = 60 - 55 = 5$

21. $7(3) + 1(1) + 7(3) + 9(1) + 5(3) + 1(1) + 0(3) + 0(1) + 4(3) + 6(1) + 1(3) + x \equiv 0 \bmod 10$
    $89 + x \equiv 0 \bmod 10$
    The next multiple of 10 is 90. $x = 90 - 89 = 1$

23. $(0 + 3 + 1 + 6 + 6 + 1 + 5 + 4 + 9 + 8) \bmod 9 \equiv 43 \bmod 9 \equiv 7$

25. $(1 + 3 + 3 + 1 + 4 + 9 + 7 + 5 + 3 + 3) \bmod 9 \equiv 39 \bmod 9 \equiv 3$

27. $(1 + 1 + 8 + 2 + 6 + 4 + 9 + 7 + 5 + 8) \bmod 7 \equiv 51 \bmod 7 \equiv 2$; valid.

29. $(2 + 0 + 2 + 6 + 1 + 7 + 8 + 9 + 1 + 4) \bmod 7 \equiv 40 \bmod 7 \equiv 5$; valid.

31. 4 4 1 7 5 4 8 6 1 7 8 5 6 4 1 1
    8 4 2 7 10 4 16 6 2 7 16 5 12 4 2 1
    $8 + 4 + 2 + 7 + (1 + 0) + 4 + (1 + 6) + 6 + 2 + 7 + (1 + 6) + 5 + (1 + 2) + 4 + 2 + 1 = 70 \equiv 0 \bmod 10$. This credit card number is valid.

33. 5 5 9 1 4 9 1 2 7 6 4 4 1 1 0 5
    10 5 18 1 8 9 2 2 14 6 8 4 2 1 0 5
    $(1 + 0) + 5 + (1 + 8) + 1 + 8 + 9 + 2 + 2 + (1 + 4) + 6 + 8 + 4 + 2 + 1 + 0 + 5 = 68$
    $68 \neq 0 \bmod 10$. This credit card number is invalid.

35. 6 0 1 1 0 4 0 8 4 9 7 7 3 1 5 8
    12 0 2 1 0 4 0 8 8 9 14 7 6 1 10 8
    $(1 + 2) + 0 + 2 + 1 + 0 + 4 + 0 + 8 + 8 + 9 + (1 + 4) + 7 + 6 + 1 + (1 + 0) + 8 = 63$
    $63 \neq 0 \bmod 10$. This credit card number is invalid.

37. 3 7 1 5 5 4 8 7 3 1 8 4 4 6 6
    3 14 1 10 5 8 8 14 3 2 8 8 4 12 6
    $3 + (1 + 4) + 1 + (1 + 0) + 5 + 8 + 8 + (1 + 4) + 3 + 2 + 8 + 8 + 4 + (1 + 2) + 6 = 70$
    $70 \equiv 0 \bmod 10$. This credit card number is valid.

39.

T	$c \equiv (20 + 8) \bmod 26 \equiv 28 \bmod 26 \equiv 2$	Code T as B.
H	$c \equiv (8 + 8) \bmod 26 \equiv 16 \bmod 26 \equiv 16$	Code H as P.
R	$c \equiv (18 + 8) \bmod 26 \equiv 26 \bmod 26 \equiv 0$	Code R as Z.
E	$c \equiv (5 + 8) \bmod 26 \equiv 13 \bmod 26 \equiv 13$	Code E as M.
E	$c \equiv (5 + 8) \bmod 26 \equiv 13 \bmod 26 \equiv 13$	Code E as M.
M	$c \equiv (13 + 8) \bmod 26 \equiv 21 \bmod 26 \equiv 21$	Code M as U.
U	$c \equiv (21 + 8) \bmod 26 \equiv 29 \bmod 26 \equiv 3$	Code U as C.
S	$c \equiv (19 + 8) \bmod 26 \equiv 27 \bmod 26 \equiv 1$	Code S as A.
K	$c \equiv (11 + 8) \bmod 26 \equiv 19 \bmod 26 \equiv 19$	Code K as S.
E	$c \equiv (5 + 8) \bmod 26 \equiv 13 \bmod 26 \equiv 13$	Code E as M.
T	$c \equiv (20 + 8) \bmod 26 \equiv 28 \bmod 26 \equiv 2$	Code T as B.
E	$c \equiv (5 + 8) \bmod 26 \equiv 13 \bmod 26 \equiv 13$	Code E as M.
E	$c \equiv (5 + 8) \bmod 26 \equiv 13 \bmod 26 \equiv 13$	Code E as M.
R	$c \equiv (18 + 8) \bmod 26 \equiv 26 \bmod 26 \equiv 0$	Code R as Z.
S	$c \equiv (19 + 8) \bmod 26 \equiv 27 \bmod 26 \equiv 1$	Code S as A.

The plaintext is coded as BPZMM UCASMBMMZA.

41.

I	$c \equiv (9 + 12) \bmod 26 \equiv 21 \bmod 26 \equiv 21$	Code I as U.
T	$c \equiv (20 + 12) \bmod 26 \equiv 32 \bmod 26 \equiv 6$	Code T as F.
S	$c \equiv (19 + 12) \bmod 26 \equiv 31 \bmod 26 \equiv 5$	Code S as E.
A	$c \equiv (1 + 12) \bmod 26 \equiv 13 \bmod 26 \equiv 13$	Code A as M.
G	$c \equiv (7 + 12) \bmod 26 \equiv 19 \bmod 26 \equiv 19$	Code G as S.
I	$c \equiv (9 + 12) \bmod 26 \equiv 21 \bmod 26 \equiv 21$	Code I as U.
R	$c \equiv (18 + 12) \bmod 26 \equiv 30 \bmod 26 \equiv 4$	Code R as D.
L	$c \equiv (12 + 12) \bmod 26 \equiv 24 \bmod 26 \equiv 24$	Code L as X.

The plaintext is coded as UF'E M SUDX.

43.

S	$c \equiv (19 + 3) \bmod 26 \equiv 22 \bmod 26 \equiv 22$	Code S as V.
T	$c \equiv (20 + 3) \bmod 26 \equiv 23 \bmod 26 \equiv 23$	Code T as W.
I	$c \equiv (9 + 3) \bmod 26 \equiv 12 \bmod 26 \equiv 12$	Code I as L.
C	$c \equiv (3 + 3) \bmod 26 \equiv 6 \bmod 26 \equiv 6$	Code C as F.
K	$c \equiv (11 + 3) \bmod 26 \equiv 14 \bmod 26 \equiv 14$	Code K as N.
S	$c \equiv (19 + 3) \bmod 26 \equiv 22 \bmod 26 \equiv 22$	Code S as V.
A	$c \equiv (1 + 3) \bmod 26 \equiv 4 \bmod 26 \equiv 4$	Code A as D.
N	$c \equiv (14 + 3) \bmod 26 \equiv 17 \bmod 26 \equiv 17$	Code N as Q.
D	$c \equiv (4 + 3) \bmod 26 \equiv 7 \bmod 26 \equiv 7$	Code D as G.
S	$c \equiv (19 + 3) \bmod 26 \equiv 22 \bmod 26 \equiv 22$	Code S as V.
T	$c \equiv (20 + 3) \bmod 26 \equiv 23 \bmod 26 \equiv 23$	Code T as W.
O	$c \equiv (15 + 3) \bmod 26 \equiv 18 \bmod 26 \equiv 18$	Code O as R.
N	$c \equiv (14 + 3) \bmod 26 \equiv 17 \bmod 26 \equiv 17$	Code N as Q.
E	$c \equiv (5 + 3) \bmod 26 \equiv 8 \bmod 26 \equiv 8$	Code E as H.
S	$c \equiv (19 + 3) \bmod 26 \equiv 22 \bmod 26 \equiv 22$	Code S as V.

The plaintext is coded as VWLFNV DQG VWRQHV.

45. $26 - 18 = 8$

Decode using the congruence $p \equiv (c + 8) \bmod 26$.

S	$p \equiv (19 + 8) \bmod 26 \equiv 27 \bmod 26 \equiv 1$	Decode S as A.
Y	$p \equiv (25 + 8) \bmod 26 \equiv 33 \bmod 26 \equiv 7$	Decode Y as G.
W	$p \equiv (23 + 8) \bmod 26 \equiv 31 \bmod 26 \equiv 5$	Decode W as E.
G	$p \equiv (7 + 8) \bmod 26 \equiv 15 \bmod 26 \equiv 15$	Decode G as O.
X	$p \equiv (24 + 8) \bmod 26 \equiv 32 \bmod 26 \equiv 6$	Decode X as F.
W	$p \equiv (23 + 8) \bmod 26 \equiv 31 \bmod 26 \equiv 5$	Decode W as E.
F	$p \equiv (6 + 8) \bmod 26 \equiv 14 \bmod 26 \equiv 14$	Decode F as N.

D	$p \equiv (4 + 8) \bmod 26 \equiv 12 \bmod 26 \equiv 12$	Decode D as L.
A	$p \equiv (1 + 8) \bmod 26 \equiv 9 \bmod 26 \equiv 9$	Decode A as I.
Y	$p \equiv (25 + 8) \bmod 26 \equiv 33 \bmod 26 \equiv 7$	Decode Y as G.
Z	$p \equiv (0 + 8) \bmod 26 \equiv 8 \bmod 26 \equiv 8$	Decode Z as H.
L	$p \equiv (12 + 8) \bmod 26 \equiv 20 \bmod 26 \equiv 20$	Decode L as T.
W	$p \equiv (23 + 8) \bmod 26 \equiv 31 \bmod 26 \equiv 5$	Decode W as E.
F	$p \equiv (6 + 8) \bmod 26 \equiv 14 \bmod 26 \equiv 14$	Decode F as N.
E	$p \equiv (5 + 8) \bmod 26 \equiv 13 \bmod 26 \equiv 13$	Decode E as M.
W	$p \equiv (23 + 8) \bmod 26 \equiv 31 \bmod 26 \equiv 5$	Decode W as E.
F	$p \equiv (6 + 8) \bmod 26 \equiv 14 \bmod 26 \equiv 14$	Decode F as N.
L	$p \equiv (12 + 8) \bmod 26 \equiv 20 \bmod 26 \equiv 20$	Decode L as T.

The ciphertext is decoded as AGE OF ENLIGHTENMENT.

47. $26 - 15 = 11$
Decode using the congruence $p \equiv (c + 11) \bmod 26$.

U	$p \equiv (21 + 11) \bmod 26 \equiv 32 \bmod 26 \equiv 6$	Decode U as F.
G	$p \equiv (7 + 11) \bmod 26 \equiv 18 \bmod 26 \equiv 18$	Decode G as R.
X	$p \equiv (24 + 11) \bmod 26 \equiv 35 \bmod 26 \equiv 9$	Decode X as I.
T	$p \equiv (20 + 11) \bmod 26 \equiv 31 \bmod 26 \equiv 5$	Decode T as E.
C	$p \equiv (3 + 11) \bmod 26 \equiv 14 \bmod 26 \equiv 14$	Decode C as N.
S	$p \equiv (19 + 11) \bmod 26 \equiv 30 \bmod 26 \equiv 4$	Decode S as D.
X	$p \equiv (24 + 11) \bmod 26 \equiv 35 \bmod 26 \equiv 9$	Decode X as I.
C	$p \equiv (3 + 11) \bmod 26 \equiv 14 \bmod 26 \equiv 14$	Decode C as N.
C	$p \equiv (3 + 11) \bmod 26 \equiv 14 \bmod 26 \equiv 14$	Decode C as N.
T	$p \equiv (20 + 11) \bmod 26 \equiv 31 \bmod 26 \equiv 5$	Decode T as E.
T	$p \equiv (20 + 11) \bmod 26 \equiv 31 \bmod 26 \equiv 5$	Decode T as E.
S	$p \equiv (19 + 11) \bmod 26 \equiv 30 \bmod 26 \equiv 4$	Decode S as D.

The ciphertext is decoded as FRIEND IN NEED.

49. Because the encoded message uses a cyclical alphabetic encrypting code, it is possible to rotate through the 26 possible codes until the ciphertext is decoded. Using just the first word:

   YVIBZM $\rightarrow$ ZWJCAN $\rightarrow$ AXKDBO $\rightarrow$ BYLECP $\rightarrow$ CZMFDQ $\rightarrow$ DANGER

It took 5 transformations to find the plaintext, so the decoding congruence is $p \equiv (c + 5) \bmod 26$. Decode the rest of the ciphertext in the usual way.

R	$p \equiv (18 + 5) \bmod 26 \equiv 23 \bmod 26 \equiv 23$	Decode R as W.
D	$p \equiv (4 + 5) \bmod 26 \equiv 9 \bmod 26 \equiv 9$	Decode D as I.
G	$p \equiv (7 + 5) \bmod 26 \equiv 12 \bmod 26 \equiv 12$	Decode G as L.
G	$p \equiv (7 + 5) \bmod 26 \equiv 12 \bmod 26 \equiv 12$	Decode G as L.
M	$p \equiv (13 + 5) \bmod 26 \equiv 18 \bmod 26 \equiv 18$	Decode M as R.
J	$p \equiv (10 + 5) \bmod 26 \equiv 15 \bmod 26 \equiv 15$	Decode J as O.
W	$p \equiv (23 + 5) \bmod 26 \equiv 28 \bmod 26 \equiv 2$	Decode W as B.
D	$p \equiv (4 + 5) \bmod 26 \equiv 9 \bmod 26 \equiv 9$	Decode D as I.
I	$p \equiv (9 + 5) \bmod 26 \equiv 14 \bmod 26 \equiv 14$	Decode I as N.
N	$p \equiv (14 + 5) \bmod 26 \equiv 19 \bmod 26 \equiv 19$	Decode N as S.
J	$p \equiv (10 + 5) \bmod 26 \equiv 15 \bmod 26 \equiv 15$	Decode J as O.
I	$p \equiv (9 + 5) \bmod 26 \equiv 14 \bmod 26 \equiv 14$	Decode I as N.

The ciphertext is decoded as DANGER WILL ROBINSON.

51. Because the encoded message uses a cyclical alphabetic encrypting code, it is possible to rotate through the 26 possible codes until the ciphertext is decoded. Using just the first word:

   UDGIJCT $\rightarrow$ VEHJKDU $\rightarrow$ WFIKLEV $\rightarrow$ XGJLMFW $\rightarrow$ YHKMNGX $\rightarrow$ ZILNOHY

   $\rightarrow$ AJMOPIZ $\rightarrow$ BKNPQJA $\rightarrow$ CLOQRKB $\rightarrow$ DMPRSLC $\rightarrow$ ENQSTMD $\rightarrow$ FORTUNE

It took 11 transformations to find the plaintext, so the decoding congruence is $p \equiv (c + 11) \bmod 26$. Decode the rest of the ciphertext in the usual way.

R	$p \equiv (18 + 11) \bmod 26 \equiv 29 \bmod 26 \equiv 3$	Decode R as C.
D	$p \equiv (4 + 11) \bmod 26 \equiv 15 \bmod 26 \equiv 15$	Decode D as O.
D	$p \equiv (4 + 11) \bmod 26 \equiv 15 \bmod 26 \equiv 15$	Decode D as O.
Z	$p \equiv (0 + 11) \bmod 26 \equiv 11 \bmod 26 \equiv 11$	Decode Z as K.
X	$p \equiv (24 + 11) \bmod 26 \equiv 35 \bmod 26 \equiv 9$	Decode X as I.
T	$p \equiv (20 + 11) \bmod 26 \equiv 31 \bmod 26 \equiv 5$	Decode T as E.

The ciphertext is decoded as FORTUNE COOKIE.

53.
M	$c \equiv (13 + 3) \bmod 26 \equiv 16 \bmod 26 \equiv 16$	Code M as P.
E	$c \equiv (5 + 3) \bmod 26 \equiv 8 \bmod 26 \equiv 8$	Code E as H.
N	$c \equiv (14 + 3) \bmod 26 \equiv 17 \bmod 26 \equiv 17$	Code N as Q.
W	$c \equiv (23 + 3) \bmod 26 \equiv 26 \bmod 26 \equiv 0$	Code W as Z.
I	$c \equiv (9 + 3) \bmod 26 \equiv 12 \bmod 26 \equiv 12$	Code I as L.
L	$c \equiv (12 + 3) \bmod 26 \equiv 15 \bmod 26 \equiv 15$	Code L as O.
L	$c \equiv (12 + 3) \bmod 26 \equiv 15 \bmod 26 \equiv 15$	Code L as O.

Continuing, the plaintext is coded as PHQ ZLOOLQJShB EHOLHYH ZKDW WKHB ZLVK.

55.
T	$c \equiv (3 \cdot 20 + 2) \bmod 26 \equiv 62 \bmod 26 \equiv 10$	Code T as J.
O	$c \equiv (3 \cdot 15 + 2) \bmod 26 \equiv 47 \bmod 26 \equiv 21$	Code O as U.
W	$c \equiv (3 \cdot 23 + 2) \bmod 26 \equiv 71 \bmod 26 \equiv 19$	Code W as S.
E	$c \equiv (3 \cdot 5 + 2) \bmod 26 \equiv 17 \bmod 26 \equiv 17$	Code E as Q.
R	$c \equiv (3 \cdot 18 + 2) \bmod 26 \equiv 56 \bmod 26 \equiv 4$	Code R as D.
O	$c \equiv (3 \cdot 15 + 2) \bmod 26 \equiv 47 \bmod 26 \equiv 21$	Code O as U.
F	$c \equiv (3 \cdot 6 + 2) \bmod 26 \equiv 20 \bmod 26 \equiv 20$	Code F as T.

Continuing, the plaintext is coded as JUSQD UT LURNUR.

57. P $\quad c \equiv (7 \cdot 16 + 8) \bmod 26 \equiv 120 \bmod 26 \equiv 16 \quad$ Code P as P.
A	$c \equiv (7 \cdot 1 + 8) \bmod 26 \equiv 15 \bmod 26 \equiv 15$	Code A as O.
R	$c \equiv (7 \cdot 18 + 8) \bmod 26 \equiv 134 \bmod 26 \equiv 4$	Code R as D.
A	$c \equiv (7 \cdot 1 + 8) \bmod 26 \equiv 15 \bmod 26 \equiv 15$	Code A as O.
L	$c \equiv (7 \cdot 12 + 8) \bmod 26 \equiv 92 \bmod 26 \equiv 14$	Code L as N.
L	$c \equiv (7 \cdot 12 + 8) \bmod 26 \equiv 92 \bmod 26 \equiv 14$	Code L as N.
E	$c \equiv (7 \cdot 5 + 8) \bmod 26 \equiv 43 \bmod 26 \equiv 17$	Code E as Q.
L	$c \equiv (7 \cdot 12 + 8) \bmod 26 \equiv 92 \bmod 26 \equiv 14$	Code L as N.

Continuing, the plaintext is coded as PODONNQN NSBQK.

59. Solve the congruence for p.

$c = 3p + 4$

$c - 4 = 3p$

The multiplicative inverse of 3 is 9.

$9(c - 4) = 9(3p)$

$[9(c - 4)] \bmod 26 \equiv p$

Decode using this congruence.

L	$[9(12 - 4)] \bmod 26 \equiv 72 \bmod 26 \equiv 20$	Decode L as T.
O	$[9(15 - 4)] \bmod 26 \equiv 99 \bmod 26 \equiv 21$	Decode O as U.
F	$[9(6 - 4)] \bmod 26 \equiv 18 \bmod 26 \equiv 18$	Decode F as R.
T	$[9(20 - 4)] \bmod 26 \equiv 144 \bmod 26 \equiv 14$	Decode T as N.
J	$[9(10 - 4)] \bmod 26 \equiv 54 \bmod 26 \equiv 2$	Decode J as B.
G	$[9(7 - 4)] \bmod 26 \equiv 27 \bmod 26 \equiv 1$	Decode G as A.
M	$[9(13 - 4)] \bmod 26 \equiv 81 \bmod 26 \equiv 3$	Decode M as C.
K	$[9(11 - 4)] \bmod 26 \equiv 63 \bmod 26 \equiv 11$	Decode K as K.

Continuing, the ciphertext is decoded as TURN BACK THE CLOCK.

61. Solve the congruence for *p*.

$$c = 5p + 9$$
$$c - 9 = 5p$$

The multiplicative inverse of 5 is 21.

$$21(c - 9) = 21(5p)$$
$$[21(c - 9)] \bmod 26 \equiv p$$

Decode using this congruence.

S	$[21(19 - 9)] \bmod 26 \equiv 210 \bmod 26 \equiv 2$	Decode S as B.
N	$[21(14 - 9)] \bmod 26 \equiv 105 \bmod 26 \equiv 1$	Decode N as A.
U	$[21(21 - 9)] \bmod 26 \equiv 252 \bmod 26 \equiv 18$	Decode U as R.
U	$[21(21 - 9)] \bmod 26 \equiv 252 \bmod 26 \equiv 18$	Decode U as R.
H	$[21(8 - 9)] \bmod 26 \equiv -21 \bmod 26 \equiv 5$	Decode H as E.
Q	$[21(17 - 9)] \bmod 26 \equiv 168 \bmod 26 \equiv 12$	Decode Q as L.
F	$[21(6 - 9)] \bmod 26 \equiv -63 \bmod 26 \equiv 15$	Decode F as O.
M	$[21(13 - 9)] \bmod 26 \equiv 84 \bmod 26 \equiv 6$	Decode M as F.

Continuing, the ciphertext is decoded as BARREL OF MONKEYS.

63. Because the check digit is simply the sum of the first 10 digits mod 9, the same digits in a different order will give the same sum and hence the same check digit.

EXERCISE SET 7.3

1. a. Yes. All possible multiplications within the set yield either -1 or 1.

   b. No. $-1 + 1 = 0$, which is not in the set.

3. a. Yes. The product of two odd integers is always an odd integer. For instance, $3 \cdot 7 = 21$ and $9 \cdot 15 = 135$.

   b. No. The sum of two odd integers is always an even integer. For instance, $3 + 7 = 10$ and $9 + 15 = 24$.

5. 1. The sum of two even integers is always an integer. Therefore, the closure property holds true.

   2. The associative property of addition holds true for the even integers.

   3. The identity element is 0, which is an even integer. Therefore, the even integers have an identity element for addition.

   4. Each element has an inverse. If *a* is an even integer, then $-a$ is the inverse of a.

All of the four properties are satisfied, so the even integers form a group with respect to addition.

7. 1. The sum of two real numbers is always a real number, so the closure property holds true.

   2. The associative property of addition holds true for the real numbers.

   3. The identity element is 0, which is a real number. Therefore the real numbers have an identity element for addition.

   4. Each element has an inverse. If *a* is a real number, then $-a$ is the inverse of *a*.

All of the four properties are satisfied, so the real numbers form a group with respect to addition.

9. 1. The product of two real numbers is always a real number. Therefore, the closure property holds true.

   2. The associative property of multiplication holds true for the real numbers.

   3. The identity element is 1, which is a real number. Therefore, the real numbers have an identity element for multiplication.

4. Not every element has an inverse. For example, there is no inverse for 0. Therefore, the inverse property fails.

Property 4 fails, so the real numbers do not form a group with respect to multiplication.

11. 1. The sum of any two rational numbers is always a rational number. Therefore, the closure property holds true.

    2. The associative property of addition holds true for the rational numbers.

    3. The identity element is 0, which is a rational number. Therefore the rational numbers have an identity element for addition.

    4. Every element has an inverse. If $\frac{a}{b}$ is a rational number, then $\frac{-a}{b}$ is the inverse of $\frac{a}{b}$.

All of the four properties are satisfied, so the rational numbers form a group with respect to addition.

13. 1. The sum modulo 4 of any two elements in the set is always a member of the set. Therefore, the closure property holds true.

    2. The associative property of addition modulo 4 holds true for the members of the set.

    3. There exists an identity element (0) in the set for addition modulo 4.

    4. Every element has an inverse. If a is an element in the set, then $(4 - a)$ mod 4 is the inverse of a.

All of the four properties are satisfied, so the set forms a group with respect to addition modulo 4.

15. 1. The product modulo 4 of any two elements in the set is always a member of the set. Therefore, the closure property holds true.

    2. The associative property of multiplication modulo 4 holds true for the members of the set.

3. There exists an identity element (1) in the set for multiplication modulo 4.

4. 0 does not have an inverse. Therefore the inverse property fails.

Property 4 fails, so the set does not form a group with respect to multiplication modulo 4.

17. 1. The product of any two elements in the set is always a member of the set. Therefore, the closure property holds true.

    2. The associative property of multiplication holds true for the members of the set.

    3. There exists an identity element (1) in the set for multiplication.

    4. Both elements have themselves as their inverse. Therefore, the inverse property holds true for the set.

All of the four properties are satisfied, so the set forms a group with respect to multiplication.

19.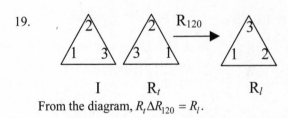

From the diagram, $R_t \Delta R_{120} = R_l$.

21.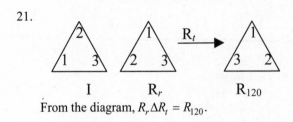

From the diagram, $R_r \Delta R_t = R_{120}$.

23. $R_{240} \Delta R_{120} = I$, so the inverse of R_{240} is R_{120}.

25.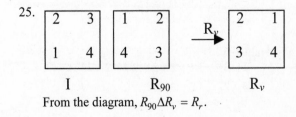

From the diagram, $R_{90} \Delta R_v = R_r$.

27.
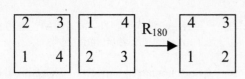

$$I \qquad R_r \qquad R_l$$

From the diagram, $R_r \Delta R_{180} = R_l$.

29. Perform all possible transformations on the identity element to determine all the elements of the group.

$$I = \begin{pmatrix} 1 & 2 & 3 & 4 \\ 1 & 2 & 3 & 4 \end{pmatrix},$$

$$R_{90} = \begin{pmatrix} 1 & 2 & 3 & 4 \\ 2 & 3 & 4 & 1 \end{pmatrix},$$

$$R_{180} = \begin{pmatrix} 1 & 2 & 3 & 4 \\ 3 & 4 & 1 & 2 \end{pmatrix},$$

$$R_{270} = \begin{pmatrix} 1 & 2 & 3 & 4 \\ 4 & 1 & 2 & 3 \end{pmatrix},$$

$$R_v = \begin{pmatrix} 1 & 2 & 3 & 4 \\ 4 & 3 & 2 & 1 \end{pmatrix}, R_h = \begin{pmatrix} 1 & 2 & 3 & 4 \\ 2 & 1 & 4 & 3 \end{pmatrix},$$

$$R_r = \begin{pmatrix} 1 & 2 & 3 & 4 \\ 3 & 2 & 1 & 4 \end{pmatrix}, R_l = \begin{pmatrix} 1 & 2 & 3 & 4 \\ 1 & 4 & 3 & 2 \end{pmatrix}.$$

31. Evaluating:

$$\begin{pmatrix} 1 & 2 & 3 & 4 \\ 2 & 1 & 4 & 3 \end{pmatrix} \Delta \begin{pmatrix} 1 & 2 & 3 & 4 \\ 2 & 3 & 4 & 1 \end{pmatrix}$$

$$= \begin{pmatrix} 1 & 2 & 3 & 4 \\ 3 & 2 & 1 & 4 \end{pmatrix} = R_r$$

33. Evaluating:

$$\begin{pmatrix} 1 & 2 & 3 & 4 \\ 4 & 1 & 2 & 3 \end{pmatrix}^{-1} = \begin{pmatrix} 1 & 2 & 3 & 4 \\ 2 & 3 & 4 & 1 \end{pmatrix} = R_{90}$$

35. Evaluating:

$$\begin{pmatrix} 1 & 2 & 3 \\ 2 & 3 & 1 \end{pmatrix} \Delta \begin{pmatrix} 1 & 2 & 3 \\ 3 & 1 & 2 \end{pmatrix} = \begin{pmatrix} 1 & 2 & 3 \\ 1 & 2 & 3 \end{pmatrix} = I$$

37. Evaluating:

$$\begin{pmatrix} 1 & 2 & 3 \\ 3 & 1 & 2 \end{pmatrix} \Delta \begin{pmatrix} 1 & 2 & 3 \\ 2 & 1 & 3 \end{pmatrix} = \begin{pmatrix} 1 & 2 & 3 \\ 3 & 2 & 1 \end{pmatrix} = D$$

39. Evaluating:

$$\begin{pmatrix} 1 & 2 & 3 \\ 1 & 3 & 2 \end{pmatrix} \Delta \begin{pmatrix} 1 & 2 & 3 \\ 2 & 3 & 1 \end{pmatrix} = \begin{pmatrix} 1 & 2 & 3 \\ 2 & 1 & 3 \end{pmatrix} = E$$

41. Evaluating:

$$\begin{pmatrix} 1 & 2 & 3 \\ 2 & 3 & 1 \end{pmatrix}^{-1} = \begin{pmatrix} 1 & 2 & 3 \\ 3 & 1 & 2 \end{pmatrix} = B$$

43.   The list follows:

$$\begin{pmatrix} 1 & 2 & 3 & 4 \\ 1 & 2 & 3 & 4 \end{pmatrix}, \begin{pmatrix} 1 & 2 & 3 & 4 \\ 1 & 2 & 4 & 3 \end{pmatrix},$$

$$\begin{pmatrix} 1 & 2 & 3 & 4 \\ 1 & 3 & 2 & 4 \end{pmatrix}, \begin{pmatrix} 1 & 2 & 3 & 4 \\ 1 & 3 & 4 & 2 \end{pmatrix},$$

$$\begin{pmatrix} 1 & 2 & 3 & 4 \\ 1 & 4 & 2 & 3 \end{pmatrix}, \begin{pmatrix} 1 & 2 & 3 & 4 \\ 1 & 4 & 3 & 2 \end{pmatrix},$$

$$\begin{pmatrix} 1 & 2 & 3 & 4 \\ 2 & 1 & 3 & 4 \end{pmatrix}, \begin{pmatrix} 1 & 2 & 3 & 4 \\ 2 & 1 & 4 & 3 \end{pmatrix},$$

$$\begin{pmatrix} 1 & 2 & 3 & 4 \\ 2 & 3 & 1 & 4 \end{pmatrix}, \begin{pmatrix} 1 & 2 & 3 & 4 \\ 2 & 3 & 4 & 1 \end{pmatrix},$$

$$\begin{pmatrix} 1 & 2 & 3 & 4 \\ 2 & 4 & 1 & 3 \end{pmatrix}, \begin{pmatrix} 1 & 2 & 3 & 4 \\ 2 & 4 & 3 & 1 \end{pmatrix},$$

$$\begin{pmatrix} 1 & 2 & 3 & 4 \\ 3 & 1 & 2 & 4 \end{pmatrix}, \begin{pmatrix} 1 & 2 & 3 & 4 \\ 3 & 1 & 4 & 2 \end{pmatrix},$$

$$\begin{pmatrix} 1 & 2 & 3 & 4 \\ 3 & 2 & 1 & 4 \end{pmatrix}, \begin{pmatrix} 1 & 2 & 3 & 4 \\ 3 & 2 & 4 & 1 \end{pmatrix},$$

$$\begin{pmatrix} 1 & 2 & 3 & 4 \\ 3 & 4 & 1 & 2 \end{pmatrix}, \begin{pmatrix} 1 & 2 & 3 & 4 \\ 3 & 4 & 2 & 1 \end{pmatrix},$$

$$\begin{pmatrix} 1 & 2 & 3 & 4 \\ 4 & 1 & 2 & 3 \end{pmatrix}, \begin{pmatrix} 1 & 2 & 3 & 4 \\ 4 & 1 & 3 & 2 \end{pmatrix},$$

$$\begin{pmatrix} 1 & 2 & 3 & 4 \\ 4 & 2 & 1 & 3 \end{pmatrix}, \begin{pmatrix} 1 & 2 & 3 & 4 \\ 4 & 2 & 3 & 1 \end{pmatrix},$$

$$\begin{pmatrix} 1 & 2 & 3 & 4 \\ 4 & 3 & 1 & 2 \end{pmatrix}, \begin{pmatrix} 1 & 2 & 3 & 4 \\ 4 & 3 & 2 & 1 \end{pmatrix}$$

45. Evaluating:

$$\begin{pmatrix} 1 & 2 & 3 & 4 \\ 1 & 4 & 2 & 3 \end{pmatrix} \Delta \begin{pmatrix} 1 & 2 & 3 & 4 \\ 2 & 3 & 4 & 1 \end{pmatrix}$$

$$= \begin{pmatrix} 1 & 2 & 3 & 4 \\ 2 & 1 & 3 & 4 \end{pmatrix}$$

47. Evaluating:

$$\begin{pmatrix} 1 & 2 & 3 & 4 \\ 2 & 4 & 1 & 3 \end{pmatrix}^{-1} = \begin{pmatrix} 1 & 2 & 3 & 4 \\ 3 & 1 & 4 & 2 \end{pmatrix}$$

49. Evaluating:

$$\begin{pmatrix} 1 & 2 & 3 & 4 \\ 4 & 3 & 2 & 1 \end{pmatrix}^{-1} = \begin{pmatrix} 1 & 2 & 3 & 4 \\ 4 & 3 & 2 & 1 \end{pmatrix}$$

51. From the table, $a \nabla a = d$.

53. From the table, $c \nabla b = c$.

55. The set satisfies the closure property because the table contains only members of the set.
From the question, the set satisfies the associative property.
From the table, note that the identity element is b, because for any element x, $x \nabla b = x$.
Therefore, the set satisfies the identity element property. Also from the table, note that every element has an inverse. The inverse of a is c, the inverse if b is b, the inverse of c is a, and the inverse of d is d. Therefore the set satisfies the inverse property.
Consequently, the set is a group with respect to the operation ∇.

57. The table is symmetrical about the diagonal from top left to bottom right, which indicates that for any two elements m and n in the set, $m \nabla n = n \nabla m$. Therefore, the group is commutative.

59. a. The product modulo 7 of any two members of the set is always within the set. Therefore the set satisfies the closure property.
Multiplication modulo 7 is associative. Therefore the set satisfies the associative property.
$1x \bmod 7 \equiv x \bmod 7$. Therefore, 1 is the identity element, and the set satisfies the identity property.
Every element of the set has an inverse with respect to multiplication modulo 7. Therefore, the set satisfies the inverse property.
The set is a group because it satisfies the four properties of a group.

   b. The set does not satisfy the closure property. For example, $2 \cdot 3 \equiv 6 \bmod 6 \equiv 0$, which is not in the set.

   c. If n is a composite number, then it can be factored into two elements x and y such that $xy = n$ and x and y are elements of the set. Then $xy \equiv n \bmod n \equiv 0$, which is not in the set. Therefore, n cannot be a composite number if the set is to be closed under multiplication. The set is a group only for prime values of n.

61. a.

$\oplus$	1	2	3	4	5
1	1	2	3	4	5
2	2	3	4	5	1
3	3	4	5	1	2
4	4	5	1	2	3
5	5	1	2	3	4

   b. The table is symmetric about the diagonal from top left to bottom right, which indicates that for any two elements a and b, $a \oplus b = b \oplus a$. Therefore, the group is commutative.

   c. The identity element is 1, because for any element a, $a \oplus 1 = a$.

   d. $1 \oplus 1 = 1$, so 1 is the inverse of 1.
$2 \oplus 5 = 1$, so 5 is the inverse of 2.
$3 \oplus 4 = 1$, so 4 is the inverse of 3.
$4 \oplus 3 = 1$, so 3 is the inverse of 4.
$5 \oplus 2 = 1$, so 2 is the inverse of 5.

CHAPTER 7 REVIEW EXERCISES

1. 2   2. 2

3. 5   4. 6

5. 9   6. 4

7. 11   8. 7

9. 3   10. 4

11. True. $17 \div 5 = 3$ remainder 2.

12. False. $14 \div 4 = 3$ remainder 2, $24 \div 4 = 6$ remainder 0.

13. False. $35 \div 10 = 3$ remainder 5, $53 \div 10 = 5$ remainder 3.

14. True. $12 \div 8 = 1$ remainder 4, $36 \div 8 = 4$ remainder 4.

15. $(8 + 12) \bmod 3 = 20 \bmod 3$
$20 \div 3 = 6$ remainder 2
So $(8 + 12) \bmod 3 \equiv 2$.

16. $(15 + 7) \bmod 6 = 22 \bmod 6$
$22 \div 6 = 3$ remainder 4
So $(15 + 7) \bmod 6 \equiv 4$.

17. $(42 - 10) \bmod 8 = 32 \bmod 8$
$32 \div 8 = 4$ remainder 0
So $(42 - 10) \bmod 8 \equiv 0$.

18. $(19 - 8) \bmod 4 = 11 \bmod 4$
$11 \div 4 = 2$ remainder 3
So $(19 - 8) \bmod 4 \equiv 3$.

19. $(7 \cdot 5) \bmod 9 = 35 \bmod 9$
$35 \div 9 = 3$ remainder 8
So $(7 \cdot 5) \bmod 9 \equiv 8$.

20. $(12 \cdot 9) \bmod 5 = 108 \bmod 5$
$108 \div 5 = 21$ remainder 3
So $(12 \cdot 9) \bmod 5 \equiv 3$.

21. $(15 \cdot 10) \bmod 11 = 150 \bmod 11$
$150 \div 11 = 13$ remainder 7
So $(15 \cdot 10) \bmod 11 \equiv 7$.

22. $(41 \cdot 13) \bmod 8 = 533 \bmod 8$
$533 \div 8 = 66$ remainder 5
So $(41 \cdot 13) \bmod 8 \equiv 5$

23. a. $(5 + 45) \bmod 12 = 50 \bmod 12$
$50 \div 12 = 4$ remainder 2
It will be 2:00.

    b. $(5 - 71) \bmod 12 = -66 \bmod 12$
$-66 + 12 = -54$
$-54 + 12 = -42$
$-42 + 12 = -30$
$-30 + 12 = -18$
$-18 + 12 = -6$
$-6 + 12 = 6$
It was 6:00.

24. There are 8 years between the two dates, two of which (2008 and 2012) are leap years. The number of days between the two dates is
$8(365) + 2 = 2922$.
$2922 \div 7 = 417$ remainder 3
April 15, 2013 will fall 3 days after Friday, which is Monday.

25. Starting with zero, substitute whole numbers less than 4 for x.
$x = 0$ $0 \not\equiv 7 \bmod 4$ NO
$x = 1$ $1 \not\equiv 7 \bmod 4$ NO
$x = 2$ $2 \not\equiv 7 \bmod 4$ NO

$x = 3$ $3 \equiv 7 \bmod 4$ YES
Adding multiples of the modulus yields other solutions.
The solutions are 3, 7, 11, 15, 19, 23, …

26. Starting with zero, substitute whole numbers less than 9 for x.
$x = 0$ $2(0) \not\equiv 5 \bmod 9$ NO
$x = 1$ $2(1) \not\equiv 5 \bmod 9$ NO
$x = 2$ $2(2) \not\equiv 5 \bmod 9$ NO
$x = 3$ $2(3) \not\equiv 5 \bmod 9$ NO
$x = 4$ $2(4) \not\equiv 5 \bmod 9$ NO
$x = 5$ $2(5) \not\equiv 5 \bmod 9$ NO
$x = 6$ $2(6) \not\equiv 5 \bmod 9$ NO
$x = 7$ $2(7) \equiv 5 \bmod 9$ YES
$x = 8$ $2(8) \not\equiv 5 \bmod 9$ NO
Adding multiples of the modulus yields other solutions.
The solutions are 7, 16, 25, 34, 43, 52, …

27. Starting with zero, substitute whole numbers less than 5 for x.
$x = 0$ $2(0) + 1 \equiv 6 \bmod 5$ YES
$x = 1$ $2(1) + 1 \not\equiv 6 \bmod 5$ NO
$x = 2$ $2(2) + 1 \not\equiv 6 \bmod 5$ NO
$x = 3$ $2(3) + 1 \not\equiv 6 \bmod 5$ NO
$x = 4$ $2(4) + 1 \not\equiv 6 \bmod 5$ NO
Adding multiples of the modulus yields other solutions.
The solution are 0, 5, 10, 15, 20, 25, 30, …

28. Starting with zero, substitute whole numbers less than 11 for x.
$x = 0$ $3(0) + 4 \not\equiv 5 \bmod 11$ NO
$x = 1$ $3(1) + 4 \not\equiv 5 \bmod 11$ NO
$x = 2$ $3(2) + 4 \not\equiv 5 \bmod 11$ NO
$x = 3$ $3(3) + 4 \not\equiv 5 \bmod 11$ NO
$x = 4$ $3(4) + 4 \equiv 5 \bmod 11$ YES
$x = 5$ $3(5) + 4 \not\equiv 5 \bmod 11$ NO
$x = 6$ $3(6) + 4 \not\equiv 5 \bmod 11$ NO
$x = 7$ $3(7) + 4 \not\equiv 5 \bmod 11$ NO
$x = 8$ $3(8) + 4 \not\equiv 5 \bmod 11$ NO
$x = 9$ $3(9) + 4 \not\equiv 5 \bmod 11$ NO
$x = 10$ $3(10) + 4 \not\equiv 5 \bmod 11$ NO
Adding multiples of the modulus yields other solutions.
The solution are 4, 15, 26, 37, 48, 59, 70, …

29. $5 + 2 = 7$, so the additive inverse is 2. Starting with zero, substitute whole numbers less than 7 for x in the congruence $5x \equiv 1 \bmod 7$.
$x = 0$ $5(0) \not\equiv 1 \bmod 7$ NO
$x = 1$ $5(1) \not\equiv 1 \bmod 7$ NO
$x = 2$ $5(2) \not\equiv 1 \bmod 7$ NO
$x = 3$ $5(3) \equiv 1 \bmod 7$ YES

$x = 4$ $5(4) \not\equiv 1 \mod 7$ NO
$x = 5$ $5(5) \not\equiv 1 \mod 7$ NO
$x = 6$ $5(6) \not\equiv 1 \mod 7$ NO
The multiplicative inverse is 3.

30. $7 + 5 = 12$, so the additive inverse is 5. Starting with zero, substitute whole numbers less than 12 for x in the congruence $7x \equiv 1 \mod 12$.
$x = 0$ $7(0) \not\equiv 1 \mod 12$ NO
$x = 1$ $7(1) \not\equiv 1 \mod 12$ NO
$x = 2$ $7(2) \not\equiv 1 \mod 12$ NO
$x = 3$ $7(3) \not\equiv 1 \mod 12$ NO
$x = 4$ $7(4) \not\equiv 1 \mod 12$ NO
$x = 5$ $7(5) \not\equiv 1 \mod 12$ NO
$x = 6$ $7(6) \not\equiv 1 \mod 12$ NO
$x = 7$ $7(7) \equiv 1 \mod 12$ YES
$x = 8$ $7(8) \not\equiv 1 \mod 12$ NO
$x = 9$ $7(9) \not\equiv 1 \mod 12$ NO
$x = 10$ $7(10) \not\equiv 1 \mod 12$ NO
$x = 11$ $7(11) \not\equiv 1 \mod 12$ NO
The multiplicative inverse is 7.

31. Assume x to be the divisor. Then $5x = 2 \mod 7$. Starting with zero, substitute whole numbers less than 7 for x into the congruence.
$x = 0$ $5(0) \not\equiv 2 \mod 7$ NO
$x = 1$ $5(1) \not\equiv 2 \mod 7$ NO
$x = 2$ $5(2) \not\equiv 2 \mod 7$ NO
$x = 3$ $5(3) \not\equiv 2 \mod 7$ NO
$x = 4$ $5(4) \not\equiv 2 \mod 7$ NO
$x = 5$ $5(5) \not\equiv 2 \mod 7$ NO
$x = 6$ $5(6) \equiv 2 \mod 7$ YES
So $(2 \div 5) \mod 7 \equiv 6$.

32. Assume x to be the divisor. Then $4x = 3 \mod 5$. Starting with zero, substitute whole numbers less than 5 for x into the congruence.
$x = 0$ $4(0) \not\equiv 3 \mod 5$ NO
$x = 1$ $4(1) \not\equiv 3 \mod 5$ NO
$x = 2$ $4(2) \equiv 3 \mod 5$ YES
$x = 3$ $4(3) \not\equiv 3 \mod 5$ NO
$x = 4$ $4(4) \not\equiv 3 \mod 5$ NO
So $(3 \div 4) \mod 5 \equiv 2$.

33. $0(10) + 8(9) + 1(8) + 2(7) + 5(6) + 8(5) + 0(4) + 0(3) + 3(2) + x \equiv 0 \mod 11$
$170 + x \equiv 0 \mod 11$
The next multiple of 11 is 176. $x = 176 - 170 = 6$.

34. $0(10) + 3(9) + 9(8) + 4(7) + 4(6) + 9(5) + 8(4) + 2(3) + 1(2) + x \equiv 0 \mod 11$
$236 + x \equiv 0 \mod 11$
The next multiple of 11 is 242. $x = 242 - 236 = 6$.

35. $0(3) + 2(1) + 9(3) + 0(1) + 0(3) + 0(1) + 0(3) + 7(1) + 0(3) + 0(1) + 4(3) + x \equiv 0 \mod 10$
$48 + x \equiv 0 \mod 10$
The next multiple of 10 is 50. $x = 50 - 48 = 2$.

36. $0(3) + 8(1) + 5(3) + 3(1) + 9(3) + 1(1) + 8(3) + 9(1) + 5(3) + 1(1) + 2(3) + x \equiv 0 \mod 10$
$109 + x \equiv 0 \mod 10$
The next multiple of 10 is 110. $x = 110 - 109 = 1$.

37. 5 1 2 6 6 9 9 3 4 2 3 1 2 9 5 6
10 1 4 6 12 9 18 3 8 2 6 1 4 9 10 6
$(1 + 0) + 1 + 4 + 6 + (1 + 2) + 9 + (1 + 8) + 3 + 8 + 2 + 6 + 1 + 4 + 9 + (1 + 0) + 6 = 73$
$73 \neq 0 \mod 10$. This credit card number is invalid.

38. 5 3 8 3 0 1 1 8 3 4 1 6 5 9 3 1
10 3 16 3 0 1 2 8 6 4 2 6 10 9 6 1
$(1 + 0) + 3 + (1 + 6) + 3 + 0 + 1 + 2 + 8 + 6 + 4 + 2 + 6 + (1 + 0) + 9 + 6 + 1 = 60$
$60 \equiv 0 \mod 10$. This credit card number is valid.

39. 3 4 1 2 4 0 8 4 3 9 8 2 5 9 4
3 8 1 4 4 0 8 8 3 18 8 4 5 18 4
$3 + 8 + 1 + 4 + 4 + 0 + 8 + 8 + 3 + (1 + 8) + 8 + 4 + 5 + (1 + 8) + 4 = 78$
$78 \neq 0 \mod 10$. This credit card number is invalid.

40. 6 0 1 1 5 1 8 5 8 2 9 5 8 3 2 8

    12 0 2 1 10 1 16 5 16 2 18 5 16 3 4 8

    $(1 + 2) + 0 + 2 + 1 + (1 + 0) + 1 + (1 + 6) + 5 + (1 + 6) + 2 + (1 + 8) + 5 + (1 + 6) + 3 + 4 + 8 = 65$

    $65 \neq 0$ mod 10. This credit card number is invalid.

41. | | | |
|---|---|---|
| M | $c \equiv (13 + 7) \bmod 26 \equiv 20 \bmod 26 \equiv 20$ | Code M as T. |
| A | $c \equiv (1 + 7) \bmod 26 \equiv 8 \bmod 26 \equiv 8$ | Code A as H. |
| Y | $c \equiv (25 + 7) \bmod 26 \equiv 32 \bmod 26 \equiv 6$ | Code Y as F. |
| T | $c \equiv (20 + 7) \bmod 26 \equiv 27 \bmod 26 \equiv 1$ | Code T as A. |
| H | $c \equiv (8 + 7) \bmod 26 \equiv 15 \bmod 26 \equiv 15$ | Code H as O. |
| E | $c \equiv (5 + 7) \bmod 26 \equiv 12 \bmod 26 \equiv 12$ | Code E as L. |
| F | $c \equiv (6 + 7) \bmod 26 \equiv 13 \bmod 26 \equiv 13$ | Code F as M. |
| O | $c \equiv (15 + 7) \bmod 26 \equiv 22 \bmod 26 \equiv 22$ | Code O as V. |
| R | $c \equiv (18 + 7) \bmod 26 \equiv 25 \bmod 26 \equiv 25$ | Code R as Y. |
| C | $c \equiv (3 + 7) \bmod 26 \equiv 10 \bmod 26 \equiv 10$ | Code C as J. |
| E | $c \equiv (5 + 7) \bmod 26 \equiv 12 \bmod 26 \equiv 12$ | Code E as L. |
| B | $c \equiv (2 + 7) \bmod 26 \equiv 9 \bmod 26 \equiv 9$ | Code B as I. |
| E | $c \equiv (5 + 7) \bmod 26 \equiv 12 \bmod 26 \equiv 12$ | Code E as L. |
| W | $c \equiv (23 + 7) \bmod 26 \equiv 30 \bmod 26 \equiv 4$ | Code W as D. |
| I | $c \equiv (9 + 7) \bmod 26 \equiv 16 \bmod 26 \equiv 16$ | Code I as P. |
| T | $c \equiv (20 + 7) \bmod 26 \equiv 27 \bmod 26 \equiv 1$ | Code T as A. |
| H | $c \equiv (8 + 7) \bmod 26 \equiv 15 \bmod 26 \equiv 15$ | Code H as O. |
| Y | $c \equiv (25 + 7) \bmod 26 \equiv 32 \bmod 26 \equiv 6$ | Code Y as F. |
| O | $c \equiv (15 + 7) \bmod 26 \equiv 22 \bmod 26 \equiv 22$ | Code O as V. |
| U | $c \equiv (21 + 7) \bmod 26 \equiv 28 \bmod 26 \equiv 2$ | Code U as B. |

The plaintext is coded as THF AOL MVYJL IL DPAO FVB.

42. | | | |
|---|---|---|
| C | $c \equiv (3 + 11) \bmod 26 \equiv 14 \bmod 26 \equiv 14$ | Code C as N. |
| A | $c \equiv (1 + 11) \bmod 26 \equiv 12 \bmod 26 \equiv 12$ | Code A as L. |
| N | $c \equiv (14 + 11) \bmod 26 \equiv 25 \bmod 26 \equiv 25$ | Code N as Y. |
| C | $c \equiv (3 + 11) \bmod 26 \equiv 14 \bmod 26 \equiv 14$ | Code C as N. |
| E | $c \equiv (5 + 11) \bmod 26 \equiv 16 \bmod 26 \equiv 16$ | Code E as P. |
| L | $c \equiv (12 + 11) \bmod 26 \equiv 23 \bmod 26 \equiv 23$ | Code L as W. |
| A | $c \equiv (1 + 11) \bmod 26 \equiv 12 \bmod 26 \equiv 12$ | Code A as L. |
| L | $c \equiv (12 + 11) \bmod 26 \equiv 23 \bmod 26 \equiv 23$ | Code L as W. |
| L | $c \equiv (12 + 11) \bmod 26 \equiv 23 \bmod 26 \equiv 23$ | Code L as W. |
| P | $c \equiv (16 + 11) \bmod 26 \equiv 27 \bmod 26 \equiv 1$ | Code P as A. |
| L | $c \equiv (12 + 11) \bmod 26 \equiv 23 \bmod 26 \equiv 23$ | Code L as W. |
| A | $c \equiv (1 + 11) \bmod 26 \equiv 12 \bmod 26 \equiv 12$ | Code A as L. |
| N | $c \equiv (14 + 11) \bmod 26 \equiv 25 \bmod 26 \equiv 25$ | Code N as Y. |
| S | $c \equiv (19 + 11) \bmod 26 \equiv 30 \bmod 26 \equiv 4$ | Code S as D. |

The plaintext is coded as NLYNPW LWW AWLYD.

43. Because the encoded message uses a cyclical alphabetic encrypting code, it is possible to rotate through the 26 possible codes until the ciphertext is decoded. Using just the first word:

    PXXM $\rightarrow$ QYYN$\rightarrow$ RZZO $\rightarrow$ SAAP $\rightarrow$ TBBQ $\rightarrow$ UCCR $\rightarrow$ VDDS $\rightarrow$ WEET $\rightarrow$ XFFU$\rightarrow$
    YGGV $\rightarrow$ ZHHW $\rightarrow$ AIIX $\rightarrow$ BJJY $\rightarrow$ CKKZ $\rightarrow$ DLLA $\rightarrow$ EMMB $\rightarrow$ FNNC $\rightarrow$ GOOD

    It took 17 transformations to find the plaintext, so the decoding congruence is $p \equiv (c + 17) \bmod 26$. Decode the rest of the ciphertext normally.

P	$p \equiv (16 + 17) \bmod 26 \equiv 33 \bmod 26 \equiv 7$	Decode P as G.
X	$p \equiv (24 + 17) \bmod 26 \equiv 41 \bmod 26 \equiv 15$	Decode X as O.
X	$p \equiv (24 + 17) \bmod 26 \equiv 41 \bmod 26 \equiv 15$	Decode X as O.
M	$p \equiv (13 + 17) \bmod 26 \equiv 30 \bmod 26 \equiv 4$	Decode M as D.
U	$p \equiv (21 + 17) \bmod 26 \equiv 38 \bmod 26 \equiv 12$	Decode U as L.

D	$p \equiv (4 + 17) \bmod 26 \equiv 21 \bmod 26 \equiv 21$	Decode D as U.
L	$p \equiv (12 + 17) \bmod 26 \equiv 29 \bmod 26 \equiv 3$	Decode L as C.
T	$p \equiv (20 + 17) \bmod 26 \equiv 37 \bmod 26 \equiv 11$	Decode T as K.
C	$p \equiv (3 + 17) \bmod 26 \equiv 20 \bmod 26 \equiv 20$	Decode C as T.
X	$p \equiv (24 + 17) \bmod 26 \equiv 41 \bmod 26 \equiv 15$	Decode X as O.
V	$p \equiv (22 + 17) \bmod 26 \equiv 39 \bmod 26 \equiv 13$	Decode V as M.
X	$p \equiv (24 + 17) \bmod 26 \equiv 41 \bmod 26 \equiv 15$	Decode X as O.
A	$p \equiv (1 + 17) \bmod 26 \equiv 18 \bmod 26 \equiv 18$	Decode A as R.
A	$p \equiv (1 + 17) \bmod 26 \equiv 18 \bmod 26 \equiv 18$	Decode A as R.
X	$p \equiv (24 + 17) \bmod 26 \equiv 41 \bmod 26 \equiv 15$	Decode X as O.
F	$p \equiv (6 + 17) \bmod 26 \equiv 23 \bmod 26 \equiv 23$	Decode F as W.

The ciphertext is decoded as GOOD LUCK TOMORROW.

44. Because the encoded message uses a cyclical alphabetic encrypting code, it is possible to rotate through the 26 possible codes until the ciphertext is decoded. Using just the first word:

HVS → IWT → JXU → KYV → LZW → MAX → NBY → OCZ → PDA → QEB→ RFC → SGD → THE

It took 12 transformations to find the plaintext, so the decoding congruence is $p \equiv (c + 17) \bmod 26$. Decode the rest of the ciphertext normally.

H	$p \equiv (8 + 12) \bmod 26 \equiv 20 \bmod 26 \equiv 20$	Decode H as T.
V	$p \equiv (22 + 12) \bmod 26 \equiv 34 \bmod 26 \equiv 8$	Decode V as H.
S	$p \equiv (19 + 12) \bmod 26 \equiv 31 \bmod 26 \equiv 5$	Decode S as E.
R	$p \equiv (18 + 12) \bmod 26 \equiv 30 \bmod 26 \equiv 4$	Decode R as D.
O	$p \equiv (15 + 12) \bmod 26 \equiv 27 \bmod 26 \equiv 1$	Decode O as A.
M	$p \equiv (13 + 12) \bmod 26 \equiv 25 \bmod 26 \equiv 25$	Decode M as Y.
V	$p \equiv (22 + 12) \bmod 26 \equiv 34 \bmod 26 \equiv 8$	Decode V as H.
O	$p \equiv (15 + 12) \bmod 26 \equiv 27 \bmod 26 \equiv 1$	Decode O as A.
G	$p \equiv (7 + 12) \bmod 26 \equiv 19 \bmod 26 \equiv 19$	Decode G as S.
O	$p \equiv (15 + 12) \bmod 26 \equiv 27 \bmod 26 \equiv 1$	Decode O as A.
F	$p \equiv (6 + 12) \bmod 26 \equiv 18 \bmod 26 \equiv 18$	Decode F as R.
F	$p \equiv (6 + 12) \bmod 26 \equiv 18 \bmod 26 \equiv 18$	Decode F as R.
W	$p \equiv (23 + 12) \bmod 26 \equiv 35 \bmod 26 \equiv 9$	Decode W as I.
J	$p \equiv (10 + 12) \bmod 26 \equiv 22 \bmod 26 \equiv 22$	Decode J as V.
S	$p \equiv (19 + 12) \bmod 26 \equiv 31 \bmod 26 \equiv 5$	Decode S as E.
R	$p \equiv (18 + 12) \bmod 26 \equiv 30 \bmod 26 \equiv 4$	Decode R as D.

The ciphertext is decoded as THE DAY HAS ARRIVED.

45.

E	$c \equiv (3 \cdot 5 + 6) \bmod 26 \equiv 21 \bmod 26 \equiv 21$	Code E as U.
N	$c \equiv (3 \cdot 14 + 6) \bmod 26 \equiv 48 \bmod 26 \equiv 22$	Code N as V.
D	$c \equiv (3 \cdot 4 + 6) \bmod 26 \equiv 18 \bmod 26 \equiv 18$	Code D as R.
O	$c \equiv (3 \cdot 15 + 6) \bmod 26 \equiv 51 \bmod 26 \equiv 25$	Code O as Y.
F	$c \equiv (3 \cdot 6 + 6) \bmod 26 \equiv 24 \bmod 26 \equiv 24$	Code F as X.
T	$c \equiv (3 \cdot 20 + 6) \bmod 26 \equiv 66 \bmod 26 \equiv 14$	Code T as N.
H	$c \equiv (3 \cdot 8 + 6) \bmod 26 \equiv 30 \bmod 26 \equiv 4$	Code H as D.
E	$c \equiv (3 \cdot 5 + 6) \bmod 26 \equiv 21 \bmod 26 \equiv 21$	Code E as U.
L	$c \equiv (3 \cdot 12 + 6) \bmod 26 \equiv 42 \bmod 26 \equiv 16$	Code L as P.
I	$c \equiv (3 \cdot 9 + 6) \bmod 26 \equiv 33 \bmod 26 \equiv 7$	Code I as G.
N	$c \equiv (3 \cdot 14 + 6) \bmod 26 \equiv 48 \bmod 26 \equiv 22$	Code N as V.
E	$c \equiv (3 \cdot 5 + 6) \bmod 26 \equiv 21 \bmod 26 \equiv 21$	Code E as U.

The plaintext is coded as UVR YX NDU PGVU.

46. Solve the congruence for p.

$c = 7p + 4$

$c - 4 = 7p$

The multiplicative inverse of 7 is 15.

$$15(c - 4) = 15(7)p$$
$$[15(c - 4)] \mod 26 \equiv p$$

Decode using this congruence.

W	$[15(23 - 4)] \mod 26 \equiv 285 \mod 26 \equiv 25$	Decode W as Y.
E	$[15(5 - 4)] \mod 26 \equiv 15 \mod 26 \equiv 15$	Decode E as O.
U	$[15(21 - 4)] \mod 26 \equiv 255 \mod 26 \equiv 21$	Decode U as U.
L	$[15(12 - 4)] \mod 26 \equiv 120 \mod 26 \equiv 16$	Decode L as P.
K	$[15(11 - 4)] \mod 26 \equiv 105 \mod 26 \equiv 1$	Decode K as A.
G	$[15(7 - 4)] \mod 26 \equiv 45 \mod 26 \equiv 19$	Decode G as S.
G	$[15(7 - 4)] \mod 26 \equiv 45 \mod 26 \equiv 19$	Decode G as S.
M	$[15(13 - 4)] \mod 26 \equiv 135 \mod 26 \equiv 5$	Decode M as E.
F	$[15(6 - 4)] \mod 26 \equiv 30 \mod 26 \equiv 4$	Decode F as D.
N	$[15(14 - 4)] \mod 26 \equiv 150 \mod 26 \equiv 20$	Decode N as T.
H	$[15(8 - 4)] \mod 26 \equiv 60 \mod 26 \equiv 8$	Decode H as H.
M	$[15(13 - 4)] \mod 26 \equiv 135 \mod 26 \equiv 5$	Decode M as E.
N	$[15(14 - 4)] \mod 26 \equiv 150 \mod 26 \equiv 20$	Decode N as T.
M	$[15(13 - 4)] \mod 26 \equiv 135 \mod 26 \equiv 5$	Decode M as E.
G	$[15(7 - 4)] \mod 26 \equiv 45 \mod 26 \equiv 19$	Decode G as S.
N	$[15(14 - 4)] \mod 26 \equiv 150 \mod 26 \equiv 20$	Decode N as T.

The ciphertext is decoded as YOU PASSED THE TEST.

47. 1. The product of two rational numbers except 0 is always another non-zero rational number. Therefore, the closure property holds true.

 2. The associative property of multiplication holds true for rational numbers except 0.

 3. There exists an identity element (1) for multiplication in the set of rational numbers except 0.

 4. If $\frac{a}{b}$ is a rational number with $a \neq 0$, then $\frac{b}{a}$ is its inverse. The inverse property is true.

Because all of the four properties of groups are satisfied, rational numbers except 0 do form a group with respect to multiplication.

48. 1. The sum of two multiples of 3 is always another multiple of 3. Therefore, the closure property holds true.

 2. The associative property of addition holds true for all multiples of 3.

 3. There exists an identity element (0) in the set of multiples of 3 for addition.

 4. If a is a multiple of 3, then $-a$ is the inverse of a. Therefore, the inverse property holds true.

Because all of the four properties of groups are satisfied, multiples of 3 do form a group with respect to addition.

49. 1. The product of two negative integers is always a positive integer. Therefore, the closure property fails.

 2. The associative property of multiplication holds true for all negative integers.

 3. Negative integers do not have an identity element in the set for multiplication. Therefore, the identity property fails.

 4. Negative integers do not have inverses in the set for multiplication. Therefore, the inverse property fails.

Properties 1, 3, and 4 fail, so the negative integers do not form a group with respect to multiplication.

50. 1. The product modulo 11 of two elements of the set is always another member of the set. Therefore, the closure property holds true.

2. The associative property of multiplication modulo 11 holds true for all members of the set.

3. There exists an identity element (1) in the set for multiplication modulo 11. Therefore, the identity property holds true.

4. If a is an element in the set, then $11 - a$ is the inverse of a. Therefore, the inverse property holds true.

Because all of the four properties of groups are satisfied, the set forms a group with respect to multiplication modulo 11.

51.

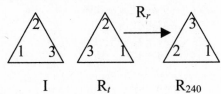

I R_t R_{240}

From the diagram, $R_t \Delta R_r = R_{240}$.

52.

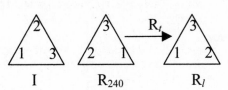

I R_{240} R_l

From the diagram, $R_{240} \Delta R_t = R_l$.

53.

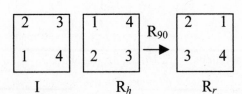

I R_h R_r

From the diagram, $R_h \Delta R_{90} = R_r$.

54.

I R_l R_{270}

From the diagram, $R_l \Delta R_v = R_{270}$.

55. $R_{180}^{-1} = \begin{pmatrix} 1 & 2 & 3 & 4 \\ 3 & 4 & 1 & 2 \end{pmatrix}^{-1} = \begin{pmatrix} 1 & 2 & 3 & 4 \\ 3 & 4 & 1 & 2 \end{pmatrix} = R_{180}$.

56. $R_r^{-1} =$

$\begin{pmatrix} 1 & 2 & 3 & 4 \\ 3 & 2 & 1 & 4 \end{pmatrix}^{-1} = \begin{pmatrix} 1 & 2 & 3 & 4 \\ 3 & 2 & 1 & 4 \end{pmatrix} = R_r$.

57. There are only four distinct ways to place the rectangle in the reference rectangle.

58. The list follows:
$\begin{pmatrix} 1 & 2 & 3 & 4 \\ 1 & 2 & 3 & 4 \end{pmatrix}, \begin{pmatrix} 1 & 2 & 3 & 4 \\ 3 & 4 & 1 & 2 \end{pmatrix},$
$\begin{pmatrix} 1 & 2 & 3 & 4 \\ 2 & 1 & 4 & 3 \end{pmatrix}, \begin{pmatrix} 1 & 2 & 3 & 4 \\ 4 & 3 & 2 & 1 \end{pmatrix}$

59. Yes.

60. The group consists of the identity, one rotation by 180° about the center, and two reflections. Two rotations of 180° about the center returns the figure to its original orientation. Thus, this rotation is its own inverse. A reflection is its own inverse since reflecting twice results in the original figure.

61. Question 59 showed that the group is commutative, so only 3 of the possible 6 combinations need to be checked.
$\begin{pmatrix} 1 & 2 & 3 & 4 \\ 3 & 4 & 1 & 2 \end{pmatrix} \Delta \begin{pmatrix} 1 & 2 & 3 & 4 \\ 2 & 1 & 4 & 3 \end{pmatrix}$
$= \begin{pmatrix} 1 & 2 & 3 & 4 \\ 4 & 3 & 2 & 1 \end{pmatrix}$
$\begin{pmatrix} 1 & 2 & 3 & 4 \\ 3 & 4 & 1 & 2 \end{pmatrix} \Delta \begin{pmatrix} 1 & 2 & 3 & 4 \\ 4 & 3 & 2 & 1 \end{pmatrix}$
$= \begin{pmatrix} 1 & 2 & 3 & 4 \\ 2 & 1 & 4 & 3 \end{pmatrix}$
$\begin{pmatrix} 1 & 2 & 3 & 4 \\ 2 & 1 & 4 & 3 \end{pmatrix} \Delta \begin{pmatrix} 1 & 2 & 3 & 4 \\ 4 & 3 & 2 & 1 \end{pmatrix}$
$= \begin{pmatrix} 1 & 2 & 3 & 4 \\ 3 & 4 & 1 & 2 \end{pmatrix}$

62. $\begin{pmatrix} 1 & 2 & 3 \\ 1 & 3 & 2 \end{pmatrix} \Delta \begin{pmatrix} 1 & 2 & 3 \\ 2 & 1 & 3 \end{pmatrix} = \begin{pmatrix} 1 & 2 & 3 \\ 2 & 3 & 1 \end{pmatrix} = A$

63. $\begin{pmatrix} 1 & 2 & 3 \\ 2 & 1 & 3 \end{pmatrix} \Delta \begin{pmatrix} 1 & 2 & 3 \\ 3 & 2 & 1 \end{pmatrix} = \begin{pmatrix} 1 & 2 & 3 \\ 2 & 3 & 1 \end{pmatrix} = A$

64. $\begin{pmatrix} 1 & 2 & 3 & 4 & 5 \\ 4 & 3 & 5 & 1 & 2 \end{pmatrix} \Delta \begin{pmatrix} 1 & 2 & 3 & 4 & 5 \\ 5 & 2 & 4 & 3 & 1 \end{pmatrix}$

$= \begin{pmatrix} 1 & 2 & 3 & 4 & 5 \\ 3 & 4 & 1 & 5 & 2 \end{pmatrix}$

65. $\begin{pmatrix} 1 & 2 & 3 & 4 & 5 \\ 3 & 5 & 1 & 4 & 2 \end{pmatrix}^{-1} = \begin{pmatrix} 1 & 2 & 3 & 4 & 5 \\ 3 & 5 & 1 & 4 & 2 \end{pmatrix}$

CHAPTER 7 TEST

1. a. 3   b. 5

2. a. 0200   b. 1300

3. a. True. $8 \div 3 = 2$ remainder 2, $20 \div 3 = 6$ remainder 2.

   b. False. $61 \div 7 = 8$ remainder 5, $38 \div 7 = 5$ remainder 3.

4. $(25 + 9) \bmod 6 = 34 \bmod 6$.
   $34 \div 6 = 5$ remainder 4
   So $(25 + 9) \bmod 6 = 4$.

5. $(31 - 11) \bmod 7 = 20 \bmod 7$.
   $20 \div 7 = 2$ remainder 6
   So $(31 - 11) \bmod 7 = 6$.

6. $(5 \cdot 16) \bmod 12 = 80 \bmod 12$.
   $80 \div 12 = 6$ remainder 8.
   So $(5 \cdot 16) \bmod 12 = 8$.

7. a. $(3 + 27) \bmod 12 = 30 \bmod 12$
   $30 \div 12 = 2$ remainder 6
   It will be 6:00.

   b. $(3 - 58) \bmod 12 = -55 \bmod 12$
   $-55 + 12 = -43$
   $-43 + 12 = -31$
   $-31 + 12 = -19$
   $-19 + 12 = -7$
   $-7 + 12 = 5$
   It was 5:00.

8. Substitute whole numbers zero through 8 for x in the congruency.

$x = 0$	$0 \not\equiv 5 \bmod 9$	NO
$x = 1$	$1 \not\equiv 5 \bmod 9$	NO
$x = 2$	$2 \not\equiv 5 \bmod 9$	NO
$x = 3$	$3 \not\equiv 5 \bmod 9$	NO
$x = 4$	$4 \not\equiv 5 \bmod 9$	NO
$x = 5$	$5 \equiv 5 \bmod 9$	YES
$x = 6$	$6 \not\equiv 5 \bmod 9$	NO
$x = 7$	$7 \not\equiv 5 \bmod 9$	NO
$x = 8$	$8 \not\equiv 5 \bmod 9$	NO

   Adding multiples of the modulus yields other solutions. The solutions are 5, 14, 23, 32, 41, 50, . . .

9. Substitute whole numbers zero through 3 for x in the congruency.

$x = 0$	$2(0) + 3 \not\equiv 1 \bmod 4$	NO
$x = 1$	$2(1) + 3 \equiv 1 \bmod 4$	YES
$x = 2$	$2(2) + 3 \not\equiv 1 \bmod 4$	NO
$x = 3$	$2(3) + 3 \equiv 1 \bmod 4$	YES

   Adding multiples of the modulus yields other solutions. The solutions are 1, 3, 5, 7, 9, 11, . . .

10. $5 + 4 = 9$, so the arithmetic inverse is 4.
    Substitute whole numbers 0 through 8 for x in the congruency $5x \equiv 1 \bmod 9$.

$x = 0$	$5(0) \not\equiv 1 \bmod 9$	NO
$x = 1$	$5(1) \not\equiv 1 \bmod 9$	NO
$x = 2$	$5(2) \equiv 1 \bmod 9$	YES
$x = 3$	$5(3) \not\equiv 1 \bmod 9$	NO
$x = 4$	$5(4) \not\equiv 1 \bmod 9$	NO

$x = 5$ $5(5) \not\equiv 1 \bmod 9$ NO
$x = 6$ $5(6) \not\equiv 1 \bmod 9$ NO
$x = 7$ $5(7) \not\equiv 1 \bmod 9$ NO
$x = 8$ $5(8) \not\equiv 1 \bmod 9$ NO
The multiplicative inverse is 2.

11. $0(10) + 4(9) + 4(8) + 1(7) + 5(6) + 6(5) + 9(4) + 5(3) + 9(2) + x \equiv 0 \bmod 11$
$204 + x \equiv 0 \bmod 11$
The next multiple of 11 is 209. $x = 209 - 204 = 5$.

12. $0(3) + 7(1) + 2(3) + 8(1) + 7(3) + 8(1) + 2(3) + 7(1) + 5(3) + 3(1) + 3(3) + x \equiv 0 \bmod 10$
$90 + x \equiv 0 \bmod 10$
The next multiple of 10 is 90. $x = 90 - 90 = 0$.

13. 4 2 3 2 8 1 8 0 5 7 3 6 4 8 7 6
8 2 6 2 16 1 16 0 10 7 6 6 8 8 14 6
$8 + 2 + 6 + 2 + (1 + 6) + 1 + (1 + 6) + 0 + (1 + 0) + 7 + 6 + 6 + 8 + 8 + (1 + 4) + 6 = 80$
$80 \equiv 0 \bmod 10$. This credit card number is valid.

14.
R	$c \equiv (18 + 10) \bmod 26 \equiv 28 \bmod 26 \equiv 2$	Code R as B.
E	$c \equiv (5 + 10) \bmod 26 \equiv 15 \bmod 26 \equiv 15$	Code E as O.
P	$c \equiv (16 + 10) \bmod 26 \equiv 26 \bmod 26 \equiv 0$	Code P as Z.
O	$c \equiv (15 + 10) \bmod 26 \equiv 25 \bmod 26 \equiv 25$	Code O as Y.
R	$c \equiv (18 + 10) \bmod 26 \equiv 28 \bmod 26 \equiv 2$	Code R as B.
T	$c \equiv (20 + 10) \bmod 26 \equiv 30 \bmod 26 \equiv 4$	Code T as D.
B	$c \equiv (2 + 10) \bmod 26 \equiv 12 \bmod 26 \equiv 12$	Code B as L.
A	$c \equiv (1 + 10) \bmod 26 \equiv 11 \bmod 26 \equiv 11$	Code A as K.
C	$c \equiv (3 + 10) \bmod 26 \equiv 13 \bmod 26 \equiv 13$	Code C as M.
K	$c \equiv (11 + 10) \bmod 26 \equiv 21 \bmod 26 \equiv 21$	Code K as U.

The plaintext is coded as BOZYBD LKMU.

15. Solve the congruence for p.
$c = 3p + 5$
$c - 5 = 3p$
The multiplicative inverse of 3 is 9.
$9(c - 5) = 9(3)p$
$[9(c - 5)] \bmod 26 \equiv p$
Decode using this congruence.

U	$[9(21 - 5)] \bmod 26 \equiv 144 \bmod 26 \equiv 14$	Decode U as N.
T	$[9(20 - 5)] \bmod 26 \equiv 135 \bmod 26 \equiv 5$	Decode T as E.
S	$[9(19 - 5)] \bmod 26 \equiv 126 \bmod 26 \equiv 22$	Decode S as V.
T	$[9(20 - 5)] \bmod 26 \equiv 135 \bmod 26 \equiv 5$	Decode T as E.
G	$[9(7 - 5)] \bmod 26 \equiv 18 \bmod 26 \equiv 18$	Decode G as R.
D	$[9(4 - 5)] \bmod 26 \equiv -9 \bmod 26 \equiv 17$	Decode D as Q.
P	$[9(16 - 5)] \bmod 26 \equiv 99 \bmod 26 \equiv 21$	Decode P as U.
F	$[9(6 - 5)] \bmod 26 \equiv 9 \bmod 26 \equiv 9$	Decode F as I.
M	$[9(13 - 5)] \bmod 26 \equiv 72 \bmod 26 \equiv 20$	Decode M as T.

The ciphertext is decoded as NEVER QUIT.

16. a. Yes. The product of any two odd integers is another odd integer.

   b. No. The sum of any two odd integers is an even integer.

17. 1. The product of any two odd integers is always another odd integer. Therefore, the closure property holds true.

    2. From the question, the associative property of multiplication holds true for odd integers.

    3. There exists an identity element (1) in the set of odd integers with respect to multiplication.

    4. Not every element has an inverse in the set. If a is an odd integer, then $\frac{1}{a}$ should be the inverse of a. However, $\frac{1}{a}$ is a non-reducible fraction for all odd numbers except for -1 and 1. Therefore, the inverse property fails.

    Property 4 fails, so the odd integers do not form a group with respect to multiplication.

18. a.

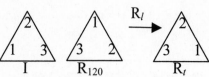

    From the diagram, $R_{120} \Delta R_l = R_t$.

    b.

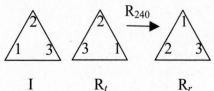

    From the diagram, $R_t \Delta R_{240} = R_r$.

19. $\begin{pmatrix} 1 & 2 & 3 \\ 3 & 1 & 2 \end{pmatrix} \Delta \begin{pmatrix} 1 & 2 & 3 \\ 3 & 2 & 1 \end{pmatrix} = \begin{pmatrix} 1 & 2 & 3 \\ 1 & 3 & 2 \end{pmatrix}$

20. Inverse $\begin{pmatrix} 1 & 2 & 3 & 4 \\ 3 & 1 & 4 & 2 \end{pmatrix} = \begin{pmatrix} 1 & 2 & 3 & 4 \\ 2 & 4 & 1 & 3 \end{pmatrix}$

Chapter 8: Geometry

EXERCISE SET 8.1

1. The three names for the angle are $\angle O$, $\angle AOB$, and $\angle BOA$.

3. The angle measures 40°. Since 40 < 90, the angle is acute.

5. The angle measures 30°. Since 30 < 90, the angle is acute.

7. The angle measures 120°. Since 120 > 90, the angle is obtuse.

9. $m\angle A + m\angle B = 51° + 39° = 90°$. Yes, they are complementary angles.

11. $m\angle C + m\angle D = 112° + 58° = 170°$. No, they are not supplementary angles.

13. Solving:
$$x + 62° = 90°$$
$$x = 28°$$
A 28° angle

15. Solving:
$$x + 162° = 180°$$
$$x = 18°$$
An 18° degree angle

17. Solving:
$$AB + BC + CD = AD$$
$$12 + BC + 9 = 35$$
$$BC = 14 \text{ cm}$$

19. Solving:
$$QR + RS = QS$$
$$QR + 3QR = QS$$
$$7 + 21 = QS$$
$$QS = 28 \text{ ft}$$

21. Solving:
$$EF + FG = EG$$
$$EF + \frac{1}{2}EF = EG$$
$$20 + 10 = EG$$
$$EG = 30 \text{ m}$$

23. Solving:
$$m\angle LOM + m\angle MON = m\angle LON$$
$$53° + m\angle MON = 139°$$
$$m\angle MON = 86°$$

25. Solving:
$$x + 74° = 145°$$
$$x = 71°$$

27. Solving:
$$x + 2x = 90°$$
$$3x = 90°$$
$$x = 30°$$

29. Solving:
$$x + x + 18° = 90°$$
$$2x + 18° = 90°$$
$$2x = 72°$$
$$x = 36°$$

31. Solving:
$$a + 53° = 180°$$
$$a = 127°$$

33. Solving:
$$a + 76° + 168° = 360°$$
$$a = 116°$$

35. Solving:
$$3x + 4x + 2x = 180°$$
$$9x = 180°$$
$$x = 20°$$

37. Solving:
$$5x + x + 20° + 2x = 180°$$
$$8x = 160°$$
$$x = 20°$$

39. Solving:
$$3x + 4x + 6x + 5x = 360°$$
$$18x = 360°$$
$$x = 20°$$

41. $\angle a$ is complementary to $\angle b$'s supplementary angle.

$$m\angle a + m\angle x = 90°$$
$$51° + m\angle x = 90°$$
$$m\angle x = 39°$$
$$m\angle b + m\angle x = 180°$$
$$m\angle b + 39° = 180°$$
$$m\angle b = 141°$$

43. Solving:
$$x + 74° = 180°$$
$$x = 106°$$

45. The two angles are vertical angles, so their measures are equal.
$$5x = 3x + 22°$$
$$2x = 22°$$
$$x = 11°$$

47. $\angle a$ is a corresponding angle with the 38° angle. Therefore, $m\angle a = 38°$.
$$m\angle a + m\angle b = 180°$$
$$38° + m\angle b = 180°$$
$$m\angle b = 142°$$

49. $\angle a$ is an alternate interior angle with the 47° angle. Therefore, $m\angle a = 47°$.
$$m\angle a + m\angle b = 180°$$
$$47° + m\angle b = 180°$$
$$m\angle b = 133°$$

51. Solving:
$$5x + 4x = 180°$$
$$9x = 180°$$
$$x = 20°$$

53. Solving:
$$2x + x + 39° = 180°$$
$$3x = 141°$$
$$x = 47°$$

55. The three lines form a triangle. The sum of the interior angles of the triangle is
$$m\angle b + (180° - m\angle a)$$
$$+ (180° - m\angle x) = 180°$$
$$70° - 95° - m\angle x = -180°$$
$$-m\angle x = -155°$$
$$m\angle x = 155°$$
$\angle y$ and $\angle b$ are vertical angles:
$$m\angle y = m\angle b = 70°$$

57. Solving:
$$m\angle a = m\angle y = 45°$$
$\angle a$ and $\angle y$ are vertical angles
The three lines form a triangle. The sum of the interior angles of the triangle is
$$m\angle a + 90° + (180° - m\angle b) = 180°$$
$$45° + 90° - m\angle b = 0°$$
$$135° = m\angle b$$

59. The three angles form a straight angle, so
$$x + m\angle AOB + m\angle BOC = 180°$$
$$x + 90° + m\angle BOC = 180°$$
$$m\angle BOC = 90° - x$$

61. Solving:
$$90° + 30° + x = 180°$$
$$x = 60°$$

63. Solving:
$$42° + 103° + x = 180°$$
$$x = 35°$$

65. Solving:
$$13° + 65° + x = 180°$$
$$x = 102°$$

67. The three angles form a straight angle. The sum of the measures of the angles of a triangle is 180°.

69. A point has zero dimensions. A line has one dimension. A line segment has one dimension. A ray has one dimension. An angle has two dimensions.

71. Solving:

$m\angle x + m\angle y + m\angle z$

$= (180° - m\angle c) + (180° - m\angle a) + (180° - m\angle b)$

$= 180° + 180° + 180° - m\angle a - m\angle b - m\angle c$

$= 540° - (m\angle a + m\angle b + m\angle c)$

The sum of the interior angles of a triangle is 180°, so

$= 540° - (180°)$

$= 360°$

73. $\angle AOC$ and $\angle BOC$ are supplementary angles; therefore, $m\angle AOC + m\angle BOC = 180°$. Because $m\angle AOC = m\angle BOC$, by substitution, $m\angle AOC + m\angle AOC = 180°$. Therefore, $2(m\angle AOC) = 180°$, and $m\angle AOC = 90°$. Hence $\overline{AB} \perp \overline{CD}$.

EXERCISE SET 8.2

1.

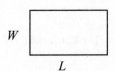

3. a. Perimeter is measured in linear units, not square units.

   b. Area is measured in square units, not linear units.

5. This is a heptagon because it has seven sides.

7. This is a quadrilateral because it has four sides.

9. The legs of the triangle are the same length. Since this triangle has two sides the same measure, it is isosceles.

11. The sides of the triangle are all different in length, so it is scalene.

13. Because this triangle has one angle that measures exactly 90 degrees, this is a right triangle.

15. Because this triangle has one angle that measures greater than 90 degrees, this is an obtuse triangle.

17. a. $P = 2l + 2w = 2(5) + 2(10) = 30$ m

    b. $A = lw = (5)(10) = 50$ m^2

19. a. $P = 4s = 4(4) = 16$ cm

    b. $A = s^2 = (4)^2 = 16$ cm^2

21. a. $P = 4s = 4(10) = 40$ km

    b. $A = s^2 = (10)^2 = 100$ km^2

23. a. $P = 2(b) + 2(s) = 2(12) + 2(8) = 40$ ft

    b. $A = bh = (12)(6) = 72$ ft^2

25. a. $C = 2\pi r = 2\pi(4) = 8\pi$ cm ≈ 25.13 cm

    b. $A = \pi r^2 = \pi(4)^2 = 16\pi$ cm ≈ 50.27 cm^2

27. a. $C = 2\pi r = 2\pi(5.5) = 11\pi$ mi ≈ 34.56 mi

    b. $A = \pi r^2 = \pi(5.5)^2 = 30.25\pi$ mi ≈ 95.03 mi^2

29. a. $C = \pi d = \pi(17) = 17\pi$ ft ≈ 53.41 ft

    b. $r = \dfrac{d}{2} = \dfrac{17}{2} = 8.5$ ft

    $A = \pi r^2 = \pi(8.5)^2 = 72.25\pi$ ft$^2 \approx 226.98$ ft^2

31. $P = a + b + c = 12\dfrac{1}{2} + 10\dfrac{3}{4} + 6\dfrac{1}{4}$

    $= 12\dfrac{2}{4} + 10\dfrac{3}{4} + 6\dfrac{1}{4}$

    $= 28\dfrac{6}{4} = 29\dfrac{2}{4} = 29\dfrac{1}{2}$ ft

33. $P = a + b + c = 4\dfrac{3}{10} + 3\dfrac{7}{10} + 2\dfrac{1}{2}$

    $= 4\dfrac{3}{10} + 3\dfrac{7}{10} + 2\dfrac{5}{10}$

    $= 9\dfrac{15}{10} = 10\dfrac{5}{10} = 10\dfrac{1}{2}$ mi

35. $P = 2l + 2w = 2(20) + 2(14) = 68$ ft

37. $P = 5s = 5(4) = 20$ in.

39. $P = 2l + 2w = 2(62) + 2(45) = 214$ yd

41. $P = 2b + 2s = 2(3) + 2(2) = 10$ mi

43. $P = 2l + 2w = 2(68) + 2(42) = 220$ in.
    15 ft = (15)(12) = 180 in.
    1 package is not enough.
    2 packages = 2(180) = 360 in.
    Two packages are needed.

45. $A = s^2 = (12)^2 = 144$ m^2

47. $P = 4s = 36$ in.
    $s = 9$ in.

49. $A = lw$
    $(40)(w) = 400$ in.^{2}
    $w = 10$ in.

51. $A = bh$
    $(b)(7) = 56$ m^2
    $b = 8$ m

53. $A = \dfrac{1}{2}bh = \dfrac{1}{2}(12)(16) = 96$ m^2

55. $A = \dfrac{(b_1 + b_2)}{2}h = \dfrac{(36 + 45)}{2}15 = 607.5$ m^2

57. $A = \dfrac{1}{2}bh = \dfrac{1}{2}(20)(24) = 240$ m^2
    $(120)(2) = 240$ m
    Two bags should be purchased.

59. $A = \dfrac{(b_1 + b_2)}{2}h = \dfrac{(10 + 12)}{2}10 = 110$ ft^2
    110 ft^2 ÷ 55 ft^2 per quart = 2 quarts

61. $A = lw$
    Floor = (10)(8) = 80 ft^2
    Tile = (2)(2) = 4 ft^2
    80 ft^2 ÷ 4 ft^2 per tile = 20 tiles

63. $A = s^2 = (80)^2 = 6{,}400$ ft^2
    6,400 ft^2 ÷ 1,500 ft^2 per bag ≈ 4.27.
    5 bags are needed.
    5 × \$8 = \$40

65. $A = lw = (16)(8) = 128$ ft^2
    $A = lw = (12)(8) = 96$ ft^2
    128 + 128 + 96 + 96 = 448 ft^2 total area
    400 ft^2 is less than 448 ft^2. Two gallons of paint
    are needed.
    2(\$17) = \$34

67. Length of drapes = 4 + 1 = 5 ft
    Width of drapes = (2)(3) = 6 ft
    Area of drapes= = lw = (5)(6) = 30 ft^2
    Area for 4 Windows = (4)(30) = 120 ft^2

69. $C = \pi d = \pi(4.2) \approx 13.19$ ft

71. $A = \pi r^2 = \pi(20)^2 = 400\pi \approx 1{,}256.6$ ft^2

73. $C = \pi d = \pi(18) \approx 56.55$ in.
    56.55 in. × 20 revolutions = 1,131.00 in.
    1,131.00 in. ÷ 12 = 94.25 ft

75. $r = \dfrac{d}{2} = \dfrac{24}{2} = 12$ in.
    $A = \pi r^2 = \pi(12)^2 = 144\pi$ in.^{2}

77. $A = \pi r^2$
    Area of Large Pizza= $\pi(10)^2 \approx 314.16$ in.^{2}
    Area of Small Pizza= $\pi(8)^2 \approx 201.06$ in.^{2}
    314.16 - 201.06 = 113.10 in.^{2}

79. $d = 2(36{,}000 + 6{,}380) = 84{,}760$ km
    $C = \pi d = \pi(84{,}760) \approx 266{,}281$ km

81. Length of Rectangle = $4r$
    Width of Rectangle = $2r$
    Area of Rectangle = $(4r)(2r) = 8r^2$
    Area of Circles = $2\pi r^2$
    Area of shaded region = $8r^2 - 2\pi r^2$

83. $A = lw$
    Double the length and the width.
    $A = (2l)(2w) = 4(lw)$
    The area is four times that of the original
    rectangle.

85. Area of shaded region = $a^2 - 16$
    Factored = $(a + 4)(a - 4)$
    The dimensions of the rectangle are $(a + 4)$ and
    $(a - 4)$.

87. a. n = number of units
        Perimeter = $2n + 2$
        2(8) + 2 = 18 cm
        Area = n = 8 cm^2

    b. n = number of units
        Perimeter = $2(n + 2)$
        2(8 + 2) = 20 cm
        Area = $2n$ = 16 cm^2

c.　n = number of units
　　Perimeter = $4n$
　　$4(8) = 32$ cm
　　Area = $n^2 = 64$ cm^2

EXERCISE SET 8.3

1.　$\dfrac{7}{14} = \dfrac{1}{2}$

3.　$\dfrac{6}{8} = \dfrac{3}{4}$

5.　$\dfrac{5}{9} = \dfrac{4}{x}$
　　$5x = 36$
　　$x = 7.2$ cm

7.　$\dfrac{3}{5} = \dfrac{2}{x}$
　　$3x = 10$
　　$x \approx 3.3$ m

9.　$\dfrac{4}{8} = \dfrac{x}{6}$
　　$8x = 24$
　　$x = 3$ m
　　$3 + 4 + 5 = 12$ m

11.　$\dfrac{4}{12} = \dfrac{x}{15}$
　　$12x = 60$
　　$x = 5$ in.
　　$3 + 4 + 5 = 12$ in.

13.　$\dfrac{15}{40} = \dfrac{x}{20}$
　　$40x = 300$
　　$x = 7.5$ cm
　　$A = \dfrac{1}{2}bh = \dfrac{1}{2}(15)(7.5) \approx 56.3$ cm^2

15.　$\dfrac{24}{8} = \dfrac{x}{6}$
　　$8x = 144$
　　$x = 18$ ft

17.　$\dfrac{8}{4} = \dfrac{x}{8}$
　　$4x = 64$
　　$x = 16$ m

19.　5 ft 9 in. = 5.75 ft
　　$\dfrac{12}{30} = \dfrac{5.75}{x}$
　　$12x = 172.5$
　　$x = 14.375 = 14\dfrac{3}{8}$ ft

21.　$\dfrac{8}{20} = \dfrac{6}{x}$
　　$8x = 120$
　　$x = 15$ m

23.　Solving:
　　$\dfrac{BC}{AC} = \dfrac{CE - DE}{CE}$
　　$AC = 3 + 3 = 6$ ft
　　$\dfrac{3}{6} = \dfrac{x - 4}{x}$
　　$3x = 6x - 24$
　　$3x = 24$
　　$x = 8$ ft

25.　$\dfrac{24}{12} = \dfrac{39 - x}{x}$
　　$24x = 468 - 12x$
　　$36x = 468$
　　$x = 13$ cm

27.　$\dfrac{8}{20} = \dfrac{14}{x}$
　　$8x = 280$
　　$x = 35$ m

29.　Yes, SAS Theorem.

31.　Yes, SSS Theorem.

33.　Yes, ASA Theorem.

35.　No.

37.　Yes, SAS Theorem.

39.　No.

41. No.

43. Using the Pythagorean Theorem:
$$a^2 + b^2 = c^2$$
$$5^2 + 12^2 = c^2$$
$$169 = c^2$$
$$13 = c$$
The side is 13 in.

45. Using the Pythagorean Theorem:
$$a^2 + b^2 = c^2$$
$$7^2 + 9^2 = c^2$$
$$130 = c^2$$
$$11.4 \approx c$$
The side is 11.4 cm.

47. Using the Pythagorean Theorem:
$$a^2 + b^2 = c^2$$
$$18^2 + b^2 = 20^2$$
$$b^2 = 400 - 324 = 76$$
$$b \approx 8.7$$
The side is 8.7 ft.

49. Using the Pythagorean Theorem:
$$a^2 + b^2 = c^2$$
$$9^2 + b^2 = 12^2$$
$$b^2 = 144 - 81 = 63$$
$$b \approx 7.9$$
The side is 7.9 m.

51. Using the Pythagorean Theorem:
$$a^2 + b^2 = c^2$$
$$3^2 + b^2 = 8^2$$
$$b^2 = 64 - 9 = 55$$
$$b \approx 7.4$$
The ladder will reach 7.4 m.

53. Using the Pythagorean Theorem:
$$a^2 + b^2 = c^2$$
$$18^2 + 12^2 = c^2$$
$$468 = c^2$$
$$21.6 \approx c$$
You are 21.6 miles from the starting point.

55. Using the Pythagorean Theorem:
$$a^2 + b^2 = c^2$$
$$6^2 + 8^2 = c^2$$
$$100 = c^2$$
$$10 = c$$
The hypotenuse has length 10 in. Adding:
$6 + 8 + 10 = 24$. The perimeter is 24 in.

57. Using the Pythagorean Theorem:
$$a^2 + b^2 = c^2$$
$$6^2 + 24^2 = c^2$$
$$612 = c^2$$
$$24.7 \approx c$$
Yes, a 25 ft. ladder will be long enough.

EXERCISE SET 8.4

1. $V = lwh = (14)(10)(6) = 840 \text{ in}^3$

3. $V = \dfrac{1}{3}s^2h = \dfrac{1}{3}(3)^2(5) = 15 \text{ ft}^3$

5. Solving:
$$r = \frac{1}{2}d = \frac{1}{2}(3) = \frac{3}{2}$$
$$V = \frac{4}{3}\pi r^3 = \frac{4}{3}\pi\left(\frac{3}{2}\right)^3 = \left(\frac{4}{3}\right)\left(\frac{27}{8}\right)\pi$$
$$V = \frac{9}{2}\pi \approx 14.14 \text{ cm}^3$$

7. Solving:
$$l = 4 \quad w = 5 \quad h = 3$$
$$S = 2lw + 2lh + 2wh$$
$$S = 2(4)(5) + 2(4)(3) + 2(5)(3)$$
$$S = 94 \text{ m}^2$$

9. Solving:
$$s = 4 \quad l = 5$$
$$S = s^2 + 4\left(\frac{1}{2}sl\right)$$
$$S = (4)^2 + 4\left(\frac{1}{2} \cdot 4 \cdot 5\right)$$
$$S = 16 + 40 = 56 \text{ m}^2$$

11. Solving:

$r = 6 \qquad h=2$

$S = 2\pi r^2 + 2\pi rh$

$S = 2\pi (6)^2 + 2\pi (6)(2)$

$S = 96\pi \approx 301.59$ in^2

13. $V = lwh = (6.8)(2.5)(2) = 34$ m^3

15. $V = s^3 = (2.5)^3 = 15.625$ in.^{3}

17. Solving:

$V = \dfrac{4}{3}\pi r^3 \qquad r = 3$

$V = \dfrac{4}{3}\pi (3)^3 = 36\pi$ ft^3

19. Solving:

$d = 24 \quad r = \dfrac{1}{2}(24) = 12 \quad h=18$

$V = \pi r^2 h$

$V = \pi (12)^2 (18) = 2592\pi \approx 8143.01$ cm^3

21. Solving:

$r = 5 \qquad h=9$

$V = \dfrac{1}{3}\pi r^2 h$

$V = \dfrac{1}{3}\pi (5)^2 (9) = 75\pi$ in.^{3}

23. Solving:

$s = 6 \qquad h=10$

$V = \dfrac{1}{3}s^2 h = \dfrac{1}{3}(6)^2 (10) = 120$ in.^{3}

25. Solving:

$C = 3.5 \quad h = 8$

If $C = 2\pi r$, then $r = \dfrac{C}{2\pi} = \dfrac{3.5}{2\pi} \approx 0.557$

$V = \pi r^2 h = \pi (.557)^2 (8) \approx 7.80$ ft^3

27. Solving:

$l = 60 \qquad w=32 \qquad h=14$

$S = 2lw + 2lh + 2wh$

$S = 2(60)(32) + 2(60)(14) + 2(32)(14)$

$S = 6416$ cm^2

29. Solving:

$s = 1.5$

$S = 6s^2 = 6(1.5)^2 = 13.5$ in.^{2}

31. Solving:

$r = 2$

$S = 4\pi r^2 = 4\pi (2)^2 = 16\pi \approx 50.27$ in.^{2}

33. Solving:

$r = \dfrac{1}{2}(1.8) = 0.9 \quad h=0.7$

$S = 2\pi r^2 + 2\pi rh = 2\pi (0.9)^2 + 2\pi (0.9)(0.7)$

$S = 1.62\pi + 1.26\pi = 2.88\pi$ m^2

35. Solving:

$r = \dfrac{1}{2}(21) = 10.5 \qquad l=16$

$S = \pi r^2 + \pi rl$

$S = \pi (10.5)^2 + \pi (10.5)(16)$

$S = 874.15$ in.^{2}

37. Solving:

$s = 16 \qquad l=18$

$S = s^2 + 2sl = (16)^2 + 2(16)(18)$

$S = 832$ m^2

39. Solving:

$l = 18 \quad w = 12 \quad V = 1836$

$V = lwh$

$h = \dfrac{V}{lw}$

$h = \dfrac{1836}{(18)(12)} = \dfrac{1836}{216} = 8.5$ in.

41. Solving:

$S = 162 \quad w = 3 \quad l = 12$

$S = 2lw + 2lh + 2wh$

$162 = 2(12)(3) + 2(12)h + 2(3)h$

$162 = 72 + 24h + 6w$

$162 = 30h + 72$

$30h = 90$

$h = 3$ ft

43. Solving:
$$r = 16$$
$$S = 4\pi r^2 = 4\pi (16)^2 \approx 3216.99 \text{ ft}^2$$
3217 ft^2 of fabric are needed.

45. Solving:
$$r = 8.25 \qquad h = 17$$
$$A = 2\pi rh$$
$$A = 2\pi (8.25)(17)$$
$$A = 280.50\pi \approx 881.22$$
The area of the label is 881.22 square centimeters.

47. Solving:
$$V_{total} = V_{prism} - V_{cylinder}$$
$$V_{prism} = lwh = (1.2)(2)(0.8) = 1.92$$
$$V_{cylinder} = \pi r^2 h = \pi (0.2)^2 (2) = 0.08\pi \approx 0.25$$
$$V_{total} = 1.92 - 0.25 = 1.67 \text{ m}^3$$

49. $$V_{total} = V_{Prism} - \frac{1}{2} V_{cylinder}$$
$$V_{Prism} = lwh = (4)(4)(8) = 128$$
$$\frac{1}{2} \cdot V_{cylinder} = \frac{1}{2} \cdot \pi r^2 h = \frac{1}{2} \pi (1)^2 (8) = 4\pi$$
$$V_{total} = V_{Prism} - \frac{1}{2} V_{cylinder} = 128 - 4\pi$$
$$V_{total} \approx 115.43 \text{ cm}^3$$

51. $$V_{total} = V_{large} - V_{small}$$
$$V_{large} = \pi (9)^2 (24) = 1944\pi$$
$$V_{small} = \pi (4.5)^2 (24) = 486\pi$$
$$V_{total} = 1944\pi - 486\pi = 1458\pi$$
$$V_{total} \approx 4580.44 \text{ cm}^3$$

53. Solving:
$$S_{front} = (1.5)(1.5) + (0.5)(0.5) = 2.5$$
$$S_{back} = (1.5)(1.5) + (0.5)(0.5) = 2.5$$
$$S_{left} = (1.5)(2) = 3$$
$$S_{bottom} = (2)(2) = 4$$
$$S_{top} = (1.5)(2) + (0.5)(2) = 4$$
$$S_{right} = (1)(2) + (0.5)(2) = 3$$
sum of all surface areas = 19 m^2

55. Solving:
$$S_{total} = S_{cylinder} + S_{prism} - 2S_{square}$$
$$S_{cylinder} = 2\pi (8)^2 + 2\pi (8)(2) = 160\pi$$
$$S_{prism} = 2(2)(2) + 2(2)(15) + 2(2)(15) = 128$$
$$S_{square} = (2)(2) = 4$$
$$S_{total} = (160\pi + 128) - 2(4) \approx 622.65 \text{ m}^2$$

57. Solving:
$$S_{total} = S_{prism} + S_{cylinder} - 4S_{circle}$$
$$S_{prism} = 2(40)(80) + 2(80)(30) + 2(40)(30)$$
$$S_{prism} = 13,600$$
$$S_{cylinder} = 2\pi (14)^2 + 2\pi (14)(80)$$
$$S_{cylinder} = 2632\pi$$
$$S_{circle} = \pi (14)^2$$
$$S_{total} = 13,600 + 2632\pi - 4(\pi (14)^2)$$
$$S_{total} \approx 19,405.66 \text{ m}^2$$

59. Solving:
$$V_{total} = V_{cylinder} + 2V_{semisphere}$$
$$V_{cylinder} = \pi (4)^2 (30) = 480\pi$$
$$2V_{semisphere} = 2\left(\frac{1}{2}\right) \cdot \frac{4}{3} \pi (4)^3 = \frac{256}{3}\pi$$
$$V_{total} = \left(480\pi + \frac{256}{3}\pi\right) \div 2 \approx 888.02 \text{ ft}^3$$

61. Solving:
$$V = (6)(2.5)(5) + (4)(1)(5) = 95 \text{ m}^3$$
$$V = (95 \text{ m}^3)\left(\frac{1000L}{1 \text{ m}^3}\right) = 95,000 \text{ L}$$

63. Solving:
$$S = \pi r^2 + 2\pi rh + \pi rl$$
$$S = \pi (5)^2 + 2\pi (5)(3) + \pi (5)(10)$$
$$S = 105\pi \approx 329.87 \text{ cm}^2$$
Find the total grams used.
$$= \left(329.87 \text{cm}^2\right)\left(\frac{0.24g}{1 \text{ cm}^2}\right) \approx 79.17 \text{ g}$$

65. Solving:

$$V_s = \frac{4}{3}r_s^{\,3}$$

$$V_l = \frac{4}{3}r_l^{\,3}$$

$$r_l = 3r_s$$

$$V_l = \frac{4}{3}(3r_s)^3 = 27V_s$$

Value of large sphere is 27 times the value of the small sphere or $4860.

67. Solving:

$$V_{\text{sphere}} = \frac{4}{3}\pi r^3 \qquad S_{\text{sphere}} = 4\pi r^2$$

$$V_{\text{hemisphere}} = \left(\frac{1}{2}\right)\left(\frac{4}{3}\right)\pi r^3 = \frac{2}{3}\pi r^3$$

$$S_{\text{hemisphere}} = \left(\frac{1}{2}\right)4\pi r^2 + \pi r^2 = 3\pi r^2$$

69. Solving:

$$S_{\text{sphere}} = 4\pi r^2$$

$$S_{\text{lateral cylinder}} = 2\pi rh$$

$$h = 2r$$

$$S_{\text{lateral cylinder}} = 2\pi r^2(2r) = 4\pi r^2$$

71. a. Solving:

$$S_{\text{original}} = 2lw + 2lh + 2wh$$

$$S_{\text{new}} = 2l(2w) + 2l(2h) + 2(2w)(2h)$$

$$S_{\text{new}} = 4lw + 4lh + 8wh$$

$$S_{\text{new}} = 2(2lw + 2lh + 2wh) + 4wh$$

$$S_{\text{new}} = 2 \cdot S_{\text{original}} + 4wh$$

The surface area is double plus 4*wh*.

b. Solving:

$$V_{\text{original}} = lwh$$

$$V_{\text{new}} = (2l)(2w)h = 4lwh$$

The volume is quadrupled.

c. Solving:

$$V_{\text{original}} = s^3$$

$$V_{\text{new}} = (2s)^3 = 8s^3$$

The volume is 8 times larger.

d. Solving:

$$S_{\text{original}} = 2\pi r^2 + 2\pi rh$$

$$S_{\text{new}} = 2\pi(2r)^2 + 2\pi(2r)(2h)$$

$$S_{\text{new}} = 8\pi r^2 + 8\pi rh = 4(2\pi r^2 + 2\pi rh)$$

The volume is quadrupled.

EXERCISE SET 8.5

1. a. $\sin A = \dfrac{\text{opp}}{\text{hyp}} = \dfrac{a}{c}$

   b. $\sin B = \dfrac{\text{opp}}{\text{hyp}} = \dfrac{b}{c}$

   c. $\cos A = \dfrac{\text{adj}}{\text{hyp}} = \dfrac{b}{c}$

   d. $\cos B = \dfrac{\text{adj}}{\text{hyp}} = \dfrac{a}{c}$

   e. $\tan A = \dfrac{\text{opp}}{\text{adj}} = \dfrac{a}{b}$

   f. $\tan B = \dfrac{\text{opp}}{\text{adj}} = \dfrac{b}{a}$

3. $\sin\theta = \dfrac{5}{13}, \cos\theta = \dfrac{12}{13}, \tan\theta = \dfrac{5}{12}$

5. $\sin\theta = \dfrac{24}{25}, \cos\theta = \dfrac{7}{25}, \tan\theta = \dfrac{24}{7}$

7. $\sin\theta = \dfrac{8}{\sqrt{113}}, \cos\theta = \dfrac{7}{\sqrt{113}}, \tan\theta = \dfrac{8}{7}$

9. $\sin\theta = \dfrac{1}{2}, \cos\theta = \dfrac{\sqrt{3}}{2}, \tan\theta = \dfrac{1}{\sqrt{3}}$

11. 0.6820

13. 1.4281

15. 0.9971

17. 1.9970

19. 0.8878

21. 0.8453

23. 0.8018

25. 0.6833

27. 38.6°

29. 41.4°

31. 21.3°

33. 38.0°

35. 72.5°

37. 0.6°

39. 66.1°

41. 29.5°

43.

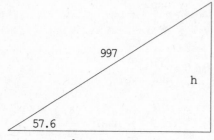

$$\sin 57.6° = \frac{h}{997}$$

$$h = 997 \sin 57.6° \approx 841.79$$

The balloon is 841.79 ft. off the ground.

45.

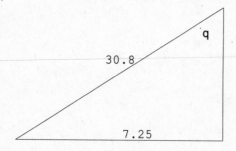

$$\sin \theta = \frac{7.25}{30.8}$$

$$\theta = \sin^{-1} \frac{7.25}{30.8} \approx 13.6$$

The angle is 13.6°

47.

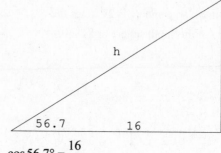

$$\cos 56.7° = \frac{16}{h}$$

$$h = \frac{16}{\cos 56.7°} \approx 29.14$$

The wire is 29.14 ft. long.

49.

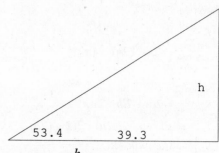

$$\tan 53.4° = \frac{h}{39.3}$$

$$h = 39.3 \tan 53.4° \approx 52.92$$

The height is 52.92 ft.

51.

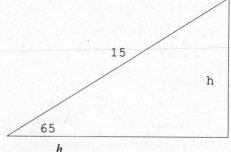

$$\sin 65° = \frac{h}{15}$$

$$h = 15 \sin 65° \approx 13.59$$

The ladder reaches 13.59 ft. up the side.

53.

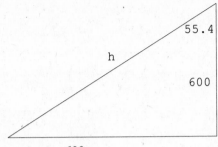

$\cos 55.4° = \dfrac{600}{h}$

$h = \dfrac{600}{\cos 55.4°} \approx 1056.63$

The wire is 1056.63 ft. long.

55.

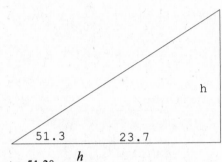

$\tan 51.3° = \dfrac{h}{23.7}$

$h = 23.7 \tan 51.3° \approx 29.58$

The tree is 29.58 yd. tall.

57. No. Explanations will vary.

59. $\sin\theta = \dfrac{2}{3} = \dfrac{o}{h}$. We wish to find $\cos\theta = \dfrac{a}{h}$.

Using the Pythagorean Theorem:

$a^2 + o^2 = h^2$

$a^2 + 2^2 = 3^2$

$a^2 = 9 - 4 = 5$

$a = \sqrt{5}$

So, $\cos\theta = \dfrac{\sqrt{5}}{3}$.

61. $\cos\theta = \dfrac{3}{4} = \dfrac{a}{h}$. We wish to find $\tan\theta = \dfrac{o}{a}$.

Using the Pythagorean Theorem:

$a^2 + o^2 = h^2$

$3^2 + o^2 = 4^2$

$o^2 = 16 - 9 = 7$

$o = \sqrt{7}$

So, $\tan\theta = \dfrac{\sqrt{7}}{3}$.

63. $\sin\theta = \dfrac{a}{1} = \dfrac{\text{opp}}{\text{hyp}}$. We wish to find $\cos\theta = \dfrac{\text{adj}}{\text{hyp}}$.

Using the Pythagorean Theorem:

$\text{adj}^2 + \text{opp}^2 = \text{hyp}^2$

$\text{adj}^2 + a^2 = 1^2$

$\text{adj}^2 = 1 - a^2$

$\text{adj} = \sqrt{1 - a^2}$

So, $\cos\theta = \dfrac{\sqrt{1-a^2}}{1} = \sqrt{1-a^2}$.

65. $\theta = \dfrac{12 \text{ cm}}{3 \text{ cm}}$ radians = 4 radians

67. $\theta = \dfrac{6 \text{ in}}{9 \text{ in}}$ radian = $\dfrac{2}{3}$ radian

69. $\left(\dfrac{180}{\pi}\right)^°$

71. Converting:

$45° \cdot \left(\dfrac{\pi \text{ radians}}{180°}\right) = \dfrac{45\pi}{180} = \dfrac{\pi}{4}$ radians

≈ 0.7854 radians

73. Converting:

$315° \cdot \left(\dfrac{\pi \text{ radians}}{180°}\right) = \dfrac{315\pi}{180} = \dfrac{7\pi}{4}$ radians

≈ 5.4978 radians

75. Converting:

$210° \cdot \left(\dfrac{\pi \text{ radians}}{180°}\right) = \dfrac{210\pi}{180} = \dfrac{7\pi}{6}$ radians

≈ 3.6652 radians

77. Converting:

$$\frac{\pi}{3} \text{ radians} \cdot \left(\frac{180°}{\pi \text{ radians}}\right) = \left(\frac{180\pi}{3\pi}\right)° = 60°$$

79. Converting:

$$\frac{4\pi}{3} \text{ radians} \cdot \left(\frac{180°}{\pi \text{ radians}}\right) = \left(\frac{720\pi}{3\pi}\right)° = 240°$$

81. Converting:

$$3 \text{ radians} \cdot \left(\frac{180°}{\pi \text{ radians}}\right) = \left(\frac{540}{\pi}\right)°$$
$$\approx 171.8873°$$

EXERCISE SET 8.6

1. a. Through a given point not on a given line, exactly one line can be drawn parallel to the given line.

    b. Through a given point not on a given line, there are at least two lines parallel to the given line.

    c. Through a given point not on a given line, there exist no lines parallel to the given line.

3. Carl Friedrich Gauss

5. a. The sum equals $180°$.

    b. The sum is less than $180°$.

    c. The sum is more than $180°$ but less than $540°$.

7. Imaginary Geometry

9. A geodesic is a curve on a surface such that for any two points of the curve the portion of the curve between the points is the shortest path on the surface that joins these points.

11. An infinite saddle surface

13. Calculating:

$$m\angle A = 150°$$
$$m\angle B = 120°$$
$$m\angle C = 90°$$
$$S = \left(m\angle A + m\angle B + m\angle C - 180°\right) \cdot \left(\frac{\pi}{180°}\right) r^2$$
$$S = \left(150° + 120° + 90° - 180°\right) \cdot \left(\frac{\pi}{180°}\right)(1)^2$$
$$S = \pi \text{ units}^2$$

15. Calculating:

$$A = \left[\sum \theta - (n-2) \cdot 180°\right] \cdot \left(\frac{\pi}{180°}\right) r^2$$
$$A = \left[380° - (4-2)180°\right] \cdot \left(\frac{\pi}{180°}\right)(1980)^2$$
$$A = \left[20°\right] \cdot \left(\frac{\pi}{180°}\right)(1980)^2 \approx 1,370,000 \text{ mi}^2$$

17. Using the formulas:

$$d_E(P,Q) = \sqrt{(x_2 - x_1)^2 + (y_2 - y_1)^2}$$
$$= \sqrt{(4-(-3))^2 + (1-1)^2}$$
$$= \sqrt{49} = 7$$

$$d_C(P,Q) = |x_2 - x_1| + |y_2 - y_1|$$
$$= |4-(-3)| + |1-1|$$
$$= |7| + |0| = 7$$

19. Using the formulas:

$$d_E(P,Q) = \sqrt{(x_2 - x_1)^2 + (y_2 - y_1)^2}$$
$$= \sqrt{(-3-2)^2 + (5-(-3))^2}$$
$$= \sqrt{25+64} = \sqrt{89} \approx 9.4$$

$$d_C(P,Q) = |x_2 - x_1| + |y_2 - y_1|$$
$$= |-3-2| + |5-(-3)|$$
$$= |-5| + |8| = 13$$

21. Using the formulas:

$$d_E(P,Q) = \sqrt{(x_2 - x_1)^2 + (y_2 - y_1)^2}$$
$$= \sqrt{(5-(-1))^2 + (-2-4)^2}$$
$$= \sqrt{36+36} = \sqrt{72} \approx 8.5$$

$$d_C(P,Q) = |x_2 - x_1| + |y_2 - y_1|$$
$$= |5-(-1)| + |-2-4|$$
$$= |6| + |-6| = 12$$

23. Using the formulas:

$$d_E(P,Q) = \sqrt{(x_2 - x_1)^2 + (y_2 - y_1)^2}$$
$$= \sqrt{(3-2)^2 + (-6-0)^2}$$
$$= \sqrt{1+36} = \sqrt{37} \approx 6.1$$

$$d_C(P,Q) = |x_2 - x_1| + |y_2 - y_1|$$
$$= |3-2| + |-6-0|$$
$$= |1| + |-6| = 7$$

25. Using the formula:

$$d_C(P,Q) = \frac{1+|m|}{\sqrt{1+m^2}} d_E(P,Q)$$

$$= \frac{1+\left|\frac{3}{4}\right|}{\sqrt{1+\left(\frac{3}{4}\right)^2}} \cdot 5 = \frac{\frac{7}{4}}{\sqrt{\frac{25}{16}}} \cdot 5$$

$$= \frac{\frac{7}{4}}{\frac{5}{4}} \cdot 5 = \frac{7}{4} \cdot \frac{4}{5} \cdot 5 = 7$$

27. Using the formula:

$$d_C(P,Q) = \frac{1+|m|}{\sqrt{1+m^2}} d_E(P,Q)$$

$$= \frac{1+\left|-\frac{2}{3}\right|}{\sqrt{1+\left(-\frac{2}{3}\right)^2}} \cdot \sqrt{13} = \frac{\frac{5}{3}}{\sqrt{\frac{13}{9}}} \cdot \sqrt{13}$$

$$= \frac{\frac{5}{3}}{\frac{\sqrt{13}}{3}} \cdot \sqrt{13} = \frac{5}{3} \cdot \frac{3}{\sqrt{13}} \cdot \sqrt{13} = 5$$

29. Using the formula:

$$d_C(P,Q) = \frac{1+|m|}{\sqrt{1+m^2}} d_E(P,Q)$$

$$= \frac{1+\left|\frac{1}{4}\right|}{\sqrt{1+\left(\frac{1}{4}\right)^2}} \cdot \sqrt{17} = \frac{\frac{5}{4}}{\sqrt{\frac{17}{16}}} \cdot \sqrt{17}$$

$$= \frac{\frac{5}{4}}{\frac{\sqrt{17}}{4}} \cdot \sqrt{17} = \frac{5}{4} \cdot \frac{4}{\sqrt{17}} \cdot \sqrt{17} = 5$$

31. A city distance may be associated with more than one Euclidean distance. For example, if $P = (0,0)$ and $Q = (2,0)$, then the city distance between the points is 2 blocks and the Euclidean distance is also 2 blocks. However, if $P = (0,0)$ and $Q = (1,1)$, then the city distance between the points is still 2 blocks, but the Euclidean distance is $\sqrt{2}$ blocks.

33.

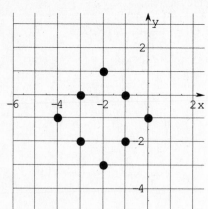

35.

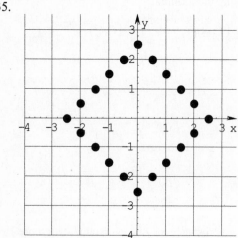

37. 4*n*

39. a.

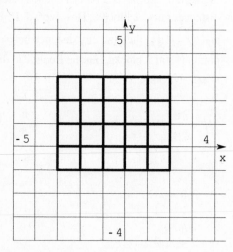

b.

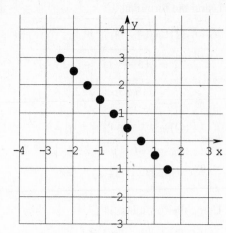

41. a.   Find all combinations of 5 points taken 2 at a time.

$$C(5,2) = \frac{5!}{3!2!} = 10$$

b.   All lines not containing points *A* or *B* are parallel to $\overline{AB}$. There are 3 such lines.

EXERCISE SET 8.7

1.   Stage 2

——   ——   ——   ——

Stage 3

– –   – –   – –   – –

3.   Stage 2

5.

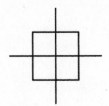

Stage 2

7. Stage 2

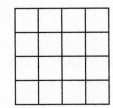

9.

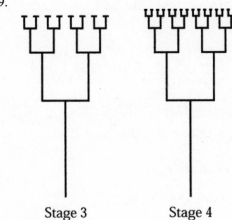

Stage 3 Stage 4

11. Replacement Ratio = 2
Scale Ratio = 3

Similarity Dimension = $\dfrac{\log 2}{\log 3} = 0.631$

13. Replacement Ratio = 5
Scale Ratio = 3

Similarity Dimension = $\dfrac{\log 5}{\log 3} = 1.465$

15. Replacement Ratio = 4
Scale Ratio = 2

Similarity Dimension = $\dfrac{\log 4}{\log 2} = 2.000$

17. Replacement Ratio = 4
Scale Ratio = 2

Similarity Dimension = $\dfrac{\log 4}{\log 2} = 2.000$

19. Replacement Ratio = 18
Scale Ratio = 6

Similarity Dimension = $\dfrac{\log 18}{\log 6} = 1.613$

21. a. Sierpinski carpet (1.893)
Variation 2 (1.771)
Variation 1 (1.465)

b. The Sierpinski carpet

23. The binary tree fractal is not a strictly self-similar fractal

CHAPTER 8 REVIEW EXERCISES

1. Calculating:
$m\angle a = 74°$ $m\angle b = 52°$
$m\angle a = m\angle b + m\angle x$
$74 = 52 + m\angle x$
$22° = m\angle x$

$180 = m\angle x + m\angle y$
$180 = 22 + m\angle y$
$158° = m\angle y$

2. $\dfrac{AC}{DF} = \dfrac{BC}{EF} \rightarrow \dfrac{AC}{12} = \dfrac{6}{9} \rightarrow 72 = 9 \cdot AC$
Perimeter $ABC = 10 + 6 + 8 = 24$ in.

3. Break figure into 2 rectangular prisms.
$V_1 = (3)(3)(8) = 72$
$V_2 = (3)(7)(8) = 168$
$V_{total} = V_1 + V_2$
$V_{total} = 72 + 168 = 240$ in^3

4. $x = 180° - 112° = 68°$

5. Calculating:
$S = 2lw + 2lh + 2wh$
$S = 2(5)(10) + 2(5)(4) + 2(10)(4)$
$S = 220$ ft^2

6. $r = \dfrac{1}{2}(4) = 2 \quad h = 8$

 $S = 2\pi r^2 + 2\pi rh$

 $S = 2\pi(2)^2 + 2\pi(2)(8)$

 $S = 8\pi + 32\pi = 40\pi \ \text{m}^2$

7. $AB = 3BC \quad BC = 11$
 $AC = AB + BC$
 $AC = 3(11) + 11 = 44 \ \text{cm}$

8. $4x + 3x + (x + 28) = 180$
 $8x + 28 = 180$
 $8x = 152$
 $x = 19°$

9. $A = bh$
 $A = (6)(4.5) = 27 \ \text{in.}^2$

10. $s = 6 \quad h = 8$

 $V = \dfrac{1}{3}s^2 h$

 $V = \dfrac{1}{3}(6)^2(8) = 96 \ \text{cm}^2$

11. $C = 2\pi r = \pi d$
 $C = \pi(4.5) \approx 14.14 \ \text{m}$

12. $\angle a$ is an alternate interior angle to the given angle. $\angle a$ and $\angle b$ are a linear pair.
 $m\angle a = 138° \quad m\angle b = 42°$

13. $180° - 32° = 148°$
 A 148° angle

14. $V = lwh = (6.5)(2)(3) = 39 \ \text{ft}^3$

15. Adding:
 $m\angle a + m\angle b + m\angle c = 180°$
 $37° + 48° + m\angle c = 180°$
 $m\angle c = 95°$

16. $A = \dfrac{1}{2}bh \quad A = 28 \quad h = 7$

 $28 = \dfrac{1}{2}b(7) = \dfrac{7}{2}b$

 $b = 8 \ \text{cm}$

17. $V = \dfrac{4}{3}\pi r^3 \quad r = \dfrac{1}{2}(12) = 6$

 $V = \dfrac{4}{3}\pi(6)^3 = 288\pi \ \text{mm}^3$

18. $P = 4s = 86$

 $s = \dfrac{86}{4} = 21.5 \ \text{cm}$

19. $r = 6 \quad h = 15$

 $S = 2\pi r^2 + 2\pi rh$

 $S = 2\pi(6)^2 + 2\pi(6)(15)$

 $S = 252\pi \approx 791.68 \ \text{ft}^2$

20. $P = 2l + 2w$
 $P = 2(56) + 2(48) = 208 \ \text{yd}$

21. $A = s^2 = (9.5)^2 = 90.25 \ \text{m}^2$

22. $A_{\text{walk}} = A_{\text{total}} - A_{\text{grass}}$
 $A_{\text{walk}} = (44)(29) + (40)(25)$
 $A_{\text{walk}} = 1276 - 1000 = 276 \ \text{m}^2$

23. Carl Friedrich Gauss

24. Yes, by the SAS Theorem.

25. Using the Pythagorean Theorem:
 $a^2 + b^2 = c^2$
 $a^2 + 7^2 = 12^2$
 $a^2 = 144 - 49 = 95$
 $a = \sqrt{95} \approx 9.75 \ \text{ft}$

26. Using the Pythagorean Theorem:

$$a^2 + b^2 = c^2$$

$$5^2 + 8^2 = c^2$$

$$89 = 95 = c^2$$

$$\sqrt{89} = c$$

$$\sin\theta = \frac{\text{opp}}{\text{hyp}} = \frac{5}{\sqrt{89}}$$

$$\cos\theta = \frac{\text{opp}}{\text{hyp}} = \frac{8}{\sqrt{89}}$$

$$\tan\theta = \frac{\text{opp}}{\text{adj}} = \frac{5}{8}$$

27. Using the Pythagorean Theorem:

$$a^2 + b^2 = c^2$$

$$a^2 + 10^2 = 20^2$$

$$a^2 = 400 - 100 = 300$$

$$a = \sqrt{300} = 10\sqrt{3}$$

$$\sin\theta = \frac{\text{opp}}{\text{hyp}} = \frac{10\sqrt{3}}{20} = \frac{\sqrt{3}}{2}$$

$$\cos\theta = \frac{\text{opp}}{\text{hyp}} = \frac{10}{20} = \frac{1}{2}$$

$$\tan\theta = \frac{\text{opp}}{\text{adj}} = \frac{10\sqrt{3}}{10} = \sqrt{3}$$

28. 25.7°

29. 29.2°

30. 53.8°

31. 1.9°

32. $\tan 50° = \dfrac{d}{84}$

$d = 84\tan 50° \approx 100.1$ ft

33. $\cos 40° = \dfrac{d}{200}$

$d = 200\cos 40° \approx 153.2$ mi.

34.

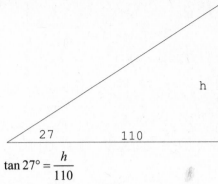

$$\tan 27° = \frac{h}{110}$$

$$h = 110\tan 27° \approx 56.0 \text{ ft}$$

35. Nikolai Lobachevsky

36. Spherical geometry or elliptical geometry

37. Hyperbolic geometry

38. Lobachevskian or hyperbolic geometry

39. Riemannian or spherical geometry

40. $m\angle A = 90°$, $m\angle B = 150°$, $m\angle C = 90°$, $r = 12$

$$S = \left(m\angle A + m\angle B + m\angle C - 180°\right)\left(\frac{\pi}{180°}\right)r^2$$

$$= \left(90° + 150° + 90° - 180°\right)\left(\frac{\pi}{180°}\right)(12)^2$$

$$= \frac{5}{6}\pi(144) = 120\pi \text{ in}^2$$

41. $m\angle A = 90°$, $m\angle B = 60°$, $m\angle C = 90°$, $r = 5$

$$S = \left(m\angle A + m\angle B + m\angle C - 180°\right)\left(\frac{\pi}{180°}\right)r^2$$

$$= \left(90° + 60° + 90° - 180°\right)\left(\frac{\pi}{180°}\right)(5)^2$$

$$= \frac{\pi}{3}(25) = \frac{25\pi}{3} \text{ ft}^2$$

42. Using the formulas:

$$d_E(P,Q) = \sqrt{(x_2 - x_1)^2 + (y_2 - y_1)^2}$$

$$= \sqrt{(3 - (-1))^2 + (4 - 1)^2}$$

$$= \sqrt{16 + 9} = \sqrt{25} = 5$$

$$d_C(P,Q) = |x_2 - x_1| + |y_2 - y_1|$$
$$= |3 - (-1)| + |4 - 1|$$
$$= |4| + |3| = 7$$

43. Using the formulas:
$$d_E(P,Q) = \sqrt{(x_2 - x_1)^2 + (y_2 - y_1)^2}$$
$$= \sqrt{(2 - (-5))^2 + (6 - (-2))^2}$$
$$= \sqrt{49 + 64} = \sqrt{113} \approx 10.6$$

$$d_C(P,Q) = |x_2 - x_1| + |y_2 - y_1|$$
$$= |2 - (-5)| + |6 - (-2)|$$
$$= |7| + |8| = 15$$

44. Using the formulas:
$$d_E(P,Q) = \sqrt{(x_2 - x_1)^2 + (y_2 - y_1)^2}$$
$$= \sqrt{(3 - 2)^2 + (2 - 8)^2}$$
$$= \sqrt{1 + 36} = \sqrt{37} \approx 6.1$$

$$d_C(P,Q) = |x_2 - x_1| + |y_2 - y_1|$$
$$= |3 - 2| + |2 - 8|$$
$$= |1| + |-6| = 7$$

45. Using the formulas:
$$d_E(P,Q) = \sqrt{(x_2 - x_1)^2 + (y_2 - y_1)^2}$$
$$= \sqrt{(5 - (-3))^2 + (-2 - 3)^2}$$
$$= \sqrt{64 + 25} = \sqrt{89} \approx 9.4$$

$$d_C(P,Q) = |x_2 - x_1| + |y_2 - y_1|$$
$$= |5 - (-3)| + |-2 - 3|$$
$$= |8| + |-5| = 13$$

46. a.   Using the formulas:
$$d_E(P,Q) = \sqrt{(x_2 - x_1)^2 + (y_2 - y_1)^2}$$
$$= \sqrt{(4 - 1)^2 + (5 - 1)^2}$$
$$= \sqrt{9 + 16} = \sqrt{25} = 5$$

$$d_E(P,R) = \sqrt{(x_2 - x_1)^2 + (y_2 - y_1)^2}$$
$$= \sqrt{(-4 - 1)^2 + (2 - 1)^2}$$
$$= \sqrt{25 + 1} = \sqrt{26} \approx 5.1$$

$$d_E(R,Q) = \sqrt{(x_2 - x_1)^2 + (y_2 - y_1)^2}$$
$$= \sqrt{(4 - (-4))^2 + (5 - 2)^2}$$
$$= \sqrt{64 + 9} = \sqrt{73} \approx 8.5$$

P and Q are closest.

b.   Using the formulas:
$$d_C(P,Q) = |x_2 - x_1| + |y_2 - y_1|$$
$$= |4 - 1| + |5 - 1|$$
$$= |3| + |4| = 7$$

$$d_C(P,R) = |x_2 - x_1| + |y_2 - y_1|$$
$$= |-4 - 1| + |2 - 1|$$
$$= |-5| + |1| = 6$$

$$d_C(R,Q) = |x_2 - x_1| + |y_2 - y_1|$$
$$= |4 - (-4)| + |5 - 2|$$
$$= |8| + |3| = 11$$

P and R are closest.

47.

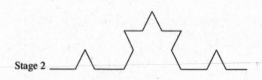

48. Stage 2

49. $\dfrac{\log 5}{\log 4} \approx 1.161$

50. Replacement Ratio = 2
    Scaling Ratio = 2
    Similarity dimension = $\dfrac{\log 2}{\log 2} = 1.000$

CHAPTER 8 TEST

1. Solving:
   $$V = \pi r^2 h$$
   $$V = \pi(3)^2(6) = 54\pi \approx 169.65 \text{ m}^3$$

2. Solving:
   $$P = 2l + 2w$$
   $$P = 2(2) + 2(1.4) = 6.8m$$

3. $90 - 32 = 58°$

4. $A = \pi r^2 = \pi(1)^2 = \pi \approx 3.14m^2$

5. Solving:
   $$m\angle x + m\angle z = 180°$$
   $$30° + m\angle z = 180°$$
   $$m\angle z = 150°$$
   $$m\angle z = m\angle y \text{ (corresponding angles)}$$
   $$m\angle y = 150°$$

6. Solving:
   $$m\angle x + m\angle b = 180°$$
   $$45 + m\angle b = 180°$$
   $$m\angle b = 135°$$
   $\angle x$ and $\angle a$ are corresponding angles, so
   $m\angle x = m\angle a$.
   $m\angle a = 45°$

7. Solving:
   $$A = s^2$$
   $$A = (2.25)^2 = 5.0625 \text{ ft}^2$$

8. Solving:
   $$V_{\text{total}} = V_{\text{large}} - V_{\text{small}}$$
   $$V_{\text{large}} = \pi(6)^2(14) = 504\pi$$
   $$V_{\text{small}} = \pi(2)^2(14) = 56\pi$$
   $$V_{\text{total}} = 504\pi - 56\pi = 448\pi \text{ cm}^3$$

9. Solving:
   $$\frac{BC}{EF} = \frac{AB}{DE} \rightarrow \frac{BC}{4} = \frac{0.75}{2.5}$$
   $$4 \cdot 0.75 = 2.5 \cdot BC$$
   $$2.5BC = 3$$
   $$BC = 1.2 = 1\frac{1}{5} \text{ ft}$$

10. By definition, one angle is $90°$, while the third will be $90 - 40 = 50°$

11. $m\angle x = 38 + 87 = 125°$

12. Solving:
    $$A = bh$$
    $$A = (8)(4) = 32m^2$$

13. Solving:
    $$\frac{5}{x} = \frac{12}{60}$$
    $$5 \cdot 60 = 12x$$
    $$12x = 300$$
    $$x = 25 \text{ ft}$$

14. Solving:
    $$A_{\text{small}} = \pi(8)^2 \qquad A_{\text{large}} = \pi(10)^2$$
    Difference $= 100\pi - 64\pi = 36\pi$
    There is about 113.10 in^2 more.

15. Yes, by the SAS Theorem.

16. Using the Pythagorean Theorem:
    $$a^2 + b^2 = c^2$$
    $$a^2 + 8^2 = 11^2$$
    $$a^2 = 121 - 64 = 57$$
    $$a = \sqrt{57} = 7.55$$
    Side BC is 7.55 cm long.

17. Using the Pythagorean Theorem:

$$a^2 + b^2 = c^2$$
$$a^2 + 8^2 = 10^2$$
$$a^2 = 100 - 64 = 36$$
$$a = \sqrt{36} = 6$$

$$\sin\theta = \frac{\text{opp}}{\text{hyp}} = \frac{8}{10} = \frac{4}{5}$$

$$\cos\theta = \frac{\text{opp}}{\text{hyp}} = \frac{6}{10} = \frac{3}{5}$$

$$\tan\theta = \frac{\text{opp}}{\text{adj}} = \frac{8}{6} = \frac{4}{3}$$

18.

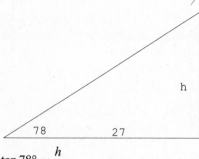

$$\tan 78° = \frac{h}{27}$$
$$h = 27\tan 78° \approx 127 \text{ ft}$$

19. Solving:

$$d = 11\text{ft }6\text{in.} = 11.5 \text{ ft}$$
$$r = \frac{1}{2}(11.5) = 5.75$$
$$A = \pi r^2$$
$$A = \pi(5.75)^2 \approx 103.87\text{ft}^2$$

20. $l = 1\text{ft }1\text{in.} = 13 \text{ in.}$ $w = 8$ in. $h = 7.5$ in.
$$V = lwh$$
$$V = (13)(8)(7.5) = 780 \text{ in.}^3$$

21. a. Through a given point not on a given line, exactly one line can be drawn parallel to the given line.

b. Through a given point not on a given line, there exist no lines parallel to the given line.

22. a. 1

b. 3

23. A great circle of a sphere is a circle on the surface of a sphere whose center is at the center of the sphere.

24. Solving:

$$m\angle A = 90°\qquad m\angle B = 100°\qquad m\angle C = 90°$$
$$r = 12$$
$$S = \left(m\angle A + m\angle B + m\angle C - 180°\right)\cdot\left(\frac{\pi}{180°}\right)r^2$$
$$S = \left(90° + 100° + 90° - 180°\right)\cdot\left(\frac{\pi}{180°}\right)(12)^2$$
$$S = \frac{5}{9}\pi(144) = 80\pi \text{ in.}^2 \approx 251.3 \text{ ft}^2$$

25. Using the formulas:

$$d_E(P,Q) = \sqrt{(x_2 - x_1)^2 + (y_2 - y_1)^2}$$
$$= \sqrt{(5-(-4))^2 + (1-2)^2}$$
$$= \sqrt{81+1} = \sqrt{82} \approx 9.1$$

$$d_C(P,Q) = |x_2 - x_1| + |y_2 - y_1|$$
$$= |5-(-4)| + |1-2|$$
$$= |9| + |-1| = 10$$

26. Using the formula: $4n = 4(4) = 16$ points.

27. Stage 2

28. Stage 2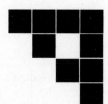

29. Replacement Ratio = 2
Scale Ratio = 2

Similarity Dimension = $\dfrac{\log 2}{\log 2} = 1.000$

30. Replacement Ratio = 3
Scale Ratio = 2

Similarity Dimension = $\dfrac{\log 3}{\log 2} \approx 1.585$

Chapter 9: The Mathematics of Graphs

EXERCISE SET 9.1

1. a. 6 b. 7

    c. 6 d. Yes

    e. No

3. a. 6 b. 4

    c. 4 d. Yes

    e. Yes

5. The graph is shown below:

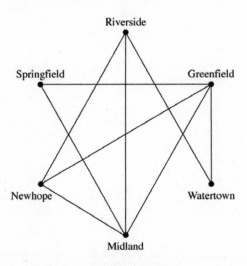

7. The graph is shown below:

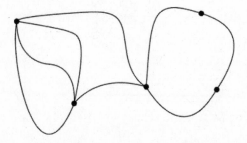

9. a. No. There is no line connecting John and Stacy.

    b. 3

c. Ada

d. A loop would correspond to a friend talking to himself or herself.

11. Equivalent

13. Not equivalent

15. The graph on the right has a vertex of degree 4 and the graph on the left does not.

17. a. Yes. D-A-E-B-D-C-E-D is an Euler circuit.

19. a. Not Eulerian. There are vertices of odd degree.

    b. Yes. A-E-A-D-E-D-C-E-C-B-E-B is an Euler walk.

21. a. Not Eulerian. There are vertices of odd degree.

    b. This graph does not have an Euler walk. More than two vertices are of odd degree.

23. a. Not Eulerian. There are vertices of odd degree.

    b. Yes. E-A-D-E-G-D-C-G-F-C-B-F-A-B-E-F is an Euler walk.

25. a. As a graph:

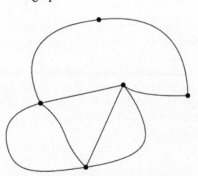

    b. It is possible to cross each bridge once and return to the starting point. Every vertex of the graph has an even degree.

27. Yes. There are exactly two vertices of odd degree, so an Euler walk is possible.

29. Yes, the hamster can travel through every tube without going through the same tube twice. Draw a graph of the Habitrail, with the vertices representing the cages and the edges representing the tubes and find an Euler circuit or an Euler walk. There is no Euler circuit; therefore the hamster cannot return to its starting point.

31. Yes. To do so would require an Euler walk or an Euler circuit. Every vertex is of even degree, so an Euler circuit is possible. You will always return to the starting room.

33. The graph is connected and has at least three vertices and no multiple edges, so the theorem applies. Every vertex has a degree of 4 or more, so the graph is Hamiltonian. A Hamiltonian circuit is A-B-C-D-E-G-F-A.

35. The graph is connected and has at least three vertices and no multiple edges, so the theorem applies. Every vertex has a degree of 4 or more, so the graph is Hamiltonian. A Hamiltonian circuit is A-B-E-C-H-D-F-G-A.

37. One such route is Springfield-Greenfield-Watertown-Riverside-Newhope-Midland-Springfield.

39. A route through the museum that visits each room once and returns to the starting point without visiting any room twice.

41. An Euler circuit would be most efficient. The officer needs to travel each street, or edge, once.

43. All of the vertices are of odd degree, so there is not an Euler circuit or an Euler walk. There are $n = 10$ vertices and each vertex has a degree of $5 = n/2$, so Dirac's Theorem says that there is a Hamiltonian circuit.

45. a. There are many possible answers.

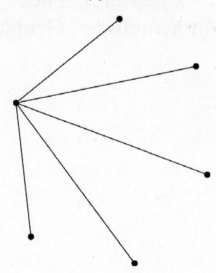

b. There are many possible answers.

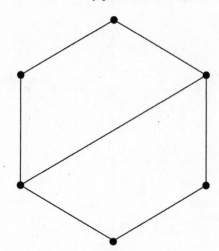

c. There are many possible answers.

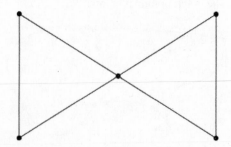

EXERCISE SET 9.2

1. Two Hamiltonian circuits are A-B-E-D-C-A, total weight 31, and A-D-E-B-C-A, total weight 32.

3. Two Hamiltonian circuits are A-D-C-E-B-F-A, total weight 114, and A-C-D-E-B-F-A, total weight 158.

5. A-D-B-C-F-E-A

7. A-C-E-B-D-A

9. A-D-B-F-E-C-A

11. A-C-E-B-D-A

13. As a graph:

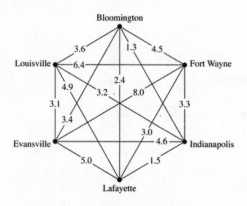

Louisville-Evansville-Bloomington-Indianapolis-Lafayette-Fort Wayne-Louisville

15. Louisville-Evansville-Bloomington-Indianapolis-Layfayette-Fort Wayne-Louisville

17. As a graph:

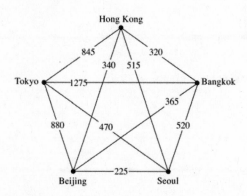

Tokyo-Seoul-Beijing-Hong Kong-Bangkok-Tokyo

19. Tokyo-Seoul-Beijing-Hong Kong-Bangkok-Tokyo

21. Represent the information as a graph:

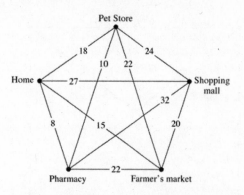

The Greedy Algorithm generates the route home-pharmacy-pet store-farmer's market-shopping mall-home. The Edge-Picking Algorithm generates the route home-pharmacy-pet store-shopping mall-farmer's market-home.

23. Represent the information as a graph:

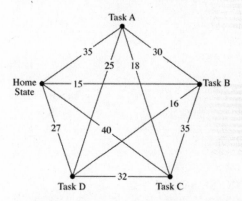

The Greedy Algorithm generates the sequence home state-task B-task D-task A-task C-home state. The Edge-Picking Algorithm gives the same sequence.

25. Assigning weights:

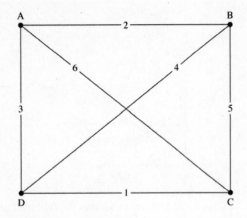

27. Assigning weights:

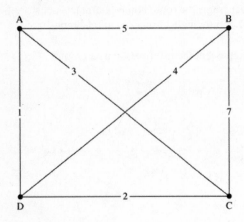

EXERCISE SET 9.3

1. As a planar graph:

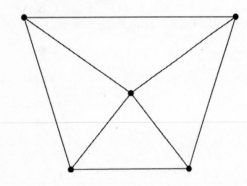

3. As a planar graph:

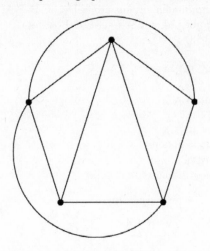

5. As a planar graph:

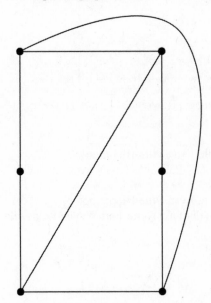

7. As a planar graph:

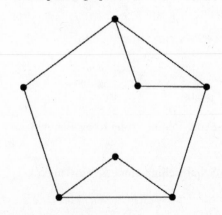

9. Finding a subgraph:

The highlighted subgraph is the Utilities Graph.

11. The graph contains K_5.

13. Starting with the original graph:

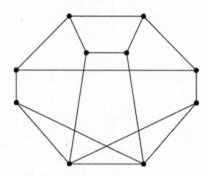

First contract the vertices on the left and right sides.

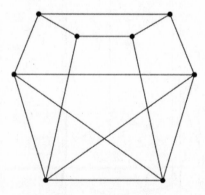

Then contract each diagonal pair of vertices at the top of the graph.

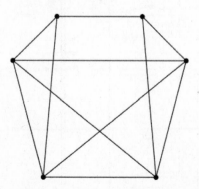

Finally contract the two vertices at the top of the graph.

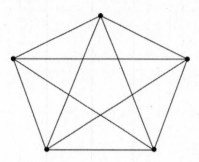

15. The graph contracts to the Utilities Graph.

17. 5 faces, 5 vertices, 8 edges
$v + f = e + 2$
$5 + 5 = 8 + 2$
$10 = 10$

19. 2 faces, 8 vertices, 8 edges
$v + f = e + 2$
$8 + 2 = 8 + 2$
$10 = 10$

21. 5 faces, 10 vertices, 13 edges
$v + f = e + 2$
$10 + 5 = 13 + 2$
$15 = 15$

23. $v + f = e + 2$
$8 + f = 15 + 2$
$f = 9$ faces

25. As a planar graph:

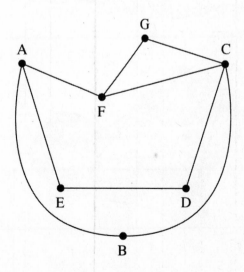

27. The graph is shown below:

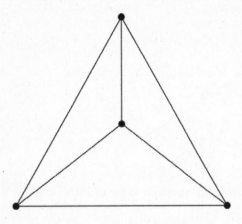

29. $v + f = e + 2$
$v + 3 = v + 2$
The above equation is never true, so Euler's Formula cannot be satisfied.

EXERCISE SET 9.4

1. Coloring the graph:

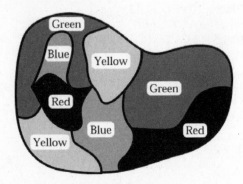

3. Coloring the graph:

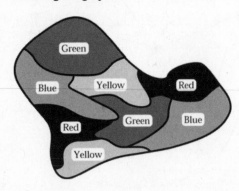

5. As a graph:

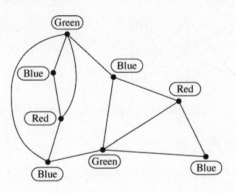

7. As a graph:

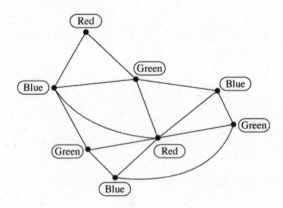

9. As a graph:

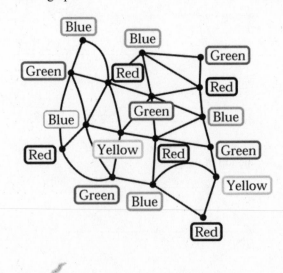

11. The graph is shown below:

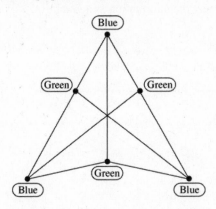

13. The graph is not two-colorable because it has a circuit that consists of an odd number of vertices.

15. The graph is shown below:

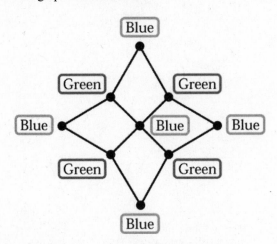

17. The graph is shown below:

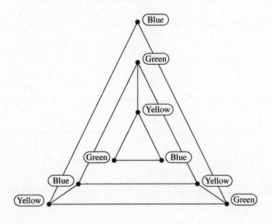

3

19. The graph is shown below:

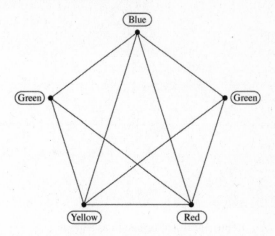

4

21. The graph is shown below:

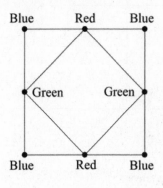

3

23. Represent the information as a graph:

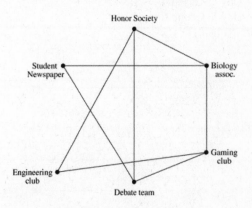

Two times slots are needed.

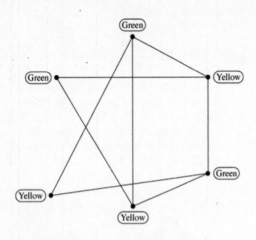

27. Represent the information as a graph:

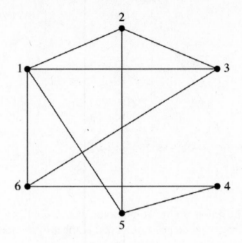

25. Represent the information as a graph:

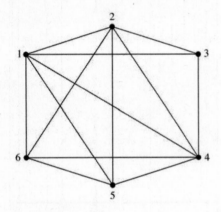

3 days are needed.

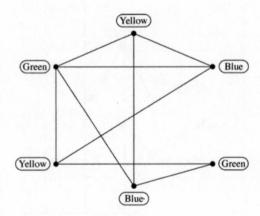

5 days are required.
Day 1: Group 1
Day 2: Group 2
Day 3: Groups 3 and 5
Day 4: Group 4
Day 5: Group 6

Day 1: Films 1 and 4
Day 2: Films 2 and 6
Day 3: Films 3 and 5

29. The graph is shown below:

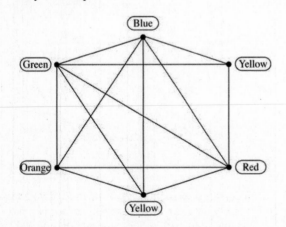

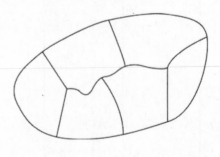

31. The graph is shown below:

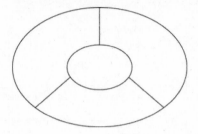

CHAPTER 9 REVIEW EXERCISES

1. a. 8

    b. 4

    c. All vertices have degree 4.

    d. Yes

2. a. 6

    b. 7

    c. 1, 1, 2, 2, 2, 2

    d. No

3. As a graph:

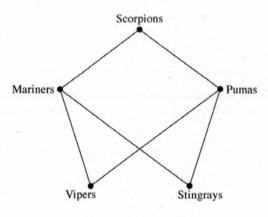

4. a. No

    b. 4

    c. 110, 405

    d. 105

5. Equivalent

6. Equivalent

7. a. E-A-B-C-D-B-E-C-A-D is an Euler walk.

    b. It is not possible to find an Euler circuit because there are vertices of odd degree.

8. a. It is not possible to find an Euler walk because there are more than 2 vertices of odd degree.

    b. It is not possible to find an Euler circuit because there are vertices of odd degree.

9. a. It is not possible to find an Euler walk because there are not exactly 2 vertices of odd degree.

    b. F-A-E-C-B-A-D-B-E-D-C-F is an Euler circuit.

10. a. B-A-E-C-A-D-F-C-B-D-E is an Euler walk.

    b. It is not possible to find an Euler circuit because there are vertices of odd degree.

11. As a graph:

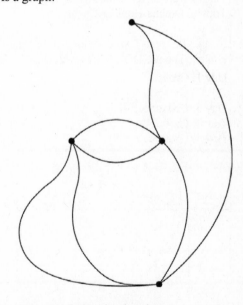

    Yes. An Euler circuit exists.

12. Yes, it is possible to walk through each doorway exactly once, i.e. an Euler walk exists. No, it is not possible to do so and return to the starting point, i.e. an Euler circuit does not exist.

13. The graph is connected and has at least three vertices and no multiple edges, so the theorem applies. Every vertex has a degree of 3 or more, so the graph is Hamiltonian. A Hamiltonian circuit is A-B-C-E-D-A.

14. The graph is connected and has at least three vertices and no multiple edges, so the theorem applies. Every vertex has a degree of 4 or more, so the graph is Hamiltonian. A Hamiltonian circuit is A-D-F-B-C-E-A.

15. As a graph:

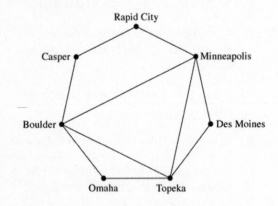

Casper-Rapid City-Minneapolis-Des Moines-Topeka-Omaha-Boulder-Casper

16. Casper-Boulder-Topeka-Minneapolis-Boulder-Omaha-Topeka-Des Moines-Minneapolis-Rapid City-Casper

17. A-D-F-E-B-C-A

18. A-B-E-C-D-A

19. A-D-F-E-C-B-A

20. A-B-E-D-C-A

21. As a graph:

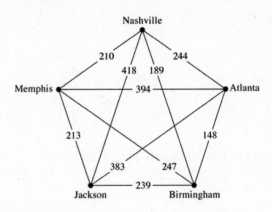

Memphis-Nashville-Birmingham-Atlanta-Jackson-Memphis

22. As a graph:

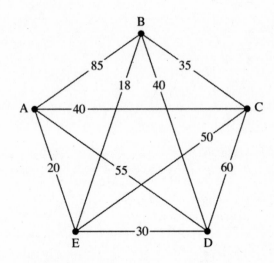

A-E-B-C-D-A

23. Redrawing:

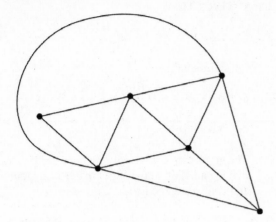

24. Redrawing:

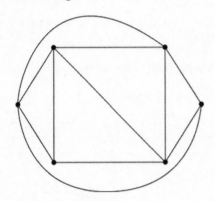

25. K_5 is a subgraph of the graph, so it is not planar.

26. The graph can be contracted into the Utilities Graph, so it is not planar.

27. 5 vertices, 8 edges, 5 faces
$v + f = e + 2$
$5 + 5 = 8 + 2$
$10 = 10$

28. 14 vertices, 16 edges, 4 faces
$v + f = e + 2$
$14 + 4 = 16 + 2$
$18 = 18$

29. As a graph:

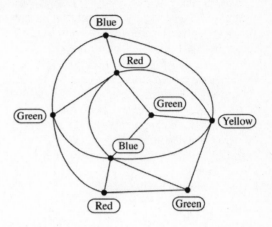

Requires 4 colors.

30. As a graph:

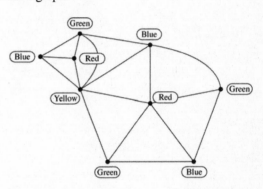

Requires 4 colors.

31. The graph has no circuits consisting of an odd number of vertices, so it is 2-colorable.

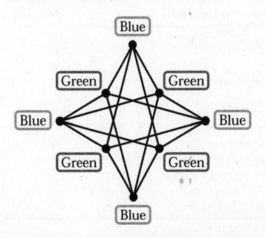

32. The graph has a circuit that consists of an odd number of vertices, so it is not 2-colorable.

33. The chromatic number is 3.

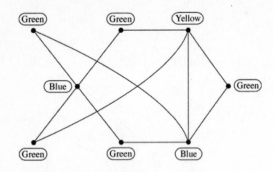

34. The chromatic number is 5.

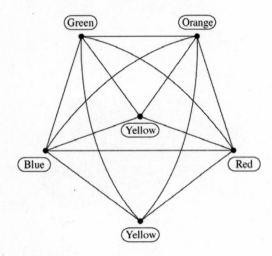

35. As a graph:

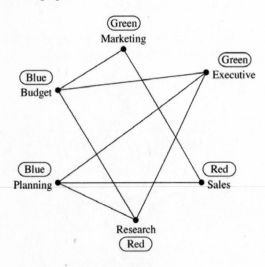

Three times slots are needed.
Time 1: Budget and Planning
Time 2: Marketing and Executive
Time 3: Sales and Research

CHAPTER 9 TEST

1. a. No

    b. Monique

    c. 0

    d. No

2. The graphs are equivalent. You can see this by making a list of all the edges in the each graph and comparing the lists.

3. a. No. The graph has vertices of odd degree, so it is not Eulerian.

    b. A-B-C-F-D-E-B-F-A-E-C-D is an Euler walk.

4. Represent the information with a graph:

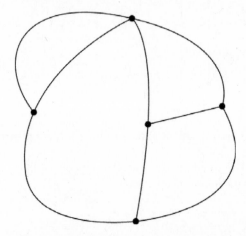

The path described corresponds to an Euler walk or an Euler circuit. The graph has more than 2 vertices of odd degree, so neither an Euler walk nor an Euler circuit exists. Thus such a walk is impossible.

5. a. A graph with more than three vertices and no multiple edges is Hamiltonian if every vertex is of degree at least $n/2$, where n is the number of vertices.
    The graph is connected and has more than three vertices and no multiple edges, so the theorem applies. All of the vertices are of degree at least 4, so the theorem is satisfied and the graph is Hamiltonian.

b.   A-G-C-D-F-B-E-A is a Hamiltonian circuit.

6.   a.   As a graph:

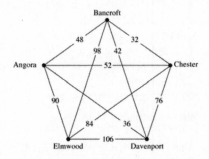

b.   Angora-Elmwood-Chester-Bancroft-
Davenport-Angora.
The total cost is 90 + 84 + 32 + 42 + 36 =
$284.

c.   Look for an Euler circuit. One exists
because each vertex has degree 4, an even
number.

7.   A-E-D-B-C-F-A

8.   Redrawing:

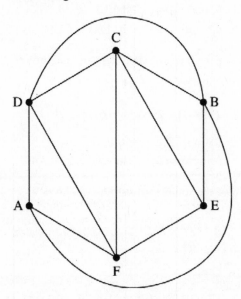

9.   Contracting vertices B and H and vertices A and
G gives the Utilities Graph, so the graph is not
planar.

10. a.   6

b.   7

c.   12 vertices, 7 faces, 17 edges
$v + f = e + 2$
$12 + 7 = 17 + 2$
$19 = 19$

11. a.   As a graph:

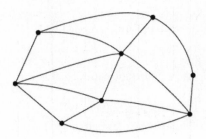

b.   3

c.   Coloring:

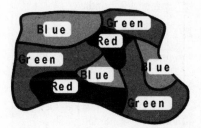

12. Coloring:

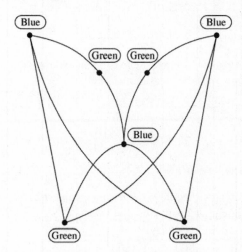

Chapter 10: The Mathematics of Finance

EXERCISE SET 10.1

1. Divide the number of months by 12.

3. I is the interest, P is the principal, r is the interest rate, and t is the time period.

5. $I = Prt = 8000(0.07)(1) = 560$
   The simple interest earned is $560.

7. Because the interest rate is an annual rate, the time must be measured in years:
   $$t = \frac{6 \text{ months}}{1 \text{ year}} = \frac{6 \text{ months}}{12 \text{ months}} = \frac{6}{12}$$
   $$I = Prt = 7000(0.065)\left(\frac{6}{12}\right) = 227.50$$
   The interest earned is $227.50.

9. Because the interest rate is an annual rate, the time must be measured in years:
   $$t = \frac{4 \text{ months}}{1 \text{ year}} = \frac{4 \text{ months}}{12 \text{ months}} = \frac{4}{12}.$$
   $$I = Prt = 9000(0.0675)\left(\frac{4}{12}\right) = 202.50$$
   The interest earned is $202.50.

11. Because the interest rate is an annual rate, the time must be measured in years:
    $$t = \frac{\text{number of days}}{360} = \frac{21}{360}.$$
    $$I = Prt = 3000(0.096)\left(\frac{21}{360}\right) = 16.80$$
    The interest earned is $16.80.

13. Because the interest rate is an annual rate, the time must be measured in years:
    $$t = \frac{\text{number of days}}{360} = \frac{114}{360}.$$
    $$I = Prt = 7000(0.072)\left(\frac{114}{360}\right) = 159.60$$
    The interest earned is $159.60.

15. Because the interest rate is per month, the time period of the loan is $t = 5$ months.
    $$I = Prt = 2000(0.0125)(5) = 125$$
    The interest earned is $125.

17. Use the simple interest formula. Because the interest rate is per month, the time period of the loan is $t = 6$ months.
    $$I = Prt = 1600(0.0175)(6) = 168$$
    The interest earned is $168.

19. Since the interest rate is per year, the time period of the loan is $t = \frac{6}{12}$ year.
    $$A = P(1+rt) = 15,000\left[1 + 0.089\left(\frac{6}{12}\right)\right]$$
    $$= 15,000(1.0445) = 15,667.50$$
    The maturity value of the loan is $15,667.50.

21. Since the interest rate is per year, the time period of the loan is $t = \frac{4}{12}$ year.
    $$A = P(1+rt) = 7200\left[1 + 0.0795\left(\frac{4}{12}\right)\right]$$
    $$= 7200(1.0265) = 7390.80$$
    The maturity value of the loan is $7390.80.

23. Since the interest rate is per year, the time period of the loan is $t = \frac{3}{12}$ year.
    $$A = P(1+rt) = 2800\left[1 + 0.092\left(\frac{3}{12}\right)\right]$$
    $$= 2800(1.023) = 2864.40$$
    The maturity value of the loan is $2864.40.

25. Solve the simple interest formula for r.
    $$I = Prt$$
    $$120 = 1600(r)(1)$$
    $$120 = 1600r$$
    $$\frac{120}{1600} = r$$
    $$0.075 = r$$
    $$r = 7.5\%$$
    The simple interest rate is 7.5%.

27. Solve the simple interest formula for r.

$$I = Prt$$

$$105 = 2000(r)\left(\frac{6}{12}\right)$$

$$105 = 1000r$$

$$\frac{105}{1000} = r$$

$$0.105 = r$$

$$r = 10.5\%$$

The simple interest rate is 10.5%.

29. Solve the simple interest formula for r.

$$I = Prt$$

$$37.20 = 1200(r)\left(\frac{4}{12}\right)$$

$$37.20 = 400r$$

$$\frac{37.20}{400} = r$$

$$0.093 = r$$

$$r = 9.3\%$$

The simple interest rate is 9.3%.

31. $I = Prt = 2300(0.07)(1) = 161$

The simple interest earned is $161.

33. $I = Prt = 1500(0.104)\left(\frac{6}{12}\right) = 78$

The simple interest earned is $78.

35. $I = Prt = 1600(0.09)\left(\frac{45}{360}\right) = 18$

The simple interest earned is $18.

37. Using the formula:

$$A = P(1+rt) = 7000\left[1 + 0.087\left(\frac{8}{12}\right)\right]$$

$$= 7000(1.058) = 7406$$

The maturity value of the loan is $7406.

39. Using the formula:

$$A = P(1+rt) = 5200[1 + 0.102(1)]$$

$$= 5200(1.102) = 5730.40$$

The maturity value of the loan is $5730.40.

41. Using the formula:

$$A = P(1+rt) = 750[1 + 0.073(1)]$$

$$= 750(1.073) = 804.75$$

The future value of the investment is $804.75.

43. Solve the simple interest formula for r.

$$I = Prt$$

$$918 = 18,000(r)\left(\frac{6}{12}\right)$$

$$918 = 9000r$$

$$\frac{918}{18,000} = r$$

$$0.102 = r$$

$$r = 10.2\%$$

The simple interest rate is 10.2%.

45. First find the amount of interest paid. Subtract the principal from the maturity value.
$$I = A - P = 5125 - 5000 = 125$$
Find the simple interest rate by solving the simple interest formula for r.

$$I = Prt$$

$$125 = 5000(r)\left(\frac{3}{12}\right)$$

$$125 = 1250r$$

$$0.10 = r$$

$$r = 10\%$$

The simple interest rate on the loan is 10%.

47. $A = P(1+rt) = 132\left[1 + 0.12\left(\frac{1}{12}\right)\right] = 133.32$

You will owe $133.32.

49. The exact method uses 365 days a year.

$$A = P(1+rt) = 10,000\left[1 + 0.08\left(\frac{140}{365}\right)\right] \approx$$

$$10,306.85$$

The exact method yields a maturity value of $10,306.85.
The ordinary method uses 360 days a year.

$$A = P(1+rt) = 10,000\left[1 + 0.08\left(\frac{140}{360}\right)\right]$$

$$\approx 10,311.11$$

The ordinary method yields a maturity value of $10,311.11.
The ordinary method yields the greater maturity value. The lender benefits from using the ordinary method.

51. There are fewer days in September than there are in August.

53. a. Use the simple interest formula $I = Prt$ to calculate each value.

Loan Amount	Interest Rate	Period	Interest
$1000	6%	1 month	$5
$2000	6%	1 month	$10
$3000	6%	1 month	$15
$4000	6%	1 month	$20
$5000	6%	1 month	$25

   b. The simple interest increases by $5 for each additional $1000 in principal. So, the interest on $6000 is $25 + $5 = $30.

   c. $30 + $5 = $35

   d. $35 + $5 = $40

   e. $40 + $5 = $45

   f. Multiply the interest due on a $1000 loan by 8.

   g. The simple interest is 2 times larger.

   h. The simple interest is 3 times larger.

EXERCISE SET 10.2

1. $r = 7\% = 0.07$, $n = 2$, $t = 12$,

$$i = \frac{r}{n} = \frac{0.07}{2} = 0.035 \text{, and}$$

$$N = nt = 2(12) = 24.$$

$$A = P(1+r)^N = 1200(1+0.035)^{24}$$
$$= 1200(1.035)^{24} \approx 2739.99$$

The compound amount after 12 years is approximately $2739.99.

3. $r = 9\% = 0.09$, $n = 4$, $t = 6$,

$$i = \frac{r}{n} = \frac{0.09}{4} = 0.0225 \text{, and}$$

$$N = nt = 4(6) = 24.$$

$$A = P(1+r)^N = 500(1+0.0225)^{24}$$
$$= 500(1.0225)^{24} \approx 852.88$$

The compound amount after 6 years is approximately $852.88.

5. $r = 9\% = 0.09$, $n = 12$, $t = 10$,

$$i = \frac{r}{n} = \frac{0.09}{12} = 0.0075 \text{, and}$$

$$N = nt = 12(10) = 120.$$

$$A = P(1+r)^N = 8500(1+0.0075)^{120}$$
$$= 8500(1.0075)^{120} \approx 20{,}836.54$$

The compound amount after 10 years is approximately $20,836.54.

7. Interest is compounded daily. Use the ordinary method. $n = 360$, $r = 9\% = 0.09$, $t = 3$,

$$i = \frac{r}{n} = \frac{0.09}{360} = 0.00025 \text{, and}$$

$$N = nt = 360(3) = 1080.$$

$$A = P(1+r)^N$$

$$A = 9600(1+0.00025)^{1080}$$

$$A = 9600(1.00025)^{1080}$$

$$A \approx 12{,}575.23$$

The compound amount after 3 years is approximately $12,575.23.

9. Use the financial mode on the calculator. Since interest is compounded quarterly, there are 4(10) = 40 payments.
The compound amount is $3532.86.

11. Use the financial mode on the calculator. Since interest is compounded monthly, there are 12(5) = 60 payments.
The compound amount is $5450.09.

13. Use the financial mode on the calculator. Since interest is compounded semiannually, there are 2(3) = 6 payments.
The compound amount is $2213.84.

15. $r = 12\% = 0.12$, $n = 12$, $t = 5$,

$$i = \frac{r}{n} = \frac{0.12}{12} = 0.01 \text{, and}$$

$$N = nt = 12(5) = 60.$$

$$A = P(1+r)^N = 7500(1+0.01)^{60}$$
$$= 7500(1.01)^{60} \approx 13{,}625.23$$

The future value is approximately $13,625.23.

$$A = P(1+i)^N \text{ ?}$$

17. $r = 10\% = 0.10$, $n = 2$, $t = 12$,

    $i = \dfrac{r}{n} = \dfrac{0.10}{2} = 0.05$, and

    $N = nt = 2(12) = 24$.

    $A = P(1+r)^N$

    $A = 4600(1+0.05)^{24}$

    $A = 4600(1.05)^{24}$

    $A \approx 14{,}835.46$

    The future value is approximately \$14,835.46.

19. $r = 9\% = 0.09$, $n = 12$, $t = 7$,

    $i = \dfrac{r}{n} = \dfrac{0.09}{12} = 0.0075$, and

    $N = nt = 12(7) = 84$.

    $A = P(1+r)^N = 22{,}000(1+0.0075)^{84}$

    $= 22{,}000(1.0075)^{84} \approx 41{,}210.44$

    The future value is approximately \$41,210.44.

21. $r = 10\% = 0.10$, $n = 4$, $t = 12$,

    $i = \dfrac{r}{n} = \dfrac{0.10}{4} = 0.025$,

    $N = nt = 4(12) = 48$.

    $P = \dfrac{A}{(1+i)^N} = \dfrac{25{,}000}{(1+0.025)^{48}}$

    $= \dfrac{25{,}000}{1.025^{48}} \approx 7641.78$

    \$7641.78 should be invested in the account in order to have \$25,000 in 12 years.

23. $r = 7.5\% = 0.075$, $n = 1$, $t = 35$,

    $i = \dfrac{r}{n} = \dfrac{0.075}{1} = 0.075$, $N = nt = 1\,(35) = 35$.

    $P = \dfrac{A}{(1+i)^N} = \dfrac{40{,}000}{(1+0.075)^{35}}$

    $= \dfrac{40{,}000}{1.075^{35}} \approx 3182.47$

    \$3182.47 should be invested in the account in order to have \$40,000 in 35 years.

25. $r = 8\% = 0.08$, $n = 4$, $t = 5$, $i = \dfrac{r}{n} = \dfrac{0.08}{4} = 0.02$,

    $N = nt = 4(5) = 20$.

$P = \dfrac{A}{(1+i)^N} = \dfrac{15{,}000}{(1+0.02)^{20}}$

$= \dfrac{15{,}000}{1.02^{20}} \approx 10{,}094.57$

\$10,094.57 should be invested in the account in order to have \$15,000 in 5 years.

27. $r = 8\% = 0.08$, $n = 4$, $t = 5$, $i = \dfrac{r}{n} = \dfrac{0.08}{4} = 0.02$,

    and $N = nt = 4(5) = 20$.

    $A = P(1+r)^N = 8000(1+0.02)^{20}$

    $= 8000(1.02)^{20} \approx 11{,}887.58$

    The compound amount after 5 years is approximately \$11,887.58.

29. Interest is compounded daily. Use the ordinary method. $n = 360$, $r = 9\% = 0.09$, $t = 4$,

    $i = \dfrac{r}{n} = \dfrac{0.09}{360} = 0.00025$, and $N = nt = 360(4) =$

    1440.

    $A = P(1+r)^N = 2500(1+0.00025)^{1440}$

    $= 2500(1.00025)^{1440} \approx 3583.16$

    The compound amount after 4 years is approximately \$3583.16.

31. $r = 12\% = 0.12$, $n = 12$, $t = 6$,

    $i = \dfrac{r}{n} = \dfrac{0.12}{12} = 0.01$, and

    $N = nt = 12(6) = 72$.

    $A = P(1+r)^N = 4000(1+0.01)^{72}$

    $= 4000(1.01)^{72} \approx 8188.40$

    The future value is approximately \$8188.40.

33. Calculate the compound amount. $r = 6\% = 0.06$,

    $n = 4$, $t = 3$, $i = \dfrac{r}{n} = \dfrac{0.06}{4} = 0.015$, and $N = nt$

    $= 4(3) = 12$.

    $A = P(1+r)^N = 2000(1+0.015)^{12}$

    $= 2000(1.015)^{12} \approx 2391.24$

    Calculate the interest earned.

    $I = A - P = 2391.24 - 2000 = 391.24$

    The amount of interest earned is approximately \$391.24.

35. Calculate the compound amount. $r = 10\% =$

    0.10, $n = 4$, $t = 8$, $i = \dfrac{r}{n} = \dfrac{0.10}{4} = 0.025$, and

    $N = nt = 4(8) = 32$.

$A = P(1+i)^N = 15,000(1+0.025)^{32}$

$ = 15,000(1.025)^{32} \approx 33,056.35$

Calculate the interest earned.

$I = A - P = 33,056.35 - 15,000 = 18,056.35$

The amount of interest earned is approximately $18,056.35.

37. $r = 6\% = 0.06$, $n = 12$, $t = 5$,

$i = \dfrac{r}{n} = \dfrac{0.06}{12} = 0.005$, and

$N = nt = 12(5) = 60$.

$P = \dfrac{A}{(1+i)^N} = \dfrac{15,000}{(1+0.005)^{60}}$

$ = \dfrac{15,000}{1.005^{60}} \approx 11,120.58$

$11,120.58 should be invested in the account in order to have $15,000 in 5 years.

39. a. $P = 1000$, $r = 0.09$, and $t = 5$.

$I = Prt$

$I = 1000(0.09)(5)$

$I = 450$

The simple interest earned is $450.

b.   Calculate the compound amount.
$r = 9\% = 0.09$, $t = 5$. Interest is compounded daily. Use the ordinary method. $n = 360$, $i = \dfrac{r}{n} = \dfrac{0.09}{360} = 0.00025$,

and $N = nt = 360(5) = 1800$.

$A = P(1+i)^N = 1000(1+0.00025)^{1800}$

$ = 1000(1.00025)^{1800} \approx 1568.22$

Calculate the interest earned.

$I = A - P = 1568.22 - 1000 = 568.22$

The amount of interest earned is approximately $568.22.

c. $568.22 - 450 = 118.22$
$118.22 more is earned when interest is compounded daily.

41. a. $r = 8\% = 0.08$, $n = 2$, $t = 4$,

$i = \dfrac{r}{n} = \dfrac{0.08}{2} = 0.04$, and

$N = nt = 2(4) = 8$.

$A = P(1+i)^N$

$A = 15,000(1+0.04)^8 = 15,000(1.04)^8$

$ \approx 20,528.54$

The future value of the investment is $20,528.54.

b. $r = 8\% = 0.08$, $n = 4$, $t = 4$,

$i = \dfrac{r}{n} = \dfrac{0.08}{4} = 0.02$, and

$N = nt = 4(4) = 16$.

$A = P(1+i)^N = 15,000(1+0.02)^{16}$

$ = 15,000(1.02)^{16} \approx 20,591.79$

The future value of the investment is $20,591.79.

c. $20,591.79 - 20,528.54 = 63.25$
The investment is $63.25 greater when interest is compounded quarterly.

43. a.   Calculate the compound amount.
$r = 8\% = 0.08$, $n = 2$, $t = 2$,

$i = \dfrac{r}{n} = \dfrac{0.08}{2} = 0.04$, and

$N = nt = 2(2) = 4$.

$A = P(1+i)^N = 10,000(1+0.04)^4$

$ = 10,000(1.04)^4 \approx 11,698.59$

Calculate the interest earned.

$I = A - P = 11,698.59 - 10,000$

$ = 1698.59$

The amount of interest earned is approximately $1698.59.

b.   Calculate the compound amount. $r = 8\% =$

0.08, $n = 4$, $t = 2$, $i = \dfrac{r}{4} = \dfrac{0.08}{4} = 0.02$, and

$N = nt = 4(2) = 8$.

$A = P(1+i)^N$

$A = 10,000(1+0.02)^8$

$A = 10,000(1.02)^8$

$A \approx 11,716.59$

Calculate the interest earned.

$I = A - P$

$I = 11,716.59 - 10,000$

$I = 1716.59$

The amount of interest earned is approximately $1716.59.

c. $1716.59 - 1698.59 = 18.00$
The investment is $18.00 greater when interest is compounded quarterly.

45. a. $r = 8\% = 0.08$, $n = 4$, $t = 5$,

$i = \dfrac{r}{n} = \dfrac{0.08}{4} = 0.02$, and

$N = nt = 4(5) = 20$.

$A = P(1+i)^N = 7000(1+0.02)^{20}$

$= 7000(1.02)^{20} \approx 10,401.63$

The maturity value of the loan is $10,401.63.

b. $10,401.63 - 7000 = 3401.63$
The interest you are paying is $3401.63.

47. $r = 8\% = 0.08$, $n = 4$, $t = 18$,

$i = \dfrac{r}{n} = \dfrac{0.08}{4} = 0.02$, $N = nt = 4(18) = 72$.

$P = \dfrac{A}{(1+i)^N} = \dfrac{40,000}{(1+0.02)^{72}}$

$= \dfrac{40,000}{1.02^{72}} \approx 9612.75$

$9612.75 should be invested in the account to have $40,000 in 18 years.

49. a. $r = 9\% = 0.09$, $n = 360$, $t = 30$,

$i = \dfrac{r}{n} = \dfrac{0.09}{360} = 0.00025$, $N = nt = 360(30) = 10,800$.

$P = \dfrac{A}{(1+i)^N}$

$= \dfrac{1,000,000}{(1+0.00025)^{10,800}}$

$= \dfrac{1,000,000}{1.00025^{10,800}} \approx 67,228.19$

$67,228.19 should be invested in the account in order to have $1,000,000 in 30 years.

b. Calculate the compound amount. Use the same values for n and i. Use $t = 1$, so $N = nt = 360(1) = 360$.

$A = P(1+i)^N$

$= 1,000,000(1+0.00025)^{360}$

$= 1,000,000(1.00025)^{360}$

$\approx 1,094,161.98$

Calculate the interest earned.

$I = A - P$

$= 1,094,161.98 - 1,000,000$

$= 94,161.98$

The amount of interest earned each year is approximately $94,161.98.

51. Calculate the compound amount for the first two years. $r = 8.1\% = 0.081$, $n = 360$, $t = 2$,

$i = \dfrac{r}{n} = \dfrac{0.081}{360} = 0.000225$, and

$N = nt = 360(2) = 720$.

$A = P(1+i)^N = 5000(1+0.000225)^{720}$

$= 5000(1.000225)^{720} \approx 5879.19$

Calculate the compound amount for the second two years. The principal is now $5879.19.
$r = 7.2\% = 0.072$, $n = 360$, $t = 2$,

$i = \dfrac{r}{n} = \dfrac{0.072}{360} = 0.0002$, and

$N = nt = 360(2) = 720$.

$A = P(1+i)^N = 5879.19(1+0.0002)^{720}$

$= 5879.19(1.0002)^{720} \approx 6789.69$

The compound amount when the second CD matures is $6789.69.

53. $r = 7\% = 0.07$, $n = 1$, $t = 15$,

$i = \dfrac{r}{n} = \dfrac{0.07}{1} = 0.07$, and

$N = nt = 1(15) = 15$.

$A = P(1+i)^N = 1686(1+0.07)^{15}$

$= 1686(1.07)^{15} \approx 4651.73$

Fifteen years from now, the average rent for an apartment will be approximately $4651.73.

55. $r = 7\% = 0.07$, $n = 1$, $t = 5$, $i = \dfrac{r}{n} = \dfrac{0.07}{1} = 0.07$,

and $N = nt = 1(5) = 5$.

$A = P(1+i)^N = 40,000(1+0.07)^5$

$= 40,000(1.07)^5 \approx 56,102.07$

In 2015, you need a salary of approximately $56,102.07 to have the same purchasing power.

57. $r = 6\% = 0.06$ and $t = 40$. The inflation rate is an

annual rate, so $n = 1$. $i = \dfrac{r}{n} = \dfrac{0.06}{1} = 0.06$,

$N = nt = 1(40) = 40$.

$$P = \frac{A}{(1+i)^N} = \frac{125{,}000}{(1+0.06)^{40}}$$

$$= \frac{125{,}000}{1.06^{40}} \approx 12{,}152.77$$

The purchasing power of the $125,000 in 2009 will be approximately $12,152.77 in 2049.

59. $r = 7\% = 0.07$ and $t = 5$. The inflation rate is an annual rate, so $n = 1$. $i = \dfrac{r}{n} = \dfrac{0.07}{1} = 0.07$,

$N = nt = 1(5) = 5$.

$$P = \frac{A}{(1+i)^N}$$

$$P = \frac{3500}{(1+0.07)^5}$$

$$P = \frac{3500}{1.07^5}$$

$P \approx 2495.45$

The purchasing power of the $3500 in will be approximately $2495.45 in 5 years.

61. Find the future value of $100 after 1 year.

$r = 7.2\%$, $n = 4$, $i = \dfrac{r}{n} = \dfrac{0.072}{4} = 0.018$ and

$N = nt = 4(1) = 4$.

$A = P(1+i)^N = 100(1+0.018)^4$

$\quad = 100(1.018)^4 \approx 107.40$

Find the interest earned on the $100.

$I = A - P = 107.40 - 100 = 7.40$

The effective interest rate is 7.40%.

63. Find the future value of $100 after 1 year.

$r = 7.5\%$, $n = 12$, $i = \dfrac{r}{n} = \dfrac{0.075}{12} = 0.00625$ and

$N = nt = 12(1) = 12$.

$A = P(1+i)^N = 100(1+0.00625)^{12}$

$\quad = 100(1.00625)^{12} \approx 107.76$

Find the interest earned on the $100.

$I = A - P$

$I = 107.76 - 100$

$I = 7.76$

The effective interest rate is 7.76%.

65. Find the future value of $100 after 1 year.

$r = 8.1\%$, $n = 360$, $i = \dfrac{r}{n} = \dfrac{0.081}{360} = 0.000225$ and

$N = nt = 360(1) = 360$.

$A = P(1+i)^N = 100(1+0.000225)^{360}$

$\quad = 100(1.000225)^{360} \approx 108.44$

Find the interest earned on the $100.

$I = A - P = 108.44 - 100 = 8.44$

The effective interest rate is 8.44%.

67. Find the future value of $100 after 1 year.

$r = 5.94\%$, $n = 12$, $i = \dfrac{r}{n} = \dfrac{0.0594}{12} = 0.00495$

and

$N = nt = 12(1) = 12$.

$A = P(1+i)^N = 100(1+0.00495)^{12}$

$\quad = 100(1.00495)^{12} \approx 106.10$.

Find the interest earned on the $100.

$I = A - P = 106.10 - 100 = 6.10$

The effective interest rate is 6.10%.

69. The inflation rate is an annual rate, so $n = 1$.

$r = 6\% = 0.06$, $i = \dfrac{r}{n} = \dfrac{0.06}{1} = 0.06$.

In 2009, $t = 5$, so $N = nt = 1(5) = 5$.

$A = P(1+r)^N = 2.00(1+0.06)^5$

$\quad = 2.00(1.06)^5 \approx 2.68$

Gasoline will cost about $2.68 in 2009.

In 2014, $t = 10$, so $N = nt = 1(10) = 10$.

$A = P(1+r)^N = 2.00(1+0.06)^{10}$

$\quad = 2.00(1.06)^{10} \approx 3.58$

Gasoline will cost about $3.58 in 2014.

In 2024, $t = 20$, so $N = nt = 1(20) = 20$.

$A = P(1+r)^N = 2.00(1+0.06)^{20}$

$\quad = 2.00(1.06)^{20} \approx 6.41$

Gasoline will cost about $6.41 in 2024.

71. The inflation rate is an annual rate, so $n = 1$.

$r = 6\% = 0.06$, $i = \dfrac{r}{n} = \dfrac{0.06}{1} = 0.06$.

In 2009, $t = 5$, so $N = nt = 1(5) = 5$.

$A = P(1+r)^N = 2.69(1+0.06)^5$

$\quad = 2.69(1.06)^5 \approx 3.60$

A loaf of bread will cost approximately $3.60 in 2009.

In 2014, $t = 10$, so $N = nt = 1(10) = 10$.

$A = P(1+r)^N = 2.69(1+0.06)^{10}$

$\quad = 2.69(1.06)^{10} \approx 4.82$

A loaf of bread will cost approximately $4.82 in 2014.

In 2024, $t = 20$, so $N = nt = 1(20) = 20$.

$A = P(1+r)^N = 2.69(1+0.06)^{20}$

$\quad = 2.69(1.06)^{20} \approx 8.63$

A loaf of bread will cost approximately \$8.63 in 2024.

73. The inflation rate is an annual rate, so $n = 1$.

$r = 6\% = 0.06, \quad i = \dfrac{r}{n} = \dfrac{0.06}{1} = 0.06$.

In 2009, $t = 5$, so $N = nt = 1(5) = 5$.

$A = P(1+r)^N = 9(1+0.06)^5$

$\quad = 9(1.06)^5 \approx 12.04$

A ticket to a movie will cost approximately \$12.04 in 2009.

In 2014, $t = 10$, so $N = nt = 1(10) = 10$.

$A = P(1+r)^N = 9(1+0.06)^{10}$

$\quad = 9(1.06)^{10} \approx 16.12$

A ticket to a movie will cost approximately \$16.12 in 2014.

In 2024, $t = 20$, so $N = nt = 1(20) = 20$.

$A = P(1+r)^N = 9(1+0.06)^{20}$

$\quad = 9(1.06)^{20} \approx 28.86$

A ticket to a movie will cost approximately \$28.86 in 2024.

75. The inflation rate is an annual rate, so $n = 1$.

$r = 6\% = 0.06, \quad i = \dfrac{r}{n} = \dfrac{0.06}{1} = 0.06$.

In 2009, $t = 5$, so $N = nt = 1(5) = 5$.

$A = P(1+r)^N = 275{,}000(1+0.06)^5$

$\quad = 275{,}000(1.06)^5 \approx 368{,}012.03$

A house will cost approximately \$368,012.03 in 2009.

In 2014, $t = 10$, so $N = nt = 1(10) = 10$.

$A = P(1+r)^N = 275{,}000(1+0.06)^{10}$

$\quad = 75{,}000(1.06)^{10} \approx 492{,}483.12$

A house will cost approximately \$492,483.12 in 2014.

In 2024, $t = 20$, so $N = nt = 1(20) = 20$.

$A = P(1+r)^N = 275{,}000(1+0.06)^{20}$

$\quad = 275{,}000(1.06)^{20} \approx 881{,}962.25$

A house will cost approximately \$881,962.25 in 2024.

77. $r = 7\% = 0.07$ and $t = 10$. The inflation rate is an annual rate, so $n = 1$. $i = \dfrac{r}{n} = \dfrac{0.07}{1} = 0.07$,

$N = nt = 1(10) = 10$.

$P = \dfrac{A}{(1+i)^N} = \dfrac{50{,}000}{(1+0.07)^{10}}$

$\quad = \dfrac{50{,}000}{1.07^{10}} \approx 25{,}417.46$

The purchasing power of the \$50,000 will be approximately \$25,417.46 in 10 years.

79. $r = 7\% = 0.07$ and $t = 20$. The inflation rate is an annual rate, so $n = 1$. $i = \dfrac{r}{n} = \dfrac{0.07}{1} = 0.07$,

$N = nt = 1(20) = 20$.

$P = \dfrac{A}{(1+i)^N} = \dfrac{100{,}000}{(1+0.07)^{20}}$

$\quad = \dfrac{100{,}000}{1.07^{20}} \approx 25{,}841.90$

The purchasing power of the \$100,000 will be approximately \$25,841.90 in 20 years.

81. $r = 7\% = 0.07$ and $t = 5$. The inflation rate is an annual rate, so $n = 1$. $i = \dfrac{r}{n} = \dfrac{0.07}{1} = 0.07$,

$N = nt = 1(5) = 5$.

$P = \dfrac{A}{(1+i)^N} = \dfrac{75{,}000}{(1+0.07)^5}$

$\quad = \dfrac{75{,}000}{1.07^5} \approx 53{,}473.96$

The purchasing power of the \$75,000 will be approximately \$53,473.96 in 5 years.

83. a. $r = 4\% = 0.04$, $P = 1$, and $t = 1$. Calculate i and N for each compounding period. Use the compound amount formula, $A = P(1+i)^N$ to find the amount after 1 year. Then use $I = A - P$ to find the effective rate. See the table below:

Nominal Rate	Effective Rate
4% annual compounding	4.0%
4% semiannual compounding	4.04%
4% quarterly compounding	4.06%
4% monthly compounding	4.07%
4% daily compounding	4.08%

b. As the number of compounding periods increases, the effective rate increases.

85. Find the future value of $100 after 1 year.

$r = 3\%$, $n = 12$, $i = \dfrac{r}{n} = \dfrac{0.03}{12} = 0.0025$ and

$N = nt = 12(1) = 12$.

$A = P(1+i)^N = 100(1+0.0025)^{12}$

$= 100(1.0025)^{12} \approx 103.04$

Find the interest earned on the $100.

$I = A - P = 103.04 - 100 = 3.04$

The effective interest rate is 3.04%.

87. Calculate $(1+i)^N$ for each investment.

6% compounded quarterly:

$i = \dfrac{r}{n} = \dfrac{0.06}{4}$

$N = nt = 4(1) = 4$

$(1+i)^N = \left(1+\dfrac{0.06}{4}\right)^4 \approx 1.061363551$

6.25% compounded semiannually:

$i = \dfrac{r}{n} = \dfrac{0.0625}{2}$

$N = nt = 12(1) = 12$

$(1+i)^N = \left(1+\dfrac{0.0625}{2}\right)^2 \approx 1.063476563$

Compare the two compound amounts.

$1.063476563 > 1.061363551$

6.25% compounded semiannually has a higher yield than 6% compounded quarterly.

89. Calculate $(1+i)^N$ for each investment.

5.8% compounded quarterly:

$i = \dfrac{r}{n} = \dfrac{0.058}{4}$

$N = nt = 4(1) = 4$

$(1+i)^N = \left(1+\dfrac{0.058}{4}\right)^4 \approx 1.05927373$

5.6% compounded monthly:

$i = \dfrac{r}{n} = \dfrac{0.056}{12}$; $N = nt = 12(1) = 12$

$(1+i)^N = \left(1+\dfrac{0.056}{12}\right)^{12} \approx 1.057459928$

Compare the two compound amounts.

$1.059273739 > 1.057459928$

5.8% compounded quarterly has a higher yield than 5.6% compounded monthly.

91. Use the compound amount formula. $n = 2$,

$t = 25$, $i = \dfrac{r}{n} = \dfrac{r}{2}$, and $N = nt = 2(25) = 50$.

$A = P(1+i)^N$

$22{,}934.80 = 2000\left(1+\dfrac{r}{2}\right)^{50}$

$\dfrac{22{,}934.80}{2000} = \left(1+\dfrac{r}{2}\right)^{50}$

$11.4674 = \left(1+\dfrac{r}{2}\right)^{50}$

$11.4674^{1/50} = 1+\dfrac{r}{2}$

$1.05 = 1+\dfrac{r}{2}$

$1.05-1 = \dfrac{r}{2}$

$0.05 = \dfrac{r}{2}$

$2(0.05) = r$

$0.10 = r$

The interest rate is 10%.

93. For the second option, make a table to find the amount in savings at the end of the year for each of the monthly payments. $r = 5\% = 0.05$, $n = 12$,

$i = \dfrac{r}{n} = \dfrac{0.05}{12} \approx 0.0041667$. The principal varies

as deposits are made and the balance increases. See the table below:

End of Month	Account Balance
1	1800
2	1800(1 + 0.05/12) + 1800 = 3607.50
3	3607.50(1 + 0.05/12) + 1800 = 5422.53
4	5422.53(1 + 0.05/12) + 1800 = 7245.12
5	7245.12(1 + 0.05/12) + 1800 = 9075.31
6	9075.30(1 + 0.05/12) + 1800 = 10913.12
7	10913.11(1 + 0.05/12) + 1800 = 12758.59
8	12758.58(1 + 0.05/12) + 1800 = 14611.75
9	14611.74(1 + 0.05/12) + 1800 = 16472.63
10	16472.62(1 + 0.05/12) + 1800 = 18341.27
11	18341.25(1 + 0.05/12) + 1800 = 20217.69
12	20217.67(1 + 0.05/12) + 1800 = 22101.93
Total in savings	**22,101.93**

Find the amount of interest. The total amount of principal is $12(1800) = 21,600$.

$I = A - P = 22,101.93 - 21,600 = 501.93$
The amount of interest is $501.93.
You must pay 28% of the interest earned in income taxes.
$0.28(501.93) = 140.54$
The amount left after taxes is
$22,101.93 - 140.54 = 21,961.39$.
Find the price of the motorcycle, after inflation.
$A = P(1 + i)^N = 20,000(1 + 0.07)^1 = 21,400$
$21,961.39 - 21,400 = 561.39$
The amount left after paying taxes and buying the motorcycle is $561.39.

EXERCISE SET 10.3

1. $I = Prt = 118.72(0.0125)(1) \approx 1.48$
    The finance charge is $1.48.

3. $I = Prt = 10,154.87(0.015)(1) \approx 152.32$
    The finance charge is $152.32.

5. Recording the balances:

Date	Payment or purchase	Balance	Days	Balance times Days
Mar. 5-11		$244	7	$1708
Mar. 12-27	$152	$396	16	$6336
Mar. 28-Apr. 4	− $100	$296	8	$2368
Total				$10,412

Average daily balance $= \dfrac{10,412}{31} \approx 335.87$.

7. Recording the balances:

Date	Payment or purchase	Balance	Days	Balance times Days
May 5-16		$944	12	$11,328
May 17-19	$255	$1199	3	$3597
May 20-Jun. 4	− $150	$1049	16	$16,784
Total				$31,709

Average daily balance $= \dfrac{31709}{31} \approx 1022.87$.

$I = Prt = (1022.87)(0.015)(1) \approx 15.34$. The finance charge is $15.34.

9. Recording the balances:

Date	Payment or purchase	Balance	Days	Balance times Days
Aug. 10-14		$345	5	$1725
Aug. 15	$56	$401	0	$0
Aug. 15-26	− $75	$326	12	$3912
Aug. 27-Sep. 9	$157	$483	14	$6762
Total				$12,399

Average daily balance $= \dfrac{12,399}{31} \approx 399.97$.

$I = Prt = (399.97)(0.0125)(1) \approx 5.00$. The finance charge is $5.00.

11. Recording the balances:

Date	Payment or purchase	Balance	Days	Balance times Days
Aug. 15		$1236.43	1	$1236.43
Aug. 16	$125	$1361.43	1	$1361.43
Aug. 17	$23.56	$1384.99	1	$1384.99
Aug. 18-19	$53.45	$1438.44	2	$2876.88
Aug. 20-21	$41.36	$1479.80	2	$2959.60
Aug. 22-24	$223.65	$1703.45	3	$5110.35
Aug. 25-29	$310	$2013.45	5	$10,067.25
Aug. 30-31	$23.36	$2036.81	2	$4073.62
Sep. 1-8	$36.45	$2073.26	8	$16,586.08
Sep. 9-11	$21.39	$2094.65	3	$6283.95
Sep. 12	$41.25	$2135.90	1	$2135.90
Sep. 13-14	− $1345	$790.90	2	$1581.80
Total				$55,658.28

Average daily balance $= \dfrac{55,658.28}{31} \approx 1795.43$.

$I = Prt = (1795.43)(0.015)(1) \approx 26.93$. The finance charge is $26.93.

13. Using the formula:

$$\text{APR} \approx \frac{2nr}{n+1}$$

$$\approx \frac{2(3)(0.09)}{3+1} = \frac{0.54}{4} = 0.135$$

The annual percentage rate on the loan is 13.5%.

15. Using the formula:

$$\text{APR} \approx \frac{2nr}{n+1}$$

$$\approx \frac{2(24)(0.10)}{24+1} = \frac{4.8}{25} = 0.192$$

The annual percentage rate on the loan is 19.2%.

17. Using the formula:

$$i = \frac{\text{annual interest rate}}{\text{number of payments per year}}$$

$$= \frac{0.069}{12}$$

Calculate the monthly payment. For a 1-year loan, $n = 12$.

$$PMT = A\left(\frac{i}{1-(1+i)^{-n}}\right)$$

$$= 400\left(\frac{0.069/12}{1-(1+0.069/12)^{-12}}\right)$$

$$\approx 34.59$$

The monthly payment is $34.59.

19. a. Sales tax = 0.0725(649) = 47.0525
The sales tax is $47.05.
Total cost = 649 + 47.05 = 696.05
The total cost is $696.05.

b. Down payment = 0.25(696.05)= 174.0125
The down payment is $174.01.

c. Loan amount = total cost − down payment
Loan amount = 696.05 − 174.01 = 522.04

$$i = \frac{\text{annual interest rate}}{\text{number of payments per year}}$$

$$= \frac{0.057}{12}$$

Calculate the monthly payment. For a 6-month loan, $n = 6$.

$$PMT = A\left(\frac{i}{1-(1+i)^{-n}}\right)$$

$$= 522.04\left(\frac{0.057/12}{1-(1+0.057/12)^{-6}}\right)$$

$$\approx 88.46$$

The monthly payment is $88.46.

21. a. Sales tax = 0.055(64,995) = 3574.725
The sales tax is $3574.73.
Total cost = 64,995 + 3574.73 = 68,569.73
The total cost is $68,569.73.

b. Down payment = 0.20(68,569.73)
= 13,713.946
The down payment is $13,713.95.

c. Loan amount = total cost − down payment
Loan amount = 68,569.73 − 13,713.95
= 54,855.78

$$i = \frac{\text{annual interest rate}}{\text{number of payments per year}}$$

$$= \frac{0.0715}{12}$$

Calculate the monthly payment. For a 10-year loan, $n = 10(12) = 120$.

$$PMT = A\left(\frac{i}{1-(1+i)^{-n}}\right)$$

$$= 54,855.78\left(\frac{0.0715/12}{1-(1+0.0715/12)^{-120}}\right)$$

$$\approx 641.17$$

The monthly payment is $641.17.

23. Sales tax = 0.06(42,600) = 2556
License fee = 0.01(42,600) = 426
Loan amount = 42,600 + 2556 + 426 = 45,582

$$i = \frac{\text{annual interest rate}}{\text{number of payments per year}}$$

$$= \frac{0.057}{12}$$

Calculate the monthly payment. For a 5-year loan, $n = 5(12) = 60$.

$$PMT = A\left(\frac{i}{1-(1+i)^{-n}}\right)$$

$$= 45,582\left(\frac{0.057/12}{1-(1+0.057/12)^{-60}}\right)$$

$$\approx 874.88$$

The monthly payment is $874.88.

25. Sales tax = 0.055(24,500) = 1347.50
Loan amount = purchase price − down payment + sales tax + license fee
Loan amount = 24,500 − 3000 + 1347.50 + 331 = 23,178.50

$$i = \frac{\text{annual interest rate}}{\text{number of payments per year}} = \frac{0.085}{12}$$

Calculate the monthly payment. For a 4-year loan, $n = 4(12) = 48$.

$$PMT = A\left(\frac{i}{1-(1+i)^{-n}}\right)$$

$$= 23,178.50\left(\frac{0.085/12}{1-(1+0.085/12)^{-48}}\right)$$

$$\approx 571.31$$

The monthly payment is $571.31.

27. a. Using the formula:

$$i = \frac{\text{annual interest rate}}{\text{number of payments per year}}$$

$$= \frac{0.08}{12}$$

Calculate the monthly payment. For a 4-year loan, $n = 4(12) = 48$.

$$PMT = A\left(\frac{i}{1-(1+i)^{-n}}\right)$$

$$= 25,455\left(\frac{0.08/12}{1-(1+0.08/12)^{-48}}\right)$$

$$\approx 621.43$$

The monthly payment is $621.43.

180   **Chapter 10: The Mathematics of Finance**

b.   There are 48 monthly payments of $621.43, so the total amount paid is 48(621.43) = $29,828.64.
Amount of interest paid = total amount paid − amount financed
= 29,828.64 − 25,455 = 4373.64
The total amount of interest paid is $4373.64.

29. She has owned the Jeep for 3 years, so she has 2 years, or 24 months, of payments remaining. Thus, $n = 24$.

$$i = \frac{APR}{12} = \frac{0.063}{12} = 0.00525$$

$$A = PMT\left(\frac{1-(1+i)^{-n}}{i}\right)$$

$$= 603.50\left(\frac{1-(1+0.00525)^{-24}}{0.00525}\right)$$

$$\approx 13,575.25$$

The loan payoff is $13,575.25.

31. You have owned the car for 3 years, so you have 1 year, or 12 months, of payments remaining. Thus, $n = 12$.

$$i = \frac{APR}{12} = \frac{0.089}{12}$$

$$A = PMT\left(\frac{1-(1+i)^{-n}}{i}\right)$$

$$= 303.52\left(\frac{1-(1+0.089/12)^{-12}}{0.089/12}\right)$$

$$\approx 3472.57$$

The loan payoff is $3472.57.

33. a.   Net capitalized cost = negotiated price − down payment − trade-in value
= 28,990 − 2400 − 3850 = 22,740
The net capitalized cost is $22,740.

b.   Average monthly finance charge = (net capitalized cost + residual value) × money factor
= (22,740 + 15,000) × 0.0027
≈ 101.90
The average monthly finance charge is $101.90.

c.   Average monthly depreciation = (net capitalized cost − residual value)/(term of the lease in months)
$$= \frac{22,740 - 15,000}{48} = 161.25$$
The average monthly depreciation is $161.25.

d.   Monthly lease payment = average monthly finance charge + average monthly depreciation
= 101.90 + 161.25 = 263.15
The monthly lease payment is $263.15.

35. a.   Net capitalized cost = negotiated price − down payment − trade-in value
= 22,100 − 1000 − 0 = 21,100
The net capitalized cost is $21,100.

b.   Money factor = (annual interest rate as a percent)/2400
= 8.1/2400 = 0.003375

c.   Average monthly finance charge = (net capitalized cost + residual value) × money factor
= (21,100 + 15,000) × 0.003375
≈ 121.84
The average monthly finance charge is $121.84.

d.   Average monthly depreciation = (net capitalized cost − residual value)/(term of the lease in months)
$$= \frac{21,100 - 15,000}{36} \approx 169.44$$
The average monthly depreciation is $169.44.

e.   Monthly lease payment = average monthly finance charge + average monthly depreciation
= 121.84 + 169.44 = 291.28
The monthly lease payment is $291.28.

37. The monthly payment for the loan is the *PMT*. The interest rate per period, *i*, is the annual interest rate divided by 12. The term of the loan, *n*, is the number of years of the loan times 12. Substitute these values into the Payment Formula for an APR loan and solve for *A*, the selling price of the car. The selling price of the car is $9775.72.

39. a. Calculate i and store the results in your calculator.

$$i = \frac{APR}{12} = \frac{0.059}{12}$$

$A = 19,788$
$V = 0.55(28,990) = 15,944.50$
$n = 4(12) = 48$

$$P = \frac{Ai(1+i)^n - Vi}{(1+i)^n - 1}$$

$= [(19,788)(0.059/12)(1 + 0.059/12)^{48}$ $-$ $(15,944.50)(0.059/12)] / [(1+0.059/12)^{48} - 1]$
$= 168.48$
The monthly lease payment is $168.48.

b. $61,432 - 87 = 61,345$
You must pay a penalty for each mile over 48,000 miles driven.
$61,345 - 48,000 = 13,345$
$13,345(0.20) = 2669$
The mileage penalty is $2669.

c. You make 48 lease payments of $168.48.
$48(168.48) = 8087.04$
Total cost = total of all lease payments + mileage penalty + cost of new tires
$= 8087.04 + 2669 + 635 = 11,391.04$
The total cost is $11,391.04.

EXERCISE SET 10.4

1. $(1.02)(375) = \$382.50$

3. $(0.63)(850) = \$535.50$

5. Solving:
$I = Prt$
$1.24 = 49.375r(1)$
$0.0251 \approx r$
The yield is 2.51%.

7. Solving:
$I = Prt$
$0.58 = 31.75r(1)$
$0.0183 \approx r$
The yield is 1.83%.

9. a. $25.72 - 16.99 = \$8.73$

b. $(0.09)(750) = \$67.50$

c. 3,750,600 shares

d. decrease

e. $22.24 + 0.35 = \$22.59$

11. Dividing:
$$\frac{5000}{98.59} \approx 50.7$$
You can buy 50 shares.

13. a. You paid:
$(3.85)(1000) = \$3850$
You sold the shares for:
$(4.14)(1000) = \$4140$
The profit is $4140 - 3850 = \$290$.

b. $(0.19)(4140) = \$78.66$

15. a. You paid:
$(14.30)(800) = \$11,440$
You sold the shares for:
$(25.67)(800) = \$20,536$
The profit is $20,536 - 11,440 = \$9096$.

b. $(0.023)(20,536) = \$472.33$

17. $I = Prt = 6000(0.042)(1) = \$252$

19. $I = Prt = 8000(0.035)(3) = \$840$

21. Using the formula:
$$NAV = \frac{50,000,000 - 5,000,000}{2,000,000} = \$22.50$$

23. First find NAV:
$$NAV = \frac{15,000,000 - 1,000,000}{2,000,000} = \$7$$
Dividing: $\frac{5000}{7} \approx 714.3$. You can buy 714 shares.

25. First find total assets:
$500,000,000 + 500,000 + 1,000,000$
$= 501,500,000$
Find NAV:

$$NAV = \frac{501,500,000 - 2,000,000}{10,000,000} = \$49.95$$

Dividing: $\frac{12,000}{49.95} \approx 240.2$. You can buy 240 shares.

27. Load fund:
    First deduct the load:
    $2500 - (0.04)(2500) = \$2400$

    At the end of the first year:
    $2400 + (0.08)(2400) = \$2592.00$

    The management fee must be deducted:
    $2592 - (0.00015)(2592) = \$2591.61$

    At the end of the second year:
    $2591.61 + (0.08)(2591.61) = \$2798.94$

    The management fee must be deducted:
    $2798.94 - (0.00015)(2798.94) = \$2598.52$

    No load fund:
    At the end of the first year:
    $2500 + (0.06)(2500) = \$2650.00$

    The management fee must be deducted:
    $2650 - (0.0015)(2650) \approx \$2650 - \$3.98 = \$2646.02$

    At the end of the second year:
    $2646.02 + (0.06)(2646.02) = \$2804.78$

    The management fee must be deducted:
    $2804.78 - (0.0015)(2804.78) = \$2800.57$

    The no-load fund's value ($2800.57) is $2.05 greater than the load fund's value ($2798.52).

EXERCISE SET 10.5

1. Down payment = 25% of $258,000
   $= 0.25(258,000)$
   $= 64,500$
   The down payment is $64,500.
   Mortgage = selling price − down payment
   $= 258,000 - 64,500 = 193,500$
   The mortgage is $193,500.

3. Points = 2.25% of $250,000
   $= 0.0225(250,000) = 5625$
   The charge for points is $5625.

5. Down payment = 30% of $309,000
   $= 0.30(309,000) = 92,700$
   Mortgage = selling price − down payment
   $= 309,000 - 92,700 = 216,300$

Points = 3% of $216,300
$= 0.03(216,300) = 6489$
Sum of down payment and closing costs
$= 92,700 + 350 + 6489 = 99539$
The total of the down payment and closing costs is $99,539.

7. Down payment = 25% of $121,500
   $= 0.25(121,500)$
   $= 30,375$
   Mortgage = selling price − down payment
   $= 121,500 - 30,375$
   $= 91,125$
   Points = $3\frac{1}{2}$% of $91,125
   $= 0.035(91,125) = 3189.38$
   Sum of down payment and closing costs
   $= 30,375 + 725 + 3189.38 = 34,289.38$
   The total of the down payment and closing costs is $34,289.38.

9. First calculate i and store the results.
   $$i = \frac{\text{annual interest rate}}{\text{number of payments per year}}$$
   $$= \frac{0.0775}{12}$$
   Calculate the monthly payment. For a 25-year loan, $n = 25(12) = 300$.
   $$PMT = A\left(\frac{i}{1-(1+i)^{-n}}\right)$$
   $$= 129,000\left(\frac{0.0775/12}{1-(1+0.0775/12)^{-300}}\right)$$
   $$\approx 974.37$$
   The monthly mortgage payment is $974.37.

11. First calculate i and store the results.
    $$i = \frac{\text{annual interest rate}}{\text{number of payments per year}}$$
    $$= \frac{0.0815}{12}$$
    Calculate the monthly payment. For a 15-year loan, $n = 15(12) = 180$.
    $$PMT = A\left(\frac{i}{1-(1+i)^{-n}}\right)$$
    $$= 223,500\left(\frac{0.0815/12}{1-(1+0.0815/12)^{-180}}\right)$$
    $$\approx 2155.28$$
    The monthly mortgage payment is $2155.28.

13. a. First calculate i and store the results.

$$i = \frac{\text{annual interest rate}}{\text{number of payments per year}}$$

$$= \frac{0.0775}{12}$$

Calculate the monthly payment. For a 30-year loan, $n = 30(12) = 360$.

$$PMT = A\left(\frac{i}{1-(1+i)^{-n}}\right)$$

$$= 152{,}000\left(\frac{0.0775/12}{1-(1+0.0775/12)^{-360}}\right)$$

$$\approx 1088.95$$

The monthly mortgage payment is $1088.95.

b. Multiply the number of payments by the monthly payment.
1088.95(360) = 392,022
The total of payments over the life of the loan is $392,022.

c. Subtract the mortgage from the total of the payments.
392,022 − 152,000 = 240,022
The amount of interest paid over the life of the loan is $240,022.

15. First calculate i and store the results.

$$i = \frac{\text{annual interest rate}}{\text{number of payments per year}}$$

$$= \frac{0.087}{12}$$

Calculate the monthly payment. For a 15-year loan, $n = 15(12) = 180$.

$$PMT = A\left(\frac{i}{1-(1+i)^{-n}}\right)$$

$$= 219{,}990\left(\frac{0.087/12}{1-(1+0.087/12)^{-180}}\right)$$

$$\approx 2192.20$$

The monthly mortgage payment is $2192.20.

Next, find the total paid over the life of the loan.
2192.20(180) = 394,596
The total of payments over the life of the loan is $394,596.

Last, find the total amount of interest paid.
394,596 − 219,990 = 174,606

The amount of interest paid over the life of the loan is $174,606.

17. Find the down payment.
0.10(208,500) = 20,850
The down payment is $20,850.
Find the mortgage.
208,500 − 20,850 = 187,650
The mortgage is $187,650.
Calculate the mortgage payment. First calculate i and store the results.

$$i = \frac{\text{annual interest rate}}{\text{number of payments per year}} = \frac{0.09}{12}$$

Calculate the monthly payment. For a 15-year loan, $n = 15(12) = 180$.

$$PMT = A\left(\frac{i}{1-(1+i)^{-n}}\right)$$

$$= 187{,}650\left(\frac{0.09/12}{1-(1+0.09/12)^{-180}}\right)$$

$$\approx 1903.27$$

The monthly mortgage payment is $1903.27.
Find the amount of interest paid on the first mortgage payment.

$$I = Prt = (187{,}650)(0.09)\left(\frac{1}{12}\right) \approx 1407.38$$

The interest paid is $1407.38.
Find the principal paid on the first mortgage payment.
1903.27 − 1407.38 = 495.89
The principal paid is $495.89.

19. Find the down payment.
0.30(185,000) = 55,500
The down payment is $55,500.
Find the mortgage.
185,000 − 55,500 = 129,500
The mortgage is $129,500.
Calculate the mortgage payment. First calculate i and store the results.

$$i = \frac{\text{annual interest rate}}{\text{number of payments per year}} = \frac{0.125}{12}$$

Calculate the monthly payment. For a 20-year loan, $n = 20(12) = 240$.

$$PMT = A\left(\frac{i}{1-(1+i)^{-n}}\right)$$

$$= 129{,}500\left(\frac{0.125/12}{1-(1+0.125/12)^{-240}}\right)$$

$$\approx 1471.30$$

The monthly mortgage payment is $1471.30.

Find the amount of interest paid on the payment for the first month.

$$I = Prt = (129,500)(0.125)\left(\frac{1}{12}\right) \approx 1348.96$$

The interest paid on the first payment is $1348.96.
Find the principal paid.
$1471.30 - 1348.96 = 122.34$
The principal paid on the first payment is $122.34.
Find the balance on the loan after the first mortgage payment.
$129,500 - 122.34 = 129,377.66$
The balance after the first payment is $129,377.66.
Find the amount of interest paid on the mortgage payment for the second month.

$$I = Prt = (129,377.66)(0.125)\left(\frac{1}{12}\right)$$

$$\approx 1347.68$$

The interest paid on the second payment is $1347.68.
Find the principal paid.
$1471.30 - 1347.68 = 123.62$
The principal paid on the second payment is $123.62.

21. Use the APR Loan Payoff Formula. You have been making payments for 6 years, or 72 months. There are 360 months in a 30-year loan, so there are $360 - 72 = 288$ remaining payments.

$$i = \frac{APR}{12} = \frac{0.085}{12}$$

$$A = PMT\left(\frac{1-(1+i)^{-n}}{i}\right)$$

$$= 913.10\left(\frac{1-(1+0.085/12)^{-288}}{0.085/12}\right)$$

$$= 112,025.49$$

The loan payoff is $112,025.49.

23. Use the APR Loan Payoff Formula. You have been making payments for 4 years, or 48 months. There are 180 months in a 15-year loan, so there are $180 - 48 = 132$ remaining payments.

$$i = \frac{APR}{12} = \frac{0.0725}{12}$$

$$A = PMT\left(\frac{1-(1+i)^{-n}}{i}\right)$$

$$= 672.39\left(\frac{1-(1+0.0725/12)^{-132}}{0.0725/12}\right)$$

$$= 61,039.75$$

The loan payoff is $61,039.75.

25. Find the monthly property tax bill.
$594 \div 12 = 49.50$
The monthly property tax bill is $49.50.
Find the monthly fire insurance bill.
$300 \div 12 = 25$
The monthly fire insurance bill is $25.
Find the total monthly payment.
$996.60 + 49.50 + 25 = 1071.10$
The total monthly payment for the mortgage, property tax, and fire insurance is $1071.10.

27. Find the monthly mortgage payment. First calculate i and store the results.

$$i = \frac{\text{annual interest rate}}{\text{number of payments per year}}$$

$$= \frac{0.0715}{12}$$

For a 25-year loan, $n = 25(12) = 300$.

$$PMT = A\left(\frac{i}{1-(1+i)^{-n}}\right)$$

$$= 259,500\left(\frac{0.0715/12}{1-(1+0.0715/12)^{-300}}\right)$$

$$\approx 1859.00$$

The monthly mortgage payment is $1859.00.
Find the monthly property tax bill.
$1320 \div 12 = 110$
The monthly property tax bill is $110.
Find the monthly fire insurance bill. $642 \div 12 = 53.50$
The monthly fire insurance bill is $53.50.
Find the total monthly payment.
$1859.00 + 110 + 53.50 = 2022.50$
The total monthly payment for the mortgage, property tax, and fire insurance is $2022.50.

29. a. Find the monthly mortgage payment for a 15-year loan. First calculate i and store the results.

$$i = \frac{\text{annual interest rate}}{\text{number of payments per year}}$$

$$= \frac{0.08125}{12}$$

For a 15-year loan, $n = 15(12) = 180$.

$$PMT = A\left(\frac{i}{1-(1+i)^{-n}}\right)$$

$$= 150,000\left(\frac{0.08125/12}{1-(1+0.08125/12)^{-180}}\right)$$

$$\approx 1444.32$$

The monthly mortgage payment on a 15-year loan is \$1444.32.

Find the monthly mortgage payment for a 30-year loan. The value of i is the same as for the 15-year loan.

For a 30-year loan, $n = 30(12) = 360$.

$$PMT = A\left(\frac{i}{1-(1+i)^{-n}}\right)$$

$$= 150,000\left(\frac{0.08125/12}{1-(1+0.08125/12)^{-360}}\right)$$

$$\approx 1113.75$$

The monthly mortgage payment on a 30-year loan is \$1113.75.

$1444.32 - 1113.75 = 330.57$

The payment on the 15-year loan is \$330.57 greater than the payment on the 30-year loan.

b. Find the amount paid over the life of each loan.

For the 15-year loan, there are 180 payments of \$1444.32.

$180(1444.32) = 259,977.60$

For the 30-year loan, there are 360 payments of \$1113.75.

$360(1113.75) = 400,950$

$400,950 - 259,977.60 = 140,972.40$

The 15-year loan costs \$140,972.40 less over the life of the loan than the 30-year loan.

31. a. Find the monthly mortgage payment for a 20-year loan. First calculate i and store the results.

$$i = \frac{\text{annual interest rate}}{\text{number of payments per year}}$$

$$= \frac{0.0675}{12}$$

For a 20-year loan, $n = 20(12) = 240$.

$$PMT = A\left(\frac{i}{1-(1+i)^{-n}}\right)$$

$$= 349,500\left(\frac{0.0675/12}{1-(1+0.0675/12)^{-240}}\right)$$

$$\approx 2657.47$$

The monthly mortgage payment on a 20-year loan is \$2657.47.

Find the monthly mortgage payment for a 30-year loan. The value of i is the same as for the 20-year loan.

For a 30-year loan, $n = 30(12) = 360$.

$$PMT = A\left(\frac{i}{1-(1+i)^{-n}}\right)$$

$$= 349,500\left(\frac{0.0675/12}{1-(1+0.0675/12)^{-360}}\right)$$

$$\approx 2266.85$$

The monthly mortgage payment on a 30-year loan is \$2266.85.

$2657.47 - 2266.85 = 390.62$

The payment on the 20-year loan is \$390.62 greater than the payment on the 30-year loan.

b. Find the amount paid over the life of each loan.

For the 20-year loan, there are 240 payments of \$2657.47.

$240(2657.47) = 637,792.80$

For the 30-year loan, there are 360 payments of \$2266.85.

$360(2266.85) = 816,066$

$816,066 - 637,792.80 = 178,273.20$

The 20-year loan costs \$178,273.20 less over the life of the loan than the 30-year loan.

33. Let x = the price of the house.

20% of the price of the house = \$25,000.

$0.20x = 25,000$

$$x = \frac{25,000}{0.20} = 125,000$$

The maximum price they can offer is \$125,000.

35. Let x = the price of the house.

Then, down payment = $0.15x$,

mortgage = $x - 0.15x$,

4 points = 4% of the mortgage =

$0.04(x - 0.15x)$.

Down payment + closing costs + points = total.

$$0.15x + 380 + 0.04(x - 0.15x) = 39,400$$
$$0.15x + 380 + 0.04(0.85x) = 39,400$$
$$0.15x + 380 + 0.034x = 39,400$$
$$0.184x + 380 = 39,400$$
$$0.184x = 39,020$$
$$x = \frac{39,020}{0.184}$$
$$x = 212,065$$

The maximum price they can offer is $212,065.

37. No. The 260th payment remains the first payment in which the principal paid exceeds the amount of interest paid, regardless of the amount of the loan.

39. You pay less total interest on a 15-year mortgage loan.

41. a. Down payment = 20% of $134,000
$$= 0.20(134,000)$$
$$= \$26,800$$
Mortgage = selling price − down payment = $134,000 − 26,800 = 107,200$
Points = 2.5% of $107,200
$$= 0.025(107,200) = 2680$$
Closing costs = points + fees
$$= 2680 + 325 = 3005$$
Down payment + closing costs = 26,800 + 3005 = 29,805
$29,805 is due at closing.
Find the amount you are willing to spend.
Savings: 36,000 − 6,000 = 30,000
Checking: 900 − 600 = 300
Willing to spend: 30,000 + 300 = 30,300
Yes. You are willing to take this $30,300 out of your accounts, so you can afford the down payment and closing costs.

b. First calculate i and store the results.
$$i = \frac{\text{annual interest rate}}{\text{number of payments per year}}$$
$$= \frac{0.09}{12}$$
Calculate the monthly payment. For a 30-year loan, $n = 30(12) = 360$.
$$PMT = A\left(\frac{i}{1-(1+i)^{-n}}\right)$$
$$= 107,200\left(\frac{0.09/12}{1-(1+0.09/12)^{-360}}\right)$$
$$\approx 862.56$$

The monthly mortgage payment is $862.56.

c. Find the monthly property tax bill.
$$1152 \div 12 = 96$$
The monthly property tax bill is $96.
The yearly utility bill is
$$8750 \times (0.07) = 612.50$$
Find the monthly utility bill.
$$612.50 \div 12 \approx 51.04$$
The monthly utility bill is $51.04.
Find the monthly insurance bill.
$$192 \div 12 = 16$$
The monthly insurance bill is $16.
Total of monthly payments = mortgage payment + maintenance fee + property tax + utilities + insurance
$$= 862.56 + 120 + 96 + 51.04 + 16$$
$$= 1145.60$$
The total of all monthly payments is $1145.60.

d. Monthly take-home pay − monthly expenses
$$= 4479.38 - 1145.60 = 3333.78$$
The difference is $3333.78.

e. $\dfrac{1145.60}{4479.38} \approx 0.256 = 25.6\%$

The condominium-related expenses are 28.9% of your take-home pay.

f. Answers will vary.

CHAPTER 10 REVIEW EXERCISES

1. Because the interest rate is an annual rate, the time must be measured in years:
$$t = \frac{4 \text{ months}}{1 \text{ year}} = \frac{4 \text{ months}}{12 \text{ months}} = \frac{4}{12}$$
$$I = Prt = 2750(0.0675)\left(\frac{4}{12}\right) \approx 61.88$$
The interest earned is $61.88.

2. The interest rate is per month and the time period is in months.
$$I = Prt = 8500(0.0115)(8) = 782$$
The interest due is $782.

3. The value for the time is
$$t = \frac{\text{number of days}}{360} = \frac{120}{360}.$$

$$I = Prt = 4000(0.0675)\left(\frac{120}{360}\right) = 90$$

The interest earned is \$90.

4. Since the interest rate is per year, the time period of the loan is $t = \frac{108}{360}$ year.

$$A = P(1+rt) = 7000\left[1 + 0.104\left(\frac{108}{360}\right)\right]$$

$$= 7000(1.0312) = 7218.40$$

The maturity value of the loan is \$7218.40.

5. Solve the simple interest formula for r.
$$I = Prt$$
$$127.50 = 6800(r)\left(\frac{3}{12}\right)$$
$$127.50 = 1700r$$
$$\frac{127.50}{1700} = r$$
$$0.075 = r$$
$$r = 7.5\%$$
The simple interest rate is 7.5%.

6. $r = 6.6\% = 0.066$, $n = 12$, $t = 3$,
$$i = \frac{r}{n} = \frac{0.066}{12} = 0.0055 \text{ , and}$$
$$N = nt = 12(3) = 36.$$
$$A = P(1+r)^N = 3000(1+0.0055)^{36}$$
$$= 3000(1.0055)^{36} \approx 3654.90$$
The compound amount after 3 years is approximately \$3654.90.

7. $r = 6\% = 0.06$, $n = 4$, $t = 10$,
$$i = \frac{r}{n} = \frac{0.06}{4} = 0.015 \text{ , and}$$
$$N = nt = 4(10) = 40.$$
$$A = P(1+r)^N = 6400(1+0.015)^{40}$$
$$= 6400(1.015)^{40} \approx 11,609.72$$
The compound amount after 10 years is approximately \$11,609.72.

8. $r = 9\% = 0.09$, $n = 360$, $t = 3$,
$$i = \frac{r}{n} = \frac{0.09}{360} = 0.00025 \text{ , and}$$
$$N = nt = 360(3) = 1080.$$
$$A = P(1+r)^N = 6000(1+0.00025)^{1080}$$
$$= 6000(1.00025)^{1080} \approx 7859.52$$
The future value is approximately \$7859.52.

9. $r = 7.2\% = 0.072$, $n = 360$, $t = 4$,
$$i = \frac{r}{n} = \frac{0.072}{360} = 0.0002 \text{ , and}$$
$$N = nt = 360(4) = 1440.$$
$$A = P(1+r)^N = 600(1+0.0002)^{1440}$$
$$= 600(1.0002)^{1440} \approx 800.23$$
Calculate the interest earned.
$$I = A - P = 800.23 - 600 = 200.23$$
The amount of interest earned is \$200.23.

10. $r = 8\% = 0.08$, $n = 2$, $t = 7$, $i = \frac{r}{n} = \frac{0.08}{2} = 0.04$,
$$N = nt = 2(7) = 14.$$
$$P = \frac{A}{(1+i)^N} = \frac{18,500}{(1+0.04)^{14}} = \frac{18,500}{1.04^{14}}$$
$$\approx 10,683.29$$
\$10,683.29 should be invested in the account in order to have \$18,500 in 7 years.

11. a. $r = 7\% = 0.07$, $n = 4$, $t = 5$,
$$i = \frac{r}{n} = \frac{0.07}{4} = 0.0175 \text{ , and}$$
$$N = nt = 4(5) = 20.$$
$$A = P(1+i)^N = 8000(1+0.0175)^{20}$$
$$= 8000(1.0175)^{20} \approx 11,318.23$$
The maturity value of the loan is \$11,318.23.

b. $11,318.23 - 8000 = 3318.23$
The interest you are paying is \$3318.23.

12. $r = 8\% = 0.08$, $n = 4$, $t = 18$,
$$i = \frac{r}{n} = \frac{0.08}{4} = 0.02 \text{ , } N = nt = 4(18) = 72.$$
$$P = \frac{A}{(1+i)^N} = \frac{80,000}{(1+0.02)^{72}} \approx 19,225.50$$
\$19,225.50 should be invested in the account in order to have \$80,000 in 18 years.

13. Using the formula:
$$I = Prt$$
$$0.66 = 60r(1)$$
$$0.66 = 60r$$
$$0.011 = r$$
$$r = 1.1\%$$
The dividend yield is 1.1%.

14. Find the annual interest payment.

$I = Prt = 20,000(0.045)(1) = 900$

Multiply the annual interest payment by the term of the bond.

$900(10) = 9000$

The total interest payments paid to the bondholder is $9000.

15. The inflation rate is an annual rate, so $n = 1$.

$r = 6\% = 0.06$, $i = \dfrac{r}{n} = \dfrac{0.06}{1} = 0.06$,

$t = 10$, so $N = nt = 1(10) = 10$.

$A = P(1+r)^N = 1.29(1+0.06)^{10}$

$= 1.29(1.06)^{10} \approx 2.31$

One pound of red delicious apples will cost approximately $2.31 in 2011.

16. $r = 7\% = 0.07$ and $t = 8$. The inflation rate is an annual rate, so $n = 1$. $i = \dfrac{r}{n} = \dfrac{0.07}{1} = 0.07$,

$N = nt = 1(8) = 8$.

$P = \dfrac{A}{(1+i)^N} = \dfrac{75,000}{(1+0.07)^8} = \dfrac{75,000}{1.07^8}$

$\approx 43,650.68$

The purchasing power of the $75,000 in will be approximately $43,650.68 in 8 years.

17. Find the future value of $100 after 1 year.

$r = 5.9\%$, $n = 12$, $i = \dfrac{r}{n} = \dfrac{0.059}{12}$ and

$N = nt = 12(1) = 12$.

$A = P(1+i)^N = 100(1+0.059/12)^{12}$

≈ 106.06

Find the interest earned on the $100.

$I = A - P = 106.06 - 100 = 6.06$. The effective interest rate is 6.06%.

18. Calculate $(1+i)^N$ for each investment.

$i = \dfrac{r}{n} = \dfrac{0.052}{4}$

$N = nt = 4(1) = 4$

$(1+i)^N = \left(1+\dfrac{0.052}{4}\right)^4 \approx 1.053022817$

$i = \dfrac{r}{n} = \dfrac{0.054}{2}$

$N = nt = 2(1) = 2$

$(1+i)^N = \left(1+\dfrac{0.054}{2}\right)^2 \approx 1.054729$

Compare the two compound amounts.
$1.054729 > 1.053022817$, so 5.4% compounded semiannually has a higher yield than 5.2% compounded quarterly.

19. Recording the balances:

Date	Payment or purchase	Balance	Days	Balance times Days
Mar. 11-17		$423.35	7	$2963.45
Mar. 18-28	$145.50	$568.85	11	$6257.35
Mar. 29-Apr. 10	$- $250	$318.85	13	$4145.05
Total				$13,365.85

Average daily balance $= \dfrac{13,365.85}{31} \approx 431.16$.

20. Recording the balances:

Date	Payment or purchase	Balance	Days	Balance times Days
Sep. 10-19		$450	10	$4500
Sep. 20-24	$47	$497	5	$2485
Sep. 25-27	$157	$654	3	$1962
Sep. 28-Oct. 9	$- $175	$479	12	$5748
Total				$14,695

Average daily balance $= \dfrac{14,695}{30} \approx 489.83$.

$I = Prt = (489.83)(0.0125)(1) \approx 6.12$.

The finance charge is $6.12.

21. a. Find the interest due.

$I = Prt = 1500(0.075)\left(\dfrac{6}{12}\right) = 56.25$

The total amount to be repaid to the bank is
$A = P + I = 1500 + 56.25 = 1556.25$

Monthly payment $= \dfrac{1556.25}{6} = 259.38$

The monthly payment is $259.38.

b. Using the formula:

$APR \approx \dfrac{2nr}{n+1}$

$\approx \dfrac{2(6)(0.075)}{6+1} = \dfrac{0.9}{7} \approx 0.129$

The annual percentage rate on the loan is 12.9%.

22. a. Find the down payment.
Down payment $= 0.10(449) = 44.9$
Find the amount financed.
$449 - 44.9 = 404.10$
Find the simple interest due.
$I = Prt = 404.10(0.07)(1) \approx 28.29$

Find the total amount to be repaid to the bank.
$A = P + I = 404.10 + 28.29 = 432.39$
Find the monthly payment.

Monthly payment $= \dfrac{432.39}{12} = 36.03$

The monthly payment is $36.03.

b. Using the formula:

$APR \approx \dfrac{2nr}{n+1}$

$\approx \dfrac{2(12)(0.07)}{12+1} = \dfrac{1.68}{13} = 0.129$

The annual percentage rate on the loan is 12.9%.

23. Using the formula:

$i = \dfrac{\text{annual interest rate}}{\text{number of payments per year}}$

$= \dfrac{0.085}{12}$

Calculate the monthly payment. For a 2-year loan, $n = 2(12) = 24$.

$PMT = A\left(\dfrac{i}{1-(1+i)^{-n}}\right)$

$= 999\left(\dfrac{0.085/12}{1-(1+0.085/12)^{-24}}\right)$

≈ 45.41

The monthly payment is $45.41.

24. a. Sales tax $= 0.0625(9499) = 593.69$
The sales tax is $593.69.
Total cost $= 9499 + 593.69 = 10{,}092.69$
The total cost is $10,092.69.

b. Down payment $= 0.20(10{,}092.69)$
$= 2018.54$
The down payment is $2018.54.

c. Loan amount $=$ total cost $-$ down payment
Loan amount $= 10{,}092.69 - 2018.54$
$= 8074.15$

$i = \dfrac{\text{annual interest rate}}{\text{number of payments per year}}$

$= \dfrac{0.08}{12}$

Calculate the monthly payment. For a 3-year loan, $n = 3(12) = 36$.

$PMT = A\left(\dfrac{i}{1-(1+i)^{-n}}\right)$

$= 8074.15\left(\dfrac{0.08/12}{1-(1+0.08/12)^{-36}}\right)$

≈ 253.01
The monthly payment is $253.01.

25. Down payment $= (0.20)(28{,}450) = 5690$
Loan amount $=$ purchase price $-$ down payment
Loan amount $= 28{,}450 - 5690 = 22{,}760$

$i = \dfrac{\text{annual interest rate}}{\text{number of payments per year}}$

$= \dfrac{0.072}{12} = 0.006$

Calculate the monthly payment. For a 3-year loan, $n = 3(12) = 36$.

$PMT = A\left(\dfrac{i}{1-(1+i)^{-n}}\right)$

$= 22{,}760\left(\dfrac{0.006}{1-(1+0.006)^{-36}}\right)$

≈ 704.85
The monthly payment is $704.85.

26. a. Using the formula:

$i = \dfrac{\text{annual interest rate}}{\text{number of payments per year}}$

$= \dfrac{0.059}{12}$

Calculate the monthly payment. For a 5-year loan, $n = 5(12) = 60$.

$$PMT = A\left(\frac{i}{1-(1+i)^{-n}}\right)$$

$$= 28,000\left(\frac{0.059/12}{1-(1+0.059/12)^{-60}}\right)$$

$$\approx 540.02$$

The monthly payment is $540.02.

b. He has owned the car for 3 years, so he has 2 years, or 24 months, of payments remaining. Thus, $n = 24$.

$$i = \frac{APR}{12} = \frac{0.059}{12}$$

$$A = PMT\left(\frac{1-(1+i)^{-n}}{i}\right)$$

$$= 540.02\left(\frac{1-(1+0.059/12)^{-24}}{0.059/12}\right)$$

$$= 12,196.80$$

The loan payoff is $12,196.80.

27. a. Net capitalized cost = negotiated price $-$ down payment $-$ trade-in value
$$= 32,450 - 3000 - 0 = 29,450$$
The net capitalized cost is $29,450.

b. Average monthly finance charge = (net capitalized cost + residual value) $\times$ money factor
$$= (29,450 + 16,000) \times 0.004$$
$$\approx 181.80$$
The average monthly finance charge is $181.80.

c. Average monthly depreciation = (net capitalized cost $-$ residual value)/(term of the lease in months)
$$= \frac{29,450 - 16,000}{60} \approx 224.17$$
The average monthly depreciation is $224.17.

d. Monthly lease payment = average monthly finance charge + average monthly depreciation
$$= 181.80 + 224.17 = 405.97$$
Sales tax $= 0.075(405.97) = 30.45$

Total payment $= 405.97 + 30.45 = 436.42$
The total monthly lease payment is $436.42.

28. a. You paid: $(500)(28.75) = 14,375$. You sold the stock for $(500)(39.40) = 19,700$. The profit is $19,700 - 14,375 = \$5,325$.

b. $(0.013)(19,700) = \$256.10$

29. First find the NAV:
$$NAV = \frac{34,000,000 - 4,000,000}{2,000,000} = 15$$
You will buy: $\frac{3000}{15} = 200$ shares.

30. Down payment = 20% of $459,000
$$= 0.20(459,000)$$
$$= 91,800$$
Mortgage = selling price $-$ down payment
$$= 459,000 - 91,800 = 367,200$$
Points = 1.75% of $367,200
$$= 0.0175(367,200) = 6426$$
Sum of down payment and closing costs
$$= 91,800 + 815 + 6426 = 99,041$$
The total of the down payment and closing costs is $99,041.

31. a. First calculate i and store the results.
$$i = \frac{\text{annual interest rate}}{\text{number of payments per year}}$$
$$= \frac{0.0675}{12}$$
Calculate the monthly payment. For a 30-year loan, $n = 30(12) = 360$.

$$PMT = A\left(\frac{i}{1-(1+i)^{-n}}\right)$$

$$= 255,800\left(\frac{0.0675/12}{1-(1+0.0675/12)^{-360}}\right)$$

$$\approx 1659.11$$
The monthly mortgage payment is $1659.11.

b. Multiply the number of payments by the monthly payment.
$$1659.11(360) = 597,279.60$$
The total of payments over the life of the loan is $597,279.60.

c. Subtract the mortgage from the total of the payments.
$$597,279.60 - 255,800 = 341,479.60$$

The amount of interest paid over the life of the loan is $341,479.60.

32. a.   First calculate i and store the results.

$$i = \frac{\text{annual interest rate}}{\text{number of payments per year}}$$

$$= \frac{0.075}{12}$$

Calculate the monthly payment. For a 25-year loan, $n = 25(12) = 300$.

$$PMT = A\left(\frac{i}{1-(1+i)^{-n}}\right)$$

$$= 189,000\left(\frac{0.075/12}{1-(1+0.075/12)^{-300}}\right)$$

$$\approx 1396.69$$

The monthly mortgage payment is $1396.69.

b.   Use the APR Loan Payoff Formula. He has been making payments for 10 years, or 120 months. There are 300 months in a 25-year loan, so there are $300 - 120 = 180$ remaining payments.

$$i = \frac{APR}{12} = \frac{0.075}{12}$$

$$A = PMT\left(\frac{1-(1+i)^{-n}}{i}\right)$$

$$= 1396.69\left(\frac{1-(1+0.075/12)^{-180}}{0.075/12}\right)$$

$$= 150,665.74$$

The loan payoff is $150,665.74.

33.   Find the monthly mortgage payment. First calculate i and store the results.

$$i = \frac{\text{annual interest rate}}{\text{number of payments per year}}$$

$$= \frac{0.07}{12}$$

Calculate the monthly payment. For a 15-year loan, $n = 15(12) = 180$.

$$PMT = A\left(\frac{i}{1-(1+i)^{-n}}\right)$$

$$= 278,950\left(\frac{0.07/12}{1-(1+0.07/12)^{-180}}\right)$$

$$\approx 2507.28$$

The monthly mortgage payment is $2507.28.
Find the monthly property tax bill.

$1134 \div 12 = 94.50$
The monthly property tax bill is $94.50.

Find the monthly fire insurance bill.
$681 \div 12 = 56.75$
The monthly fire insurance bill is $56.75.

Find the total monthly payment.
$2507.28 + 94.50 + 56.75 = 2658.53$
The total monthly payment for the mortgage, property tax, and fire insurance is $2658.53.

CHAPTER 10 TEST

1.   Because the interest rate is an annual rate, the time must be measured in years:

$$t = \frac{3 \text{ months}}{1 \text{ year}} = \frac{3 \text{ months}}{12 \text{ months}} = \frac{3}{12}$$

$$I = Prt$$

$$= 5250(0.0825)\left(\frac{3}{12}\right)$$

$$\approx 108.28$$

The interest earned is $108.28.

2.   The value for the time is

$$t = \frac{\text{number of days}}{360} = \frac{180}{360}.$$

$$I = Prt$$

$$= 6000(0.0675)\left(\frac{180}{360}\right)$$

$$= 202.50$$

The interest earned is $202.50.

3.   Since the interest rate is per year, the time period of the loan is $t = \frac{200}{360}$ year.

$$A = P(1+rt)$$

$$= 8000\left[1+0.092\left(\frac{200}{360}\right)\right]$$

$$\approx 8000(1.051111) = 8408.89$$

The maturity value of the loan is $8408.89.

4. Solve the simple interest formula for r.

$$I = Prt$$

$$114 = 7600(r)\left(\frac{2}{12}\right)$$

$$114 = \frac{7600}{6}r$$

$$\frac{6}{7600}(114) = r$$

$$0.09 = r$$

$$r = 9\%$$

The simple interest rate is 9%.

5. $r = 7\% = 0.07$, $n = 12$, $t = 8$, $i = \frac{r}{n} = \frac{0.07}{12}$, and

$N = nt = 12(8) = 96$.

$$A = P(1+r)^N$$

$$A = 4200\left(1 + \frac{0.07}{12}\right)^{96}$$

$$A \approx 7340.87$$

The compound amount after 8 years is $7340.87.

6. $r = 6.3\% = 0.063$, $n = 360$, $t = 3$,

$i = \frac{r}{n} = \frac{0.063}{360} = 0.000175$, and

$N = nt = 360(3) = 1080$.

$$A = P(1+r)^N = 1500(1 + 0.000175)^{1080}$$

$$= 1500(1.000175)^{1080} \approx 1812.03$$

Calculate the interest earned.

$I = A - P = 1812.03 - 1500 = 312.03$

The amount of interest earned is $312.03.

7. a. $r = 9.5\% = 0.095$, $n = 12$, $t = 4$,

$i = \frac{r}{n} = \frac{0.095}{12} \approx 0.00791667$, and

$N = nt = 12(4) = 48$.

$$A = P(1+i)^N$$

$$A = 10,500\left(1 + \frac{0.095}{12}\right)^{48}$$

$$A \approx 15,331.03$$

The maturity value of the loan is $15,331.03.

   b. $15,331.03 - 10,500 = 4831.03$

The interest you are paying is $4831.03.

8. $r = 6.25\% = 0.0625$, $n = 360$, $t = 5$,

$i = \frac{r}{n} = \frac{0.0625}{360}$, $N = nt = 360(5) = 1800$.

$$P = \frac{A}{(1+i)^N}$$

$$P = \frac{30,000}{(1 + 0.0625/360)^{1800}}$$

$$P \approx 21,949.06$$

$21,949.06 should be invested in the account in order to have $30,000 in 5 years.

9. Using the formula:

$$I = Prt$$

$$0.48 = 40r(1)$$

$$0.48 = 40r$$

$$0.012 = r$$

$$r = 1.2\%$$

The dividend yield is 1.2%.

10. Find the annual interest payment.

$I = Prt = 5000(0.038)(1) = 190$

Multiply the annual interest payment by the term of the bond.

$190(10) = 1900$

The total interest payments paid to the bondholder is $1900.

11. The inflation rate is an annual rate, so $n = 1$.

$r = 7\% = 0.07$, $i = \frac{r}{n} = \frac{0.07}{1} = 0.07$,

$t = 2022 - 2002 = 20$, so $N = nt = 1(20) = 20$.

$$A = P(1+r)^N = 158,200(1 + 0.07)^{20}$$

$$= 158,200(1.07)^{20} \approx 612,184.08$$

The median value of a single family house will be approximately $612,184.08 in 2022.

12. Find the future value of $100 after 1 year.

$r = 6.25\%$, $n = 4$, $i = \frac{r}{n} = \frac{0.0625}{4}$ and

$N = nt = 4(1) = 4$.

$$A = P(1+i)^N = 100(1 + 0.0625/4)^4$$

$$\approx 106.40$$

Find the interest earned on the $100.

$I = A - P = 106.40 - 100 = 6.40$

The effective interest rate is 6.40%.

13. Calculate $(1+i)^N$ for each investment.

$i = \dfrac{r}{n} = \dfrac{0.044}{12}$

0.4% compounded monthly:

$N = nt = 12\,(1) = 12$

$(1+i)^N = \left(1 + \dfrac{0.044}{12}\right)^{12} \approx 1.044898269$

$i = \dfrac{r}{n} = \dfrac{0.046}{2}$

4.6% compounded semiannually:

$N = nt = 2(1) = 2$

$(1+i)^N = \left(1 + \dfrac{0.046}{2}\right)^{2} \approx 1.046529$

Compare the two compound amounts.
1.046529 > 1.044898269
4.6% compounded semiannually has a higher yield than 4.4% compounded monthly.

14. Recording the balances:

Date	Payment or purchase	Balance	Days	Balance times Days
Oct. 15-19		$515	5	$2575
Oct. 20-27	$75	$590	8	$4720
Oct. 28-Nov. 14	– $250	$340	18	$6120
Total				$13,415

Average daily balance $= \dfrac{13,415}{31} \approx 432.74$.

$I = Prt = (432.74)(0.018)(1) \approx 7.79$

The finance charge is $7.79.

15. a. Find the down payment.
   Down payment = 0.15(629) = 94.35
   Find the amount financed.
   629 – 94.35 = 534.65
   Find the simple interest due.
   $I = Prt = 534.65(0.09)(1) \approx 48.12$

   Find the total amount to be repaid to the bank.
   $A = P + I = 534.65 + 48.12 = 582.77$
   Find the monthly payment.

Monthly payment $= \dfrac{582.77}{12} = 48.56$

The monthly payment is $48.56.

b. Using the formula:

$\text{APR} \approx \dfrac{2nr}{n+1}$

$\approx \dfrac{2(12)(0.09)}{12+1} = \dfrac{2.16}{13} = 0.166$

The annual percentage rate on the loan is 16.6%.

16. Calculate the payment. $n = 3(12) = 36$.

$PMT = A\left(\dfrac{i}{1-(1+i)^{-n}}\right)$

$= 1899\left(\dfrac{0.0925/12}{1-(1+0.0925/12)^{-36}}\right)$

≈ 60.61

The monthly payment is $60.61.

17. a. You paid: $(800)(31.82) = 25,456$. You sold the stock for $(800)(25.70) = 20,560$.
   The loss is $25,456 - 20,560 = \$4,896$.

   b. $(0.011)(20560) = \$226.16$

18. First find the NAV:

$NAV = \dfrac{42,000,000 - 6,000,000}{3,000,000} = 12$

Dividing: $\dfrac{2500}{12} \approx 208.3$ You will buy 208 shares.

19. a. Sales tax = 0.0625(6575) = 410.94
   The sales tax is $410.94.
   Total cost = 6575 + 410.94 = 6985.94
   The total cost is $6985.94.

   b. Down payment = 0.20(6985.94) = 1397.19
   The down payment is $1397.19.

   c. Loan amount = total cost – down payment
   Loan amount = 6985.94 – 1397.19
   = 5588.75

   $i = \dfrac{\text{annual interest rate}}{\text{number of payments per year}}$

   $= \dfrac{0.078}{12}$

Calculate the monthly payment. For a 3-year loan, $n = 3(12) = 36$.

$$PMT = A\left(\frac{i}{1-(1+i)^{-n}}\right)$$

$$= 5588.75\left(\frac{0.078/12}{1-(1+0.078/12)^{-36}}\right)$$

$$\approx 174.62$$

The monthly payment is $174.62

20. Down payment = 20% of $262,250
$$= 0.20(262,250)$$
$$= 52,450$$
Mortgage = selling price − down payment
$$= 262,250 − 52,450$$
$$= 209,800$$
Points = 3.25% of $209,800
$$= 0.0325(209,800) = 6818.50$$
Sum of down payment and closing costs
$$= 52,450 + 815 + 6818.50 = 60,083.50$$
The total of the down payment and closing costs is $60,083.50.

21. a. First calculate i and store the results.
$$i = \frac{\text{annual interest rate}}{\text{number of payments per year}}$$
$$= \frac{0.0675}{12}$$
Calculate the monthly payment. For a 30-year loan, $n = 30(12) = 360$.

$$PMT = A\left(\frac{i}{1-(1+i)^{-n}}\right)$$

$$= 236,000\left(\frac{0.0675/12}{1-(1+0.0675/12)^{-360}}\right)$$

$$\approx 1530.69$$

The monthly mortgage payment is $1530.69.

b. Use the APR Loan Payoff Formula. He has been making payments for 5 years, or 60 months. There are 360 months in a 30-year loan, so there are $360 − 60 = 300$ remaining payments.

$$i = \frac{APR}{12} = \frac{0.0675}{12}$$

$$A = PMT\left(\frac{1-(1+i)^{-n}}{i}\right)$$

$$= 1530.69\left(\frac{1-(1+0.0675/12)^{-300}}{0.0675/12}\right)$$

$$= 221,546.46$$

The loan payoff is $221,546.46.

22. Find the monthly mortgage payment. First calculate i and store the results.
$$i = \frac{\text{annual interest rate}}{\text{number of payments per year}}$$
$$= \frac{0.0725}{12}$$
Calculate the monthly payment. For a 20-year loan, $n = 20(12) = 240$.

$$PMT = A\left(\frac{i}{1-(1+i)^{-n}}\right)$$

$$= 312,000\left(\frac{0.0725/12}{1-(1+0.0725/12)^{-240}}\right)$$

$$\approx 2465.97$$

The monthly mortgage payment is $2465.97.
Find the monthly property tax bill.
$$1044 \div 12 = 87$$
The monthly property tax bill is $87.
Find the monthly fire insurance bill.
$$516 \div 12 = 43$$
The monthly fire insurance bill is $43.
Find the total monthly payment.
$$2465.97 + 87 + 43 = 2595.97$$
The total monthly payment for the mortgage, property tax, and fire insurance is $2595.97.

Chapter 11: Combinatorics and Probability

EXERCISE SET 11.1

1. $\{0, 2, 4, 6, 8\}$

3. {Monday, Tuesday, Wednesday, Thursday, Friday, Saturday, Sunday}

5.

	H	T
H	HH	HT
T	TH	TT

{HH, TT, HT, TH}

7.

	H	T
1	1H	1T
2	2H	2T
3	3H	3T
4	4H	4T
5	5H	5T
6	6H	6T

{1H, 2H, 3H, 4H, 5H, 6H, 1T, 2T, 3T, 4T, 5T, 6T}

9.

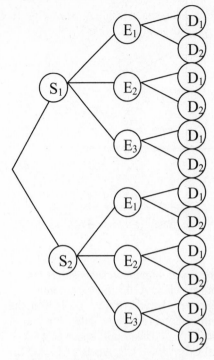

$\{S_1E_1D_1, S_1E_1D_2, S_1E_2D_1, S_1E_2D_2,$
$S_1E_3D_1, S_1E_3D_2, S_2E_1D_1, S_2E_1D_2, S_2E_2D_1,$
$S_2E_2D_2, S_2E_3D_1, S_2E_3D_2\}$

11.

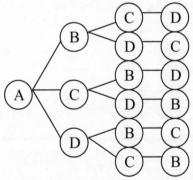

{ABCD, ABDC, ACBD, ACDB, ADBC, ADCB}

13.

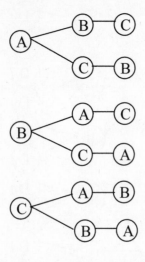

The first column represents letters put into envelope A, the second column represents letters put into envelope B, the third column represents letters put into envelope C. The sample space is: {AA BB CC, AA BC CB, AB BA CC, AC BA CB, AC BB CA, AB BC CA}

15. Any of the 4 numbers can be used in the first digit, so $n_1 = 4$. Because each number can be used only once, $n_2 = 3$. There are $4 \cdot 3 = 12$ two digit numbers.

17. There are four choices on each question, so there
$$4 \cdot 4 \cdot 4 \cdot 4 \cdot 4 \cdot 4 \cdot 4 \cdot 4 \cdot 4 \cdot 4 \cdot 4 \cdot 4 \cdot$$
are
$$4 \cdot 4 \cdot 4 \cdot 4 \cdot 4 \cdot 4 \cdot 4 \cdot 4 = 4^{20}$$
possible ways to complete the test.

19. There are 7 numbers that can be used in the first digit, so $n_1 = 7$. There are 10 numbers that can be used in the last three digits, so $n_2 = 10$, $n_3 = 10$ and $n_4 = 10$. By the counting principle, there are $7 \cdot 10 \cdot 10 \cdot 10 = 7000$ possible four-digit numbers.

21. Because a number starting with 0 is not two-digits, there are 9 numbers that can be used in the first digit, so $n_1 = 9$. There are 10 numbers that can be used in the last digit, so $n_2 = 10$. By the counting principle, there are $9 \cdot 10 = 90$ possible two-digit numbers.

23. Because a number starting with 0 is not two-digits, there are 9 numbers that can be used in the first digit, so $n_1 = 9$. Because a number divisible by 5 must end in 0 or 5, there are 2 numbers that can be used in the last digit, so $n_2 = 2$. By the

counting principle, there are $9 \cdot 2 = 18$ possible two-digit numbers divisible by 5.

25. Because a number with 4 or greater in the first place is greater than 37, there are 6 numbers that can be used in the first digit, so $n_1 = 6$. For numbers starting with 4 or greater, any number may be in the second digit, so $n_2 = 10$. Additionally, some numbers may start with 3, so for these cases $n_1 = 1$. When paired with 3, 8 and 9 are the only numbers that make a number greater than 37, so $n_2 = 2$. There are $6 \cdot 10 = 60$ numbers and $1 \cdot 2 = 2$ numbers or $60 + 2 = 62$ possible two-digit numbers greater than 37.

27. Any one of the cards can be placed in box A, so $n_1 = 4$. Because a card cannot be placed in more than one box, there are 3 cards that can be placed in box B, so $n_2 = 3$. Similarly, there are 2 cards that can be placed in box C and 1 card in box D. By the counting principle, there are $4 \cdot 3 \cdot 2 \cdot 1 = 24$ possible placements.

29. Using the sample space found in Exercise 28, there are 15 elements where at least one card is in the box with the corresponding letter.

31. There are thirteen choices in each position, so there are $13 \cdot 13 \cdot 13 \cdot 13 = 13^4$ possible ways to choose the cards.

33. There is 1 choice for each position, so n_1, n_2, n_3, and $n_4 = 1$. By the counting principle, there are $1 \cdot 1 \cdot 1 \cdot 1 = 1$ possible ways to choose the cards.

35. Answers may vary.

37. There are 10 numbers that I can equal, so $n_1 = 10$. There are 15 numbers that J can equal, so $n_2 = 15$. By the counting principle, the instruction will have been executed $10 \cdot 15 = 150$ times.
The last number for I is 10 and the last number for J is 15. Therefore, the final value will be $10 + 15 = 25$.

EXERCISE SET 11.2

1. $8! = 8 \cdot 7 \cdot 6 \cdot 5 \cdot 4 \cdot 3 \cdot 2 \cdot 1 = 40,320$

3. Evaluating:
$$9!-5! = (9\cdot8\cdot7\cdot6\cdot5\cdot4\cdot3\cdot2\cdot1)-(5\cdot4\cdot3\cdot2\cdot1)$$
$$= 362,880-120 = 362,760$$

5. $(8-3)! = 5! = 5\cdot4\cdot3\cdot2\cdot1 = 120$

7. $8! = 8\cdot7! = 8\cdot5040 = 40,320$

9. $5! = \dfrac{6!}{6} = \dfrac{720}{6} = 120$

11. Using the formula:
$$P(n,2) = 42$$
$$\frac{n!}{(n-2)!} = 42$$
$$\frac{n(n-1)(n-2)!}{(n-2)!} = 42$$
$$n(n-1) = 42$$
$$n^2 - n = 42$$
$$n^2 - n - 42 = 0$$
$$n = 7, n = -6$$
Since n cannot be negative, $n = 7$.

13. Evaluating:
$$P(8,5) = \frac{8!}{(8-5)!} = \frac{8!}{3!} = \frac{8\cdot7\cdot6\cdot5\cdot4\cdot3!}{3!}$$
$$= 8\cdot7\cdot6\cdot5\cdot4 = 6720$$

15. Evaluating:
$$P(9,7) = \frac{9!}{(9-7)!} = \frac{9!}{2!} = \frac{9\cdot8\cdot7\cdot6\cdot5\cdot4\cdot3\cdot2!}{2!}$$
$$= 9\cdot8\cdot7\cdot6\cdot5\cdot4\cdot3 = 181,440$$

17. $P(8,0) = \dfrac{8!}{(8-0)!} = \dfrac{8!}{8!} = 1$

19. Evaluating:
$$P(8,8) = \frac{8!}{(8-8)!} = \frac{8!}{0!} = \frac{8\cdot7\cdot6\cdot5\cdot4\cdot3\cdot2\cdot1}{1}$$
$$= 8\cdot7\cdot6\cdot5\cdot4\cdot3\cdot2\cdot1 = 40,320$$

21. Evaluating:
$$P(8,2)\cdot P(5,3) = \frac{8!}{(8-2)!}\cdot\frac{5!}{(5-3)!}$$
$$= \frac{8\cdot7\cdot6!}{6!}\cdot\frac{5\cdot4\cdot3\cdot2!}{2!}$$
$$= 56\cdot60 = 3360$$

23. Evaluating:
$$\frac{P(6,0)}{P(6,6)} = \frac{\dfrac{6!}{(6-0)!}}{\dfrac{6!}{(6-6)!}} = \frac{\dfrac{6!}{6!}}{\dfrac{6!}{0!}} = \frac{1}{\dfrac{6\cdot5\cdot4\cdot3\cdot2\cdot1}{1}}$$
$$= \frac{1}{720}$$

25. Evaluating:
$$C(9,2) = \frac{9!}{2!(9-2)!} = \frac{9!}{2!7!}$$
$$= \frac{9\cdot8\cdot7!}{2!7!} = \frac{9\cdot8}{2\cdot1} = 36$$

27. Evaluating:
$$C(12,0) = \frac{12!}{0!(12-0)!} = \frac{12!}{0!12!}$$
$$= \frac{1}{1} = 1$$

29. Evaluating:
$$C(6,2)\cdot C(7,3) = \frac{6!}{2!(6-2)!}\cdot\frac{7!}{3!(7-3)!}$$
$$= \frac{6!}{2!4!}\cdot\frac{7!}{3!4!}$$
$$= \frac{6\cdot5\cdot4!}{2!4!}\cdot\frac{7\cdot6\cdot5\cdot4!}{3!4!}$$
$$= \frac{6\cdot5}{2\cdot1}\cdot\frac{7\cdot6\cdot5}{3\cdot2\cdot1}$$
$$= 15\cdot35 = 525$$

31. Evaluating:

$$\frac{C(10,4)\cdot C(5,2)}{C(15,6)}=\frac{\dfrac{10!}{4!(10-4)!}\cdot\dfrac{5!}{2!(5-2)!}}{\dfrac{15!}{6!(15-6)!}}$$

$$=\frac{\dfrac{10!}{4!6!}\cdot\dfrac{5!}{2!3!}}{\dfrac{15!}{6!9!}}$$

$$=\frac{\dfrac{10\cdot9\cdot8\cdot7\cdot6!}{4!6!}\cdot\dfrac{5\cdot4\cdot3!}{2!3!}}{\dfrac{15\cdot14\cdot13\cdot12\cdot11\cdot10\cdot9!}{6!9!}}$$

$$=\frac{\dfrac{10\cdot9\cdot8\cdot7}{4\cdot3\cdot2\cdot1}\cdot\dfrac{5\cdot4}{2\cdot1}}{\dfrac{15\cdot14\cdot13\cdot12\cdot11\cdot10}{6\cdot5\cdot4\cdot3\cdot2\cdot1}}$$

$$=\frac{210\cdot10}{5005}=\frac{60}{143}$$

33. Evaluating:

$$\frac{C(9,7)\cdot C(5,3)}{C(14,10)}=\frac{\dfrac{9!}{7!(9-7)!}\cdot\dfrac{5!}{3!(5-3)!}}{\dfrac{14!}{10!(14-10)!}}$$

$$=\frac{\dfrac{9!}{7!2!}\cdot\dfrac{5!}{3!2!}}{\dfrac{14!}{10!4!}}$$

$$=\frac{\dfrac{9\cdot8\cdot7!}{7!2!}\cdot\dfrac{5\cdot4\cdot3!}{3!2!}}{\dfrac{14\cdot13\cdot12\cdot11\cdot10!}{10!4!}}$$

$$=\frac{\dfrac{9\cdot8}{2\cdot1}\cdot\dfrac{5\cdot4}{2\cdot1}}{\dfrac{14\cdot13\cdot12\cdot11}{4\cdot3\cdot2\cdot1}}$$

$$=\frac{36\cdot10}{1001}=\frac{360}{1001}$$

35. Evaluating:

$$4!\cdot C(10,3)=4!\cdot\frac{10!}{3!(10-3)!}=4!\cdot\frac{10!}{3!7!}$$

$$=4\cdot3!\cdot\frac{10\cdot9\cdot8\cdot7!}{3!7!}$$

$$=4\cdot10\cdot9\cdot8=2880$$

37. Using the formula:

$$C(7,5)=\frac{7!}{5!(7-5)!}=\frac{7!}{5!2!}$$

$$=\frac{7\cdot6\cdot5!}{5!2!}=\frac{7\cdot6}{2\cdot1}=21$$

39. Using the formula:

$$C(12,7)=\frac{12!}{7!(12-7)!}=\frac{12!}{7!5!}$$

$$=\frac{12\cdot11\cdot10\cdot9\cdot8\cdot7!}{7!5!}$$

$$=\frac{12\cdot11\cdot10\cdot9\cdot8}{5\cdot4\cdot3\cdot2\cdot1}=792$$

41. $C(11,11)=\dfrac{11!}{11!(11-11)!}=\dfrac{11!}{11!0!}=\dfrac{1}{1}=1$

43. No. If there are only 7 items, there is no way to choose 9 of them.

45. The order in which the songs are played is important, so the number of ways to order the songs is

$$P(5,5)=\frac{5!}{(5-5)!}=\frac{5!}{0!}=\frac{5\cdot4\cdot3\cdot2\cdot1}{1}$$
$$=5\cdot4\cdot3\cdot2\cdot1=120$$

47. Because each officer position is different, the order of the selection is important, so the number of ways to make the selection is

$$P(16,4)=\frac{16!}{(16-4)!}=\frac{16!}{12!}$$
$$=\frac{16\cdot15\cdot14\cdot13\cdot12!}{12!}$$
$$=16\cdot15\cdot14\cdot13=43{,}680$$

49. The order in which the student reads the books is not important, so the number of ways to choose the books is

$$C(7,3)=\frac{7!}{3!(7-3)!}=\frac{7!}{3!4!}$$
$$=\frac{7\cdot6\cdot5\cdot4!}{3!4!}=\frac{7\cdot6\cdot5}{3\cdot2\cdot1}=35$$

51. The order in which the players are selected for testing is not important, so the number of groups of players that can be chosen is:

$$C(15,3) = \frac{15!}{3!(15-3)!} = \frac{15!}{3!12!}$$

$$= \frac{15 \cdot 14 \cdot 13 \cdot 12!}{3!12!}$$

$$= \frac{15 \cdot 14 \cdot 13}{3 \cdot 2 \cdot 1} = 455$$

53. 16!

55. Because the choice of team A playing team B is the same as the choice of team B playing team A, the order of the selection is not important, so the number of different games is

$$C(8,2) = \frac{8!}{2!(8-2)!} = \frac{8!}{2!6!}$$

$$= \frac{8 \cdot 7 \cdot 6!}{2!6!} = \frac{8 \cdot 7}{2 \cdot 1} = 28$$

57. Because the choice of team A playing team B is the same as the choice of team B playing team A, the order of the selection is not important, so the number of different ways to match the teams is

$$C(6,2) = \frac{6!}{2!(6-2)!} = \frac{6!}{2!4!}$$

$$= \frac{6 \cdot 5 \cdot 4!}{2!4!} = \frac{6 \cdot 5}{2 \cdot 1} = 15$$

Since the teams must play each other twice, the number of games is $15 \cdot 2 = 30$

59. The order in which the heads and tails are chosen is not important, so the number of distinct arrangements is

$$C(10,5) = \frac{10!}{5!(10-5)!} = \frac{10!}{5!5!}$$

$$= \frac{10 \cdot 9 \cdot 8 \cdot 7 \cdot 6 \cdot 5!}{5!5!}$$

$$= \frac{10 \cdot 9 \cdot 8 \cdot 7 \cdot 6}{5 \cdot 4 \cdot 3 \cdot 2 \cdot 1} = 252$$

61. Because it takes 2 points to draw a line, and because the choice of points A and B is the same as the choice of points B and A, the order of the selection is not important, so the number of different lines is

$$C(7,2) = \frac{7!}{2!(7-2)!} = \frac{7!}{2!5!}$$

$$= \frac{7 \cdot 6 \cdot 5!}{2!5!} = \frac{7 \cdot 6}{2 \cdot 1} = 21$$

63. There are 6 points and we must choose 2 to form a diagonal. Because order is not important, we find C(6,2).

$$C(6,2) = \frac{6!}{2!(6-2)!} = \frac{6!}{2!4!}$$

$$= \frac{6 \cdot 5 \cdot 4!}{2!4!} = \frac{6 \cdot 5}{2 \cdot 1} = 15$$

Because 6 of the lines made are the sides of the hexagon, the number of diagonals is $15 - 6 = 9$.

65. The order in which the players are selected is not important, so the number of different teams is

$$C(18,9) = \frac{18!}{9!(18-9)!} = \frac{18!}{9!9!}$$

$$= \frac{18 \cdot 17 \cdot 16 \cdot 15 \cdot 14 \cdot 13 \cdot 12 \cdot 11 \cdot 10 \cdot 9!}{9!9!}$$

$$= \frac{18 \cdot 17 \cdot 16 \cdot 15 \cdot 14 \cdot 13 \cdot 12 \cdot 11 \cdot 10}{9 \cdot 8 \cdot 7 \cdot 6 \cdot 5 \cdot 4 \cdot 3 \cdot 2 \cdot 1}$$

$$= 48,620$$

67. There are C(12,2) ways to choose the first group, C(10, 2) ways to choose the second group, C(8, 2) ways to choose the third group, C(6, 2) ways to choose the fourth group, C(4, 2) ways to choose the fifth group and C(2, 2) ways to choose the last group.
By the counting principle, the number of ways to divide the people is

$$C(12,2) \cdot C(10,2) \cdot C(8,2) \cdot C(6,2)$$
$$\cdot C(4,2) \cdot C(2,2)$$

$$= \frac{12!}{2!(12-2)} \cdot \frac{10!}{2!(10-2)!} \cdot \frac{8!}{2!(8-2)!}$$
$$\cdot \frac{6!}{2!(6-2)!} \cdot \frac{4!}{2!(4-2)!} \cdot \frac{2!}{2!(2-2)!}$$

$$= \frac{12!}{2!10!} \cdot \frac{10!}{2!8!} \cdot \frac{8!}{2!6!} \cdot \frac{6!}{2!4!} \cdot \frac{4!}{2!2!} \cdot \frac{2!}{2!0!}$$

$$= \frac{12 \cdot 11 \cdot 10!}{2!10!} \cdot \frac{10 \cdot 9 \cdot 8!}{2!8!} \cdot \frac{8 \cdot 7 \cdot 6!}{2!6!} \cdot \frac{6 \cdot 5 \cdot 4!}{2!4!}$$
$$\cdot \frac{4 \cdot 3 \cdot 2!}{2!2!} \cdot \frac{2!}{2!0!}$$

$$= \frac{12 \cdot 11}{2 \cdot 1} \cdot \frac{10 \cdot 9}{2 \cdot 1} \cdot \frac{8 \cdot 7}{2 \cdot 1} \cdot \frac{6 \cdot 5}{2 \cdot 1} \cdot \frac{4 \cdot 3}{2 \cdot 1} \cdot \frac{1}{1}$$

$$= 66 \cdot 45 \cdot 28 \cdot 15 \cdot 6 \cdot 1 = 7,484,400$$

69. We are looking for the number of permutations of the letters *committee*. With $n = 9$ (number of letters), $k_1 = 1$ (number of c's), $k_2 = 1$ (number of o's), $k_3 = 2$ (number of m's), $k_4 = 1$ (number of i's), $k_5 = 2$ (number of t's) and $k_6 = 2$ (number of e's), we have

$$\frac{9!}{1! \cdot 1! \cdot 2! \cdot 1! \cdot 2! \cdot 2!} = 45,360$$

There are 45,360 different arrangements possible.

71. The order in which the firefighters are selected is not important, so the number of different teams is

$$C(24,8) = \frac{24!}{8!(24-8)!} = \frac{24!}{8!16!}$$

$$= \frac{24 \cdot 23 \cdot 22 \cdot 21 \cdot 20 \cdot 19 \cdot 18 \cdot 17 \cdot 16!}{8!16!}$$

$$= \frac{24 \cdot 23 \cdot 22 \cdot 21 \cdot 20 \cdot 19 \cdot 18 \cdot 17}{8 \cdot 7 \cdot 6 \cdot 5 \cdot 4 \cdot 3 \cdot 2 \cdot 1}$$

$$= 735,471$$

73. There are $C(4, 4)$ ways of choosing four aces from four aces and 48 choices for the fifth card in the hand. By the counting principle, the number of hands containing 4 aces is

$$C(4,4) \cdot 48 = \frac{4!}{4!(4-4)!} \cdot 48$$

$$= \frac{4!}{4!0!} \cdot 48 = \frac{1}{1} \cdot 48 = 48$$

75. There are $C(4, 3)$ ways of choosing three jacks from four jacks and $C(48, 2)$ ways of choosing the other 2 cards from the remaining 48 cards. By the counting principle, the number of hands containing exactly three jacks is

$$C(4,3) \cdot C(48,2) = \frac{4!}{3!(4-3)!} \cdot \frac{48!}{2!(48-2)!}$$

$$= \frac{4!}{3!1!} \cdot \frac{48!}{2!46!}$$

$$= \frac{4 \cdot 3!}{3!1!} \cdot \frac{48 \cdot 47 \cdot 46!}{2!46!}$$

$$= \frac{4}{1} \cdot \frac{48 \cdot 47}{2 \cdot 1}$$

$$= 4 \cdot 1128 = 4512$$

77. There are $C(4, 2)$ ways of choosing two sevens from four sevens and $C(48, 3)$ ways of choosing the other 3 cards from the remaining 48 cards.

By the counting principle, the number of hands containing exactly two sevens is

$$C(4,2) \cdot C(48,3) = \frac{4!}{2!(4-2)!} \cdot \frac{48!}{3!(48-3)!}$$

$$= \frac{4!}{2!2!} \cdot \frac{48!}{3!45!}$$

$$= \frac{4 \cdot 3 \cdot 2!}{2!2!} \cdot \frac{48 \cdot 47 \cdot 46 \cdot 45!}{3!45!}$$

$$= \frac{4 \cdot 3}{2 \cdot 1} \cdot \frac{48 \cdot 47 \cdot 46}{3 \cdot 2 \cdot 1}$$

$$= 6 \cdot 17,296 = 103,776$$

79. Answers may vary.

81. The coefficients are

$$C(5,0) = \frac{5!}{0!(5-0)!} = \frac{5!}{0!5!} = 1$$

$$C(5,1) = \frac{5!}{1!(5-1)!} = \frac{5!}{1!4!}$$

$$= \frac{5 \cdot 4!}{1!4!} = \frac{5}{1} = 5$$

$$C(5,2) = \frac{5!}{2!(5-2)!} = \frac{5!}{2!3!}$$

$$= \frac{5 \cdot 4 \cdot 3!}{2!3!} = \frac{5 \cdot 4}{2 \cdot 1} = 10$$

$$C(5,3) = \frac{5!}{3!(5-3)!} = \frac{5!}{3!2!}$$

$$= \frac{5 \cdot 4 \cdot 3!}{3!2!} = \frac{5 \cdot 4}{2 \cdot 1} = 10$$

$$C(5,4) = \frac{5!}{4!(5-4)!} = \frac{5!}{4!1!}$$

$$= \frac{5 \cdot 4!}{4!1!} = \frac{5}{1} = 5$$

$$C(5,5) = \frac{5!}{5!(5-5)!} = \frac{5!}{5!0!} = 1$$

Therefore, the expansion of $(x + y)^5$ is $x^5 + 5x^4y + 10x^3y^2 + 10x^2y^3 + 5xy^4 + y^5$.

EXERCISE SET 11.3

1. $S = \{$HHH, HHT, HTH, HTT, THH, THT, TTH, TTT$\}$

3. $S = \{$Nov. 1, Nov. 2, Nov. 3, Nov. 4, Nov. 5, Nov. 6, Nov. 7, Nov. 8, Nov. 9, Nov. 10, Nov. 11, Nov. 12, Nov. 13, Nov. 14$\}$

5. $S = \{$Alaska, Alabama, Arizona, Arkansas$\}$

7. $S = \{$BBB, BBG, BGB, BGG, GBB, GBG, GGB, GGG$\}$

9. $S = \{$BGG, GBG, GGB, GGG$\}$

11. $S = \{$BBG, BGB, BGG, GBB, GBG, GGB, GGG$\}$

13. The sample space was found in Exercise 7. The elements in the event that the Lins will have at least two girls was found in Exercise 9. Then the probability is $P(E) = \dfrac{n(E)}{n(S)} = \dfrac{4}{8} = \dfrac{1}{2}$.

15. The sample space was found in Exercise 7. The elements in the event that the Lins will have at least one girl are $\{$BBG, BGB, GBB, BGG, GBG, GGB, GGG$\}$. Then the probability is $P(E) = \dfrac{n(E)}{n(S)} = \dfrac{7}{8}$.

17. The sample space for tossing a coin four times is $\{$HHHH, HHHT, HHTH, HHTT, HTHH, HTHT, HTTH, HTTT, THHH, THHT, THTH, THTT, TTHH, TTHT, TTTH, TTTT$\}$. The elements in the event that one head and three tails are tossed are $\{$HTTT, THTT, TTHT, TTTH$\}$. Then the probability is $P(E) = \dfrac{n(E)}{n(S)} = \dfrac{4}{16} = \dfrac{1}{4}$.

19. The sample space for tossing a coin four times was found in Exercise 17. The elements in the event that all four coins tossed are identical are $\{$HHHH, TTTT$\}$. Then the probability is $P(E) = \dfrac{n(E)}{n(S)} = \dfrac{2}{16} = \dfrac{1}{8}$.

21. The sample space for tossing a coin four times was found in Exercise 17. The elements in the event that at least two tails are tossed are $\{$HHTT, HTHT, HTTH, HTTT, THHT, THTH, THTT, TTHH, TTHT, TTTH, TTTT$\}$. Then the probability is $P(E) = \dfrac{n(E)}{n(S)} = \dfrac{11}{16}$.

23. The sample space for tossing a dodecahedral die is $\{1, 2, 3, 4, 5, 6, 7, 8, 9, 10, 11, 12\}$. The element in the event that the number on the upward face is 12 is $\{12\}$. Then the probability is

$P(E) = \dfrac{n(E)}{n(S)} = \dfrac{1}{12}$.

25. The sample space for tossing a dodecahedral die was found in Exercise 23. The elements in the event that the number on the upward face is divisible by 4 are $\{4, 8, 12\}$. Then the probability is

$P(E) = \dfrac{n(E)}{n(S)} = \dfrac{3}{12} = \dfrac{1}{4}$.

27. The sample space for tossing two six-sided dice was found in Figure 11.6. The dice must be considered as distinct, so there are 36 possible outcomes. The elements in the event that the sum of the pips on the upward faces is 6 are

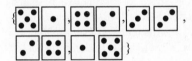

Then the probability is

$P(E) = \dfrac{n(E)}{n(S)} = \dfrac{5}{36}$.

29. The sample space for tossing two six-sided dice was found in Figure 11.6. The dice must be considered as distinct, so there are 36 possible outcomes. The element in the event that the sum of the pips on the upward faces is 2 is

Then the probability is $P(E) = \dfrac{n(E)}{n(S)} = \dfrac{1}{36}$.

31. The sample space for tossing two six-sided dice was found in Figure 11.6. The dice must be considered as distinct, so there are 36 possible outcomes. There are no elements in the event that the sum of the pips on the upward faces is 1.

Then the probability is $P(E) = \dfrac{n(E)}{n(S)} = \dfrac{0}{36} = 0$.

33. The sample space for tossing two six-sided dice was found in Figure 11.6. The dice must be considered as distinct, so there are 36 possible outcomes. The elements in the event that the sum of the pips on the upward faces is at least 10 are

(6 and 6, 6 and 5, 6 and 4, 5 and 6, 5 and 5, 4 and 6}. Then the probability is

$$P(E) = \frac{n(E)}{n(S)} = \frac{6}{36} = \frac{1}{6}.$$

35. The sample space for tossing two six-sided dice was found in Figure 11.6. The dice must be considered as distinct, so there are 36 possible outcomes. The elements in the event that the sum of the pips on the upward faces is an even number are {6 and 6, 6 and 4, 6 and 2, 5 and 5, 5 and 3, 5 and 1, 4 and 6, 4 and 4, 4 and 2, 3 and 5, 3 and 3, 3 and 1, 2 and 6, 2 and 4, 2 and 2, 1 and 5, 1 and 3, 1 and 1}.

Then the probability is $P(E) = \dfrac{n(E)}{n(S)} = \dfrac{18}{36} = \dfrac{1}{2}.$

37. The sample space for tossing two six-sided dice was found in Figure 11.6. The dice must be considered as distinct, so there are 36 possible outcomes. The elements in the event that both dice show the same number of pips are {6 and 6, 5 and 5, 4 and 4, 3 and 3, 2 and 2, 1 and 1}. Then

the probability is $P(E) = \dfrac{n(E)}{n(S)} = \dfrac{6}{36} = \dfrac{1}{6}.$

39. Because there are 52 cards in a standard deck of playing cards, there are 52 items in the sample space. The elements in the event that a red card is drawn include 13 cards from the heart suit and 13 cards from the diamonds suit. Then the

probability is $P(E) = \dfrac{n(E)}{n(S)} = \dfrac{26}{52} = \dfrac{1}{2}.$

41. Because there are 52 cards in a standard deck of playing cards, there are 52 items in the sample space. The elements in the event that a nine is drawn are {9 hearts, 9 diamonds, 9 spades, 9 clubs}. Then the probability is

$$P(E) = \frac{n(E)}{n(S)} = \frac{4}{52} = \frac{1}{13}.$$

43. Because there are 52 cards in a standard deck of playing cards, there are 52 items in the sample space. The elements in the event that a card between 5 and 9 is drawn include the 6, 7, and 8 in each of four suits. Then the probability is

$$P(E) = \frac{n(E)}{n(S)} = \frac{12}{52} = \frac{3}{13}.$$

45. Let E be the event that the voter is a Democrat. Then $P(E) = \dfrac{1267}{3228}.$

47. Let E be the event that the voter is 50 years old or older. Then $P(E) = \dfrac{804}{3228} = \dfrac{67}{269}.$

49. Let E be the event that the voter is an Independent. Then $P(E) = \dfrac{150}{3228} = \dfrac{25}{538}.$

51. Let E be the event that the respondent did not complete high school. Then

$$P(E) = \frac{52}{850} = \frac{26}{425}.$$

53. Let E be the event that the respondent has a Ph.D. or professional degree. Then

$$P(E) = \frac{36}{850} = \frac{18}{425}.$$

55. Let E be the event that the respondent earns from $36,000 to $45,999 annually. Then $P(E) = \dfrac{58}{293}.$

57. Let E be the event that the respondent earns at least $80,000 annually. There are $22 + 14 = 36$ people in the survey that earn at least $80,000 annually. Then

$$P(E) = \frac{36}{293}.$$

59. There are 30 answers total. Two are desirable and the other 28 are undesirable. The odds of choosing a Daily Double are 2:28 or 1:14.

61. To have white flowers, the plant must be *rr*. From the table, only one of the four possible genotypes is *rr*, so the probability that an offspring will have white flowers is $\dfrac{1}{4}.$

63. Make a Punnett square.

Parents	*E*	*E*
e	*Ee*	*Ee*
e	*Ee*	*Ee*

To have red eyes, the mouse must be *ee*. From the table, none of the four possible genotypes is *ee*, so the probability that an offspring will have red eyes is 0.

65. Answers may vary.

67. Use the formula for calculating the probability of an event when the odds in favor are known.
$$P(E) = \frac{a}{a+b} = \frac{1}{1+2} = \frac{1}{3}$$

69. Use the formula for calculating the probability of an event when the odds in favor are known.
$$P(E) = \frac{a}{a+b} = \frac{3}{3+7} = \frac{3}{10}$$

71. Use the formula for calculating the probability of an event when the odds in favor are known.
$$P(E) = \frac{a}{a+b} = \frac{8}{8+5} = \frac{8}{13}$$

73. The odds in favor are
$$\frac{P(E)}{1-P(E)} = \frac{0.2}{1-0.2} = \frac{0.2}{0.8} = \frac{1}{4}$$
The odds in favor are 1 to 4.

75. The odds in favor are
$$\frac{P(E)}{1-P(E)} = \frac{0.375}{1-0.375} = \frac{0.375}{0.625} = \frac{3}{5}$$
The odds in favor are 3 to 5.

77. The odds in favor are
$$\frac{P(E)}{1-P(E)} = \frac{0.55}{1-0.55} = \frac{0.55}{0.45} = \frac{11}{9}$$
The odds in favor are 11 to 9.

79. Let E be the event of rolling a sum of 9. Refer to Figure 11.6. Because there are four ways to roll a sum of 9, there are four favorable outcomes, leaving 32 unfavorable outcomes.
Odds in favor of $E = \frac{4}{32} = \frac{1}{8}$
The odds in favor are 1 to 8.

81. Let E be the event of pulling a heart. Because there are thirteen ways to draw a heart, there are thirteen favorable outcomes, leaving 39 unfavorable outcomes.
Odds in favor of $E = \frac{13}{39} = \frac{1}{3}$
The odds in favor are 1 to 3.

83. Because the odds against are 8 to 3, the odds in favor are 3 to 8. Use the formula for calculating the probability of an event when the odds in favor are known.
$$P(E) = \frac{a}{a+b} = \frac{3}{3+8} = \frac{3}{11}$$

85. a. Let E be the event of getting a green M&M. Because there are 7 green M&M's, there are 7 favorable outcomes, leaving 50 unfavorable outcomes.
Odds in favor of $E = \frac{7}{50}$
The odds in favor are 7 to 50.

   b. Because there are 57 M&M's in the bag, there are 57 items in the sample space. Let E be the event of choosing a green M&M. There are 7 green M&M's so there are 7 items in E. Then the probability is
$$P(E) = \frac{n(E)}{n(S)} = \frac{7}{57}.$$

87. By making a tree diagram, we can find the sample space for the possible arrangements of the cards to be {ABCD, ABDC, ACBD, ACDB, ADBC, ADCB, BACD, BADC, BCAD, BCDA, BDAC, BDCA, CABD, CADB, CBAD, CBDA, CDAB, CDBA, DABC, DACB, DBAC, DBCA, DCAB, DCBA}. Assuming the boxes are in the order ABCD, the elements in the event that no card will be in a box with the same letter are {BADC, BCDA, BDAC, CADB, CDAB, CDBA, DABC, DCAB, DCBA}. Then the probability is
$$P(E) = \frac{n(E)}{n(S)} = \frac{9}{24} = \frac{3}{8}.$$

89. Answers may vary.

91. There are C(52, 5) different hands of five cards possible.
$$C(52,5) = \frac{52!}{5!(52-5)!} = \frac{52!}{5!47!}$$
$$= \frac{52\cdot51\cdot50\cdot49\cdot48\cdot47!}{5!47!}$$
$$= \frac{52\cdot51\cdot50\cdot49\cdot48}{5\cdot4\cdot3\cdot2\cdot1} = 2,598,960$$
There are $C(4, 3)$ ways of choosing three jacks from four jacks and $C(4, 2)$ ways of choosing two queens from four queens. By the counting

principle, the number of hands containing three jacks and two queens is

$$C(4,3) \cdot C(4,2) = \frac{4!}{3!(4-3)!} \cdot \frac{4!}{2!(4-2)!}$$

$$= \frac{4!}{3! 1!} \cdot \frac{4!}{2! 2!}$$

$$= \frac{4 \cdot 3!}{3! 1!} \cdot \frac{4 \cdot 3 \cdot 2!}{2! 2!}$$

$$= \frac{4}{1} \cdot \frac{4 \cdot 3}{2 \cdot 1}$$

$$= 4 \cdot 6 = 24$$

Then the probability is

$$P(E) = \frac{n(E)}{n(S)} = \frac{24}{2,598,960} = \frac{1}{108,290}.$$

EXERCISE SET 11.4

1. Answers may vary.

3. Let A = {♥4, ♥4, ♦4, ♣4} and B = {♠A, ♥A, ♦A, ♣A}. The events in A and B cannot occur at the same time, so they are mutually exclusive. There are 52 cards in a standard deck, thus $n(S)$ = 52. The probability is

$$P(A \text{ or } B) = P(A) + P(B) = \frac{1}{13} + \frac{1}{13} = \frac{2}{13}.$$

5. Let A be the event of rolling a 2, and let B be the event of rolling a 10. The events in A and B cannot occur at the same time, so they are mutually exclusive. From the sample space on page 721,

$P(A) = \frac{1}{36}$ and $P(B) = \frac{3}{36} = \frac{1}{12}$. Because A and B are mutually exclusive events,

$$P(A \text{ or } B) = P(A) + P(B) = \frac{1}{36} + \frac{1}{12} = \frac{4}{36} = \frac{1}{9}.$$

7. $P(A \text{ or } B) = P(A) + P(B) - P(A \text{ and } B)$
   $= 0.2 + 0.5 - 0.1$
   $= 0.6$

9. $P(A \cap B) = P(A) + P(B) - P(A \cup B)$
   $= 0.3 + 0.8 - 0.9$
   $= 0.2$

11. Let A be the event of the number being more than 6, and let B be the event of the number being odd. The sample space consists of 10 numbers. $n(A) = 4$, $n(B) = 5$ and $n(A \text{ and } B) = 2$.

$$P(A \text{ or } B) = P(A) + P(B) - P(A \text{ and } B)$$
$$= \frac{4}{10} + \frac{5}{10} - \frac{2}{10} = \frac{7}{10}$$

13. Let A be the event of the number being even, and let B be the event of the number being prime. The sample space consists of 10 numbers. $n(A) = 5$, $n(B) = 4$ and $n(A \text{ and } B) = 1$.
$$P(A \text{ or } B) = P(A) + P(B) - P(A \text{ and } B)$$
$$= \frac{4}{10} + \frac{5}{10} - \frac{1}{10} = \frac{8}{10} = \frac{4}{5}$$

15. Let A be the event of rolling a 6, and let B be the event of rolling doubles. From the figure on page 721, there are 36 elements in the sample space. $n(A) = 5$, $n(B) = 6$ and $n(A \text{ and } B) = 1$.
$$P(A \text{ or } B) = P(A) + P(B) - P(A \text{ and } B)$$
$$= \frac{5}{36} + \frac{6}{36} - \frac{1}{36} = \frac{10}{36} = \frac{5}{18}$$

17. Let A be the event of rolling an even number, and let B be the event of rolling doubles. From the figure on page 721, there are 36 elements in the sample space. $n(A) = 18$, $n(B) = 6$ and $n(A$ and $B) = 6$.
$$P(A \text{ or } B) = P(A) + P(B) - P(A \text{ and } B)$$
$$= \frac{18}{36} + \frac{6}{36} - \frac{6}{36} = \frac{18}{36} = \frac{1}{2}$$

19. Let A be the event of rolling an odd number, and let B be the event of rolling a number less than 6. From the figure on page 721, there are 36 elements in the sample space. $n(A) = 18$, $n(B) = 10$ and $n(A \text{ and } B) = 6$.
$$P(A \text{ or } B) = P(A) + P(B) - P(A \text{ and } B)$$
$$= \frac{18}{36} + \frac{10}{36} - \frac{6}{36} = \frac{22}{36} = \frac{11}{18}$$

21. Let A be the event of drawing an 8, and let B be the event of drawing a spade. The sample space consists of 52 cards. $n(A) = 4$, $n(B) = 13$ and $n(A \text{ and } B) = 1$.
$$P(A \text{ or } B) = P(A) + P(B) - P(A \text{ and } B)$$
$$= \frac{4}{52} + \frac{13}{52} - \frac{1}{52} = \frac{16}{52} = \frac{4}{13}$$

23. Let A be the event of drawing a jack, and let B be the event of drawing a face card. The sample space consists of 52 cards. $n(A) = 4$, $n(B) = 12$ and $n(A \text{ and } B) = 4$. So

$$P(A \text{ or } B) = P(A) + P(B) - P(A \text{ and } B)$$
$$= \frac{4}{52} + \frac{12}{52} - \frac{4}{52} = \frac{12}{52} = \frac{3}{13}$$

25. Let A be the event of drawing a diamond, and let B be the event of drawing a black card. The sample space consists of 52 cards. $n(A) = 13$, $n(B) = 26$ and $n(A \text{ and } B) = 0$.
$$P(A \text{ or } B) = P(A) + P(B)$$
$$= \frac{13}{52} + \frac{26}{52} = \frac{39}{52} = \frac{3}{4}$$

27. Let $A =$ {people aged 26-34}, and let $B =$ {people employed part-time}. Then, from the table, $n(A) = 348 + 67 + 27 = 442$, $n(B) = 164 + 203 + 67 + 179 + 162 = 775$ and $n(A \text{ and } B) = 67$. The total number of people represented in the table is 3179.
$$P(A \text{ or } B) = P(A) + P(B) - P(A \text{ and } B)$$
$$= \frac{442}{3179} + \frac{775}{3179} - \frac{67}{3179} = \frac{1150}{3179}$$

29. Let $A =$ {people under 18}, and let $B =$ {people employed part-time}. Then, from the table, $n(A) = 24 + 164 + 371 = 559$, $n(B) = 164 + 203 + 67 + 179 + 162 = 775$ and $n(A \text{ and } B) = 164$. The total number of people represented in the table is 3179.
$$P(A \text{ or } B) = P(A) + P(B) - P(A \text{ and } B)$$
$$= \frac{559}{3179} + \frac{775}{3179} - \frac{164}{3179} = \frac{1170}{3179}$$

31. If E is the event of winning the contest, then E^C is the event of not winning the contest, and
$$P(E^C) = 1 - P(E) = 1 - 0.04 = 0.96$$

33. If E is the event of an individual being involved in a car accident, then E^C is the event of not being involved in a car accident, and
$$P(E^C) = 1 - P(E) = 1 - \frac{1}{2500} = \frac{2499}{2500}$$

35. If E is the event of tossing a 7, then E^C is the event of not tossing a 7, and
$$P(E^C) = 1 - P(E) = 1 - \frac{6}{36} = \frac{30}{36} = \frac{5}{6}$$

37. If E is the event of tossing a sum less than 4, then E^C is the event of tossing a sum of at least four, and

$$P(E^C) = 1 - P(E) = 1 - \frac{3}{36} = \frac{33}{36} = \frac{11}{12}$$

39. If E is the event of drawing an ace, then E^C is the event of not drawing an ace, and
$$P(E^C) = 1 - P(E) = 1 - \frac{4}{52} = \frac{48}{52} = \frac{12}{13}$$

41. If E is the event of getting no tails, then E^C is the event of getting at least one tail, and
$$P(E^C) = 1 - P(E) = 1 - \frac{1}{16} = \frac{15}{16}$$

43. Let $E =$ {at least one 1}; then $E^C =$ {no 1's}. Because on each toss of the die there are six possible outcomes, $n(S) = 6 \cdot 6 \cdot 6 = 216$. On each toss of the die there are five numbers that are not 1's. Therefore, $n(E^C) = 5 \cdot 5 \cdot 5 = 125$.
$$P(E) = 1 - P(E^C) = 1 - \frac{125}{216} = \frac{91}{216} \approx 0.421$$
The probability is 42.1%.

45. Let $E =$ {at least one roll of sum of 8}; then $E^C =$ {no sum of 8 is rolled}. Using the table on page 721, there are 36 possibilities for each toss of the dice. Thus, $n(S) = 36 \cdot 36 \cdot 36 = 46,656$. On each toss of the dice there are 31 numbers that do not total 8, so $n(E^C) = 31 \cdot 31 \cdot 31 = 29,791$.
$$P(E) = 1 - P(E^C) = 1 - \frac{29,791}{46,656} = \frac{16,865}{46,656} \approx 0.361$$
The probability is 36.1%.

47. Let $E =$ {at least one face card is drawn}; then $E^C =$ {no face card is drawn}. There are 52 possibilities for each card drawn. Thus, $n(S) = 52 \cdot 52 \cdot 52 = 140,608$. Each time a card is drawn, there are 40 cards that are not face cards, so $n(E^C) = 40 \cdot 40 \cdot 40 = 64,000$.
$$P(E) = 1 - P(E^C)$$
$$= 1 - \frac{64,000}{140,608} = \frac{76,608}{140,608} \approx 0.545$$
The probability is 54.5%.

49. Let $E =$ {at least one face card is drawn}; then $E^C =$ {no face card is drawn}. There are 52 possibilities for the first card drawn, 51 possibilities for the second card drawn and 50 for the third. Thus, $n(S) = 52 \cdot 51 \cdot 50 = 132,600$. The first time a card is drawn, there are 40 cards that are not face cards, the second time there are 39

cards that are not face cards and 38 cards the third time, so $n(E^C) = 40 \cdot 39 \cdot 38 = 59,280$.

$P(E) = 1 - P(E^C)$

$= 1 - \dfrac{59,280}{132,600} = \dfrac{73,320}{132,600} \approx 0.553$

The probability is 55.3%.

51. Let $E = \{$at least one tape is defective$\}$; then $E^C = \{$no tapes are defective$\}$. The number of elements in the sample space is

$n(S) = C(30,12) = \dfrac{30!}{12!(30-12)!}$

$= \dfrac{30!}{12!18!} = 86,493,225.$

To find the number of outcomes that contain no defective tapes, all of them must come from the 26 nondefective tapes. The number of elements is

$n(E^C) = C(26,12) = \dfrac{26!}{12!(26-12)!}$

$= \dfrac{26!}{12!14!} = 9,657,700$

$P(E) = 1 - P(E^C)$

$= 1 - \dfrac{9,657,700}{86,493,225} = \dfrac{76,835,525}{86,493,225} \approx 0.888.$

The probability is 88.8%.

53. Let $E = \{$at least one employee wins a prize$\}$; then $E^C = \{$no employee wins a prize$\}$. The number of elements in the sample space is

$n(S) = C(42,3) = \dfrac{42!}{3!(42-3)!}$

$= \dfrac{42!}{3!39!} = 11,480.$

To find the number of outcomes where no employee wins, all 3 must come from the 39 non-employee cards. The number of elements is

$n(E^C) = C(39,3) = \dfrac{39!}{3!(39-3)!}$

$= \dfrac{39!}{3!36!} = 9139.$

$P(E) = 1 - P(E^C)$

$= 1 - \dfrac{9139}{11,480} = \dfrac{2341}{11,480} \approx 0.204$

The probability is 20.4%.

55. No, the ratio of the cards is still the same.

57. There are 2 possible settings for each of the 8 digits, so by the counting principle, there are $2 \cdot 2 \cdot 2 \cdot 2 \cdot 2 \cdot 2 \cdot 2 \cdot 2 = 256$ possible settings. Because there are less than 300 settings possible, the probability that at least two homeowners will set their switches to the same code is 100%.

59. Answers may vary.

EXERCISE SET 11.5

1. Answers may vary.

3. $P(A|B) = \dfrac{P(A \cap B)}{P(B)} = \dfrac{0.25}{0.4} = 0.625$

$P(B|A) = \dfrac{P(A \cap B)}{P(A)} = \dfrac{0.25}{0.7} \approx 0.357$

5. $P(A|B) = \dfrac{P(A \cap B)}{P(B)} = \dfrac{0.07}{0.18} \approx 0.389$

$P(B|A) = \dfrac{P(A \cap B)}{P(A)} = \dfrac{0.07}{0.61} \approx 0.115$

7. Let $B = \{$people employed part-time$\}$ and $A = \{$people between 35 and 49$\}$. From the table, $n(A \cap B) = 179$, $n(A) = 581 + 179 + 104 = 864$, and $n(S) = 3179$. We have

$P(B|A) = \dfrac{P(A \cap B)}{P(A)} = \dfrac{\frac{179}{3179}}{\frac{864}{3179}} = \dfrac{179}{864}.$

9. Let $B = \{$people employed full-time$\}$ and $A = \{$people between 18 and 49$\}$. From the table, $n(A \cap B) = 185 + 348 + 581 = 1114$, $n(A) = 185 + 203 + 148 + 348 + 67 + 27 + 581 + 179 + 104 = 1842$, and $n(S) = 3179$. We have

$P(B|A) = \dfrac{P(A \cap B)}{P(A)} = \dfrac{\frac{1114}{3179}}{\frac{1842}{3179}} = \dfrac{1114}{1842} = \dfrac{557}{921}.$

11. There are 2000 people in the sample space. The number of elements in the event that a participant prefers the Playstation 2 system is 607. Then the probability is

$P(E) = \dfrac{n(E)}{n(S)} = \dfrac{607}{2000} \approx 0.30.$

13. Let B = {participants who prefer Nintendo GameCube} and A = {participants between 13 and 24}. From the table, $n(A \cap B) = 92 + 83 = 175$, $n(A) = 105 + 139 + 92 + 113 + 248 + 217 + 83 + 169 = 1166$, and $n(S) = 2000$. We have

$$P(B|A) = \frac{P(A \cap B)}{P(A)} = \frac{\frac{175}{2000}}{\frac{1166}{2000}} \approx 0.15.$$

15. Let B = {the sum is 8} and A = {the sum is even}. From the table on page 721, there are five possible rolls of the dice for which the sum is 8 and the sum is even. So $P(A \cap B) = \frac{5}{36}$. There are 18 possibilities for which the sum is even, so $P(A) = \frac{18}{36} = \frac{1}{2}$. We have

$$P(B|A) = \frac{P(A \cap B)}{P(A)} = \frac{\frac{5}{36}}{\frac{1}{2}} \approx 0.28.$$

17. Let B = {doubles} and A = {sum less than 7}. From the table on page 721, there are three possible rolls of the dice for which the roll is doubles and the sum is less than 7. So $P(A \cap B) = \frac{3}{36} = \frac{1}{12}$. There are 15 possibilities for which the sum is less than 7, so $P(A) = \frac{15}{36} = \frac{5}{12}$. We have

$$P(B|A) = \frac{P(A \cap B)}{P(A)} = \frac{\frac{1}{12}}{\frac{5}{12}} = 0.20.$$

19. Let A = {a face card is drawn first} and B = {a face card is drawn second}. On the first draw, there are 12 face cards in the deck of 52 cards. Therefore, $P(A) = \frac{12}{52} = \frac{3}{13}$. On the second draw, there are only 51 cards remaining and only 11 face cards. Therefore, $P(B|A) = \frac{11}{51}$. We have

$$P(A \cap B) = P(A) \cdot P(B | A)$$
$$= \frac{3}{13} \cdot \frac{11}{51} = \frac{33}{663} \approx 0.05.$$

21. Let A = {a spade is drawn first} and B = {a red card is drawn second}. On the first draw, there are 13 spades in the deck of 52 cards. Therefore, $P(A) = \frac{13}{52} = \frac{1}{4}$. On the second draw, there are only 51 cards remaining and 26 red cards. Therefore, $P(B|A) = \frac{26}{51}$. We have

$$P(A \cap B) = P(A) \cdot P(B | A)$$
$$= \frac{1}{4} \cdot \frac{26}{51} = \frac{26}{204} \approx 0.127.$$

23. Let A = {a blue is selected first}, B = {a red is selected second} and C = {a green is selected third}. Then
$P(A$ followed by B followed by $C)$
$= P(A) \cdot P(B | A) \cdot P(C | A$ and $B)$
$$= \frac{12}{57} \cdot \frac{12}{56} \cdot \frac{7}{55} = \frac{6}{1045}.$$

25. Let A = {a green is selected first}, B = {a green is selected second} and C = {a red is selected third}. Then
$P(A$ followed by B followed by $C)$
$= P(A) \cdot P(B | A) \cdot P(C | A$ and $B)$
$$= \frac{7}{57} \cdot \frac{6}{56} \cdot \frac{12}{55} = \frac{3}{1045}.$$

27. Let A = {a red card is dealt first}, B = {a black card is dealt second} and C = {a red card is dealt third}. Then
$P(A$ followed by B followed by $C)$
$= P(A) \cdot P(B | A) \cdot P(C | A$ and $B)$
$$= \frac{26}{52} \cdot \frac{26}{51} \cdot \frac{25}{50} = \frac{13}{102}.$$

29. Let A = {an ace is dealt first}, B = {a face card is dealt second} and C = {an 8 is dealt third}. Then
$P(A$ followed by B followed by $C)$
$= P(A) \cdot P(B | A) \cdot P(C | A$ and $B)$
$$= \frac{4}{52} \cdot \frac{12}{51} \cdot \frac{4}{50} = \frac{8}{5525}.$$

31. Let A = {absent the first day}, B = {absent the second day} and C = {absent the third day}. Then
$P(A$ followed by B followed by $C)$
$= P(A) \cdot P(B | A) \cdot P(C | A$ and $B)$
$= 0.04 \cdot 0.11 \cdot 0.11 = 0.000484.$

33. Let A = {absent the first day}, B = {not absent the second day} and C = {absent the third day}. Then
$P(A$ followed by B followed by $C)$
$= P(A) \cdot P(B | A) \cdot P(C | A$ and $B)$
$= 0.04 \cdot 0.89 \cdot 0.04 = 0.001424.$

35. The outcome of the first roll has no effect on the outcome of the second roll so the events are independent.

37. The probability of numbers on the second slip of paper depends on the result of the first slip of paper so the events are not independent.

39. The rolls of a pair of dice are independent. Let A = {sum of 8 on the first roll} and B = {sum of 8 on the second roll}.
$$P(A \cap B) = P(A) \cdot P(B)$$
$$= \frac{5}{36} \cdot \frac{5}{36} = \frac{25}{1296}$$

41. The rolls of a pair of dice are independent. Let A = {sum of at least 10 on the first roll} and B = {sum of at least 11 on the second roll}.
$$P(A \cap B) = P(A) \cdot P(B)$$
$$= \frac{6}{36} \cdot \frac{3}{36} = \frac{18}{1296} = \frac{1}{72}$$

43. The rolls of a pair of dice are independent. Let A = {sum of even number on the first roll} and B = {sum of even number on the second roll}.
$$P(A \cap B) = P(A) \cdot P(B)$$
$$= \frac{18}{36} \cdot \frac{18}{36} = \frac{324}{1296} = \frac{1}{4}$$

45. Each flip of a coin is independent. Let A = {heads}, B = {heads}, C = {tails} and D = {tails}.
$$P(A \cap B \cap C \cap D) = P(A) \cdot P(B) \cdot P(C) \cdot P(D)$$
$$= \frac{1}{2} \cdot \frac{1}{2} \cdot \frac{1}{2} \cdot \frac{1}{2} = \frac{1}{16}$$

47. The rolls of a pair of dice are independent. Let A = {sum of at least 10 on the first roll}, B = {sum of at least 10 on the second roll} and C = {sum of at least 10 on the third roll}.
$$P(A \cap B \cap C) = P(A) \cdot P(B) \cdot P(C)$$
$$= \frac{6}{36} \cdot \frac{6}{36} \cdot \frac{6}{36}$$
$$= \frac{216}{46,656} = \frac{1}{216}$$

49. The choices of a card are independent. Let A = {ace on the first draw} and B = {ace on the second draw}.

$$P(A \cap B) = P(A) \cdot P(B)$$
$$= \frac{4}{52} \cdot \frac{4}{52} = \frac{16}{2704} = \frac{1}{169}$$

51. The choices of a card are independent. Let A = {spade on the first draw} and B = {diamond on the second draw}.
$$P(A \cap B) = P(A) \cdot P(B)$$
$$= \frac{13}{52} \cdot \frac{13}{52} = \frac{169}{2704} = \frac{1}{16}$$

53. The choices of a card are independent. Let A = {heart on the first draw} and B = {spade on the second draw}.
$$P(A \cap B) = P(A) \cdot P(B)$$
$$= \frac{13}{52} \cdot \frac{13}{52} = \frac{169}{2704} = \frac{1}{16}$$

55. a.   The choices of a card with replacement are independent. Let
A = {diamond on the first draw},
B = {diamond on the second draw} and
C = {black card on the third draw}.
$$P(A \cap B \cap C) = P(A) \cdot P(B) \cdot P(C)$$
$$= \frac{13}{52} \cdot \frac{13}{52} \cdot \frac{26}{52}$$
$$= \frac{4394}{140,608} = \frac{1}{32}$$

b.   The choices of a card without replacement are not independent Let A = {diamond on the first draw}, B = {diamond on the second draw} and C = {black card on the third draw}.
$P(A$ followed by B followed by $C)$
$= P(A) \cdot P(B \mid A) \cdot P(C \mid A$ and $B)$
$$= \frac{13}{52} \cdot \frac{12}{51} \cdot \frac{26}{50} = \frac{4056}{132,600} = \frac{13}{425}$$

$P(A$ followed by B followed by $C)$
$= P(A) \cdot P(B \mid A) \cdot P(C \mid A$ and $B)$
$$= \frac{12}{52} \cdot \frac{11}{51} \cdot \frac{10}{50} = \frac{1320}{132,600} = \frac{11}{1105}$$

57. a.   The choices of a marble with replacement are independent. Let
A = {red marble chosen},
B = {red marble chosen} and
C = {green marble chosen}.

$$P(A \cap B \cap C) = P(A) \cdot P(B) \cdot P(C)$$
$$= \frac{5}{17} \cdot \frac{5}{17} \cdot \frac{4}{17} = \frac{100}{4913}$$

b.   The choices of a marble without
replacement are not independent. Let
A = {red marble chosen },
B = {red marble chosen} and
C = {green marble chosen}.
$P(A$ followed by B followed by $C)$
   $= P(A) \cdot P(B \mid A) \cdot P(C \mid A$ and $B)$

$$= \frac{5}{17} \cdot \frac{4}{16} \cdot \frac{4}{15} = \frac{80}{4080} = \frac{1}{51}$$

$$= P(A) \cdot P(B \mid A) \cdot P(C \mid A \text{ and } B)$$

$$= \frac{5}{12} \cdot \frac{3}{11} \cdot \frac{4}{10} = \frac{60}{1320} = \frac{1}{22}$$

59. Let D be the event that a person has been using
the drug and let T be the event that the test is
positive. The probability we wish to determine is
$P(D|T)$.

$$P(D \mid T) = \frac{P(D \text{ and } T)}{P(T)}$$

A tree diagram will help us compute the
needed probabilities.

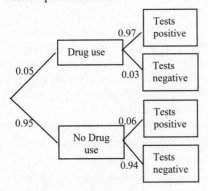

From the diagram, $P(D$ and $T) =$
$P(D) \cdot P(T \mid D) = (0.05)(0.97)$. To compute $P(T)$
we need to combine two branches from the
diagram, one corresponding to a correct positive
test result when the person has drug use, and one
corresponding to a false positive result when the
person has no drug use: $P(T) = (0.05)(0.97) +$
(0.95)(0.06).

$$P(D \mid T) = \frac{P(D \text{ and } T)}{P(T)}$$
$$= \frac{(0.05)(0.97)}{(0.05)(0.97) + (0.95)(0.06)} \approx 0.46$$

61. Let D be the event that a person has the
disease and let T be the event that the test
is positive. The probability we wish to
determine is $P(D|T)$.

$$P(D \mid T) = \frac{P(D \text{ and } T)}{P(T)}$$

A tree diagram will help us compute the
needed probabilities.

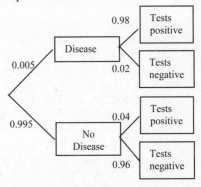

From the diagram, $P(D$ and $T) =$
$P(D) \cdot P(T \mid D) = (0.005)(0.98)$. To compute
$P(T)$ we need to combine two branches from the
diagram, one corresponding to a correct positive
test result when the person has the disease, and
one corresponding to a false positive result when
the person does not have the disease:
$P(T) = (0.005)(0.98) + (0.995)(0.04)$.

$$P(D \mid T) = \frac{P(D \text{ and } T)}{P(T)}$$
$$= \frac{(0.005)(0.98)}{(0.005)(0.98) + (0.995)(0.04)} \approx 0.11$$

63. a.   To be 3 blocks north, the coin would have
the same outcome 3 times, such as heads.
The sample space for tossing a coin three
times is {HHH, HHT, HTH, HTT, THH,
THT, TTH, TTT}. The element in the event
that three heads are tossed is {HHH}. Then
the probability is

$$P(E) = \frac{n(E)}{n(S)} = \frac{1}{8}.$$

b. To be 2 blocks north, the coin would have the same outcome 3 times and the other outcome 1 time, such as heads 3 times and tails once. The sample space for tossing a coin four times is {HHHH, HHHT, HHTH, HHTT, HTHH, HTHT, HTTH, HTTT, THHH, THHT, THTH, THTT, TTHH, TTHT, TTTH, TTTT}. The elements in the event that three heads and one tails are tossed are {HHHT, HHTH, HTHH, THHH}. Then the probability is

$$P(E) = \frac{n(E)}{n(S)} = \frac{4}{16} = \frac{1}{4}.$$

c. To be back at the original position, the coin would have the same outcome 2 times and the other outcome 2 times, such as heads 2 times and tails twice. The sample space for tossing a coin four times is {HHHH, HHHT, HHTH, HHTT, HTHH, HTHT, HTTH, HTTT, THHH, THHT, THTH, THTT, TTHH, TTHT, TTTH, TTTT}. The elements in the event that two heads and two tails are tossed are {HHTT, THTH, HTTH, THHT, HTHT, TTHH}. Then the probability is

$$P(E) = \frac{n(E)}{n(S)} = \frac{6}{16} = \frac{3}{8}.$$

EXERCISE SET 11.6

1. Expectation
$$= P(S_1) \cdot S_1 + P(S_2) \cdot S_2 + P(S_3) \cdot S_3$$
$$+ P(S_4) \cdot S_4 + P(S_5) \cdot S_5$$
$$= 30 \cdot 0.15 + 40 \cdot 0.2 + 50 \cdot 0.4$$
$$+ 60 \cdot 0.05 + 70 \cdot 0.2 = 49.5$$

3. Let S_1 be the event that the roulette ball lands on a black number, in which case the player wins $1. There are 38 possible numbers, so $P(S_1) = \frac{18}{38}$. Let S_2 be the event that a black number does not come up, in which case the player loses $1. Then $P(S_2) = 1 - \frac{18}{38} = \frac{20}{38}$.
Expectation = $P(S_1) \cdot S_1 + P(S_2) \cdot S_2$
$$= \frac{18}{38}(1) + \frac{20}{38}(-1) \approx -0.05$$
Expectation is − 5 cents.

5. Let S_1 be the event that the wheel stops on $40, in which case the player wins $40. There are 54 possible numbers, so $P(S_1) = \frac{2}{54}$. Let S_2 be the event that the wheel does not stop on $40, in which case the player loses $1. Then
$$P(S_2) = 1 - \frac{2}{54} = \frac{52}{54}.$$
Expectation = $P(S_1) \cdot S_1 + P(S_2) \cdot S_2$
$$= \frac{2}{54}(40) + \frac{52}{54}(-1) \approx 0.52$$
Expectation is 52 cents.

7. Let S_1 be the event that the wheel stops on $5, in which case the player wins $5. There are 54 possible numbers, so $P(S_1) = \frac{8}{54}$. Let S_2 be the event that the wheel does not stop on $5, in which case the player loses $1. Then
$$P(S_2) = 1 - \frac{8}{54} = \frac{46}{54}.$$
Expectation = $P(S_1) \cdot S_1 + P(S_2) \cdot S_2$
$$= \frac{8}{54}(5) + \frac{46}{54}(-1) \approx -0.11$$
Expectation is − 11 cents.

9. Let S_1 be the event that the person dies within one year. Then $P(S_1) = 0.002$ and the company must pay out $25,000. Since the company charged $75 for the policy, the company's actual loss is $24,925. Let S_2 be the event that the policy holder does not die during the year of the policy. Then $P(S_2) = 0.998$ and the company keeps the premium of $75.
Expectation = $P(S_1) \cdot S_1 + P(S_2) \cdot S_2$
$$= 0.002(-24925) + 0.998(75)$$
$$= 25$$
Expectation is $25.

11. Let S_1 be the event that the person dies within one year. Then $P(S_1) = 0.073$ and the company must pay out $10,000. Since the company charged $495 for the policy, the company's actual loss is $9505. Let S_2 be the event that the policy holder does not die during the year of the policy. Then $P(S_2) = 0.927$ and the company keeps the premium of $495.

Expectation = $P(S_1) \cdot S_1 + P(S_2) \cdot S_2$

$\qquad = 0.073(-9505) + 0.927(495)$

$\qquad = -235$

Expectation is $-\$235$.

13. Let S_1 be the event that the person dies within one year. Then $P(S_1) = 0.02$ and the company must pay out $\$30,000$. If the company charges $\$x$ for the policy, the company's actual loss is $\$(30,000 - x)$. Let S_2 be the event that the policyholder does not die during the year of the policy. Then $P(S_2) = 0.98$ and the company keeps the premium of $\$x$.

Expectation = $P(S_1) \cdot S_1 + P(S_2) \cdot S_2$

$\quad 0 = 0.02(-(30,000 - x)) + 0.98(x)$

$\quad 0 = -600 + 0.02x + 0.98x$

$\quad 600 = x$

The company should charge more than $\$600$.

15. Expectation

$= 0.10(100,000) + 0.40(60,000) +$
$0.25(30,000) + 0.15(0) +$
$0.08(-20,000) + 0.02(-40,000)$
$= 10,000 + 24,000 + 7500 + 0 - 1600 -$
$800 = 39,100$

The expected profit is $\$39,100$.

17. Expectation

$= 0.05(40,000) + 0.2(30,000) +$
$0.5(20,000) + 0.2(10,000) + 0.05(5,000)$
$= 2000 + 6000 + 10,000 + 2000 + 250$
$= 20,250$

The expected profit is $\$20,250$.

19. Let S_1 be the event that the player wins the Match 4 + Mega in which case the player wins $\$1,999,999$. $P(S_1) = \dfrac{1}{2,986,522}$. Let S_2 be the event that the player wins the Match 4 in which case the player wins $\$2,163$. $P(S_2) = \dfrac{3}{426,646}$.

Let S_3 be the event that the player wins the Match 3 + Mega in which case the player wins $\$439$. $P(S_3) = \dfrac{80}{1,493,261}$. Let S_4 be the event that the player wins the Match 3 in which case the player wins $\$69$. $P(S_4) = \dfrac{240}{213,323}$. Let S_5 be the event that the player wins the Match 2 + Mega in which case the player wins $\$26.50$.

$P(S_5) = \dfrac{2340}{1,493,261}$. Let S_6 be the event that the player wins the Match 2 in which case the player wins $\$1.50$. $P(S_6) = \dfrac{7020}{213,323}$. Let S_7 be the event that the player wins the Match 1 + Mega in which case the player wins $\$2$.

$P(S_7) = \dfrac{19,760}{1,493,261}$. Let S_8 be the event that the player wins the Mega in which case the player wins $\$0$. $P(S_8) = \dfrac{45,695}{1,493,261}$. Let S_9 be the event that no numbers are chosen correctly, in which case the player loses $\$1$.

$P(S_9) = 1 - \dfrac{1}{2,986,522} - \dfrac{3}{426,646} - \dfrac{80}{1,493,261} -$

$\qquad \dfrac{240}{213,323} - \dfrac{2340}{1,493,261} - \dfrac{7020}{213,323} -$

$\qquad \dfrac{19,760}{1,493,261} - \dfrac{45,695}{1,493,261}$

$\quad = \dfrac{196,365}{213,323}$

Expectation =

$\quad P(S_1) \cdot S_1 + P(S_2) \cdot S_2 + P(S_3) \cdot S_3 +$

$\quad P(S_4) \cdot S_4 + P(S_5) \cdot S_5 + P(S_6) \cdot S_6 +$

$\quad P(S_7) \cdot S_7 + P(S_8) \cdot S_8 + P(S_9) \cdot S_9$

$\dfrac{1}{2,986,522}(1,999,999) + \dfrac{3}{426,646}(2,163) +$

$\dfrac{80}{1,493,261}(439) + \dfrac{240}{213,323}(69) +$

$\dfrac{2340}{1,493,261}(26.50) + \dfrac{7020}{213,323}(1.50) +$

$\dfrac{19,760}{1,493,261}(2) + \dfrac{45,695}{1,493,261}(0) + \dfrac{196,365}{213,323}(-1)$

≈ -0.017

Expectation is -1.7 cents.

21. Possible outcomes can be found on the table on page 721. Let $S_1 = \{\text{sum of } 2\}$, $S_2 = \{\text{sum of } 3\}$, $S_3 = \{\text{sum of } 4\}$, $S_4 = \{\text{sum of } 5\}$, $S_5 = \{\text{sum of } 6\}$, $S_6 = \{\text{sum of } 7\}$, $S_7 = \{\text{sum of } 8\}$, $S_8 = \{\text{sum of } 9\}$, $S_9 = \{\text{sum of } 10\}$, $S_{10} = \{\text{sum of } 11\}$, and $S_{11} = \{\text{sum of } 12\}$. Then probabilities are:

$P(S_1) = \frac{1}{36}, P(S_2) = \frac{2}{36}, P(S_3) = \frac{3}{36},$

$P(S_4) = \frac{4}{36}, P(S_5) = \frac{5}{36}, P(S_6) = \frac{6}{36},$

$P(S_7) = \frac{5}{36}, P(S_8) = \frac{4}{36}, P(S_9) = \frac{3}{36},$

$P(S_{10}) = \frac{2}{36}, P(S_{11}) = \frac{1}{36}$

Expectation =

$P(S_1) \cdot S_1 + P(S_2) \cdot S_2 + P(S_3) \cdot S_3 +$

$P(S_4) \cdot S_4 + P(S_5) \cdot S_5 + P(S_6) \cdot S_6 +$

$P(S_7) \cdot S_7 + P(S_8) \cdot S_8 + P(S_9) \cdot S_9 +$

$P(S_{10}) \cdot S_{10} + P(S_{11}) \cdot S_{11}$

$= \frac{1}{36}(2) + \frac{2}{36}(3) + \frac{3}{36}(4) + \frac{4}{36}(5)$

$+ \frac{5}{36}(6) + \frac{6}{36}(7) + \frac{5}{36}(8) + \frac{4}{36}(9)$

$+ \frac{3}{36}(10) + \frac{2}{36}(11) + \frac{1}{36}(12) = 7$

The expected sum is 7.

23. The sample space can be found using a table.

	1	**2**	**2**	**3**	**3**	**4**
1	11	12	12	13	13	14
3	31	32	32	33	33	34
4	41	42	42	43	43	44
5	51	52	52	53	53	54
6	61	62	62	63	63	64
8	81	82	82	83	83	84

Possible sums are: {2, 3, 4, 5, 6, 7, 8, 9, 10, 11, 12}

Probabilities are:

$P(2) = \frac{1}{36}, P(3) = \frac{2}{36}, P(4) = \frac{3}{36},$

$P(5) = \frac{4}{36}, P(6) = \frac{5}{36}, P(7) = \frac{6}{36},$

$P(8) = \frac{5}{36}, P(9) = \frac{4}{36}, P(10) = \frac{3}{36},$

$P(11) = \frac{2}{36}, P(12) = \frac{1}{36}.$

Expectation =

$= \frac{1}{36}(2) + \frac{2}{36}(3) + \frac{3}{36}(4) + \frac{4}{36}(5)$

$+ \frac{5}{36}(6) + \frac{6}{36}(7) + \frac{5}{36}(8) + \frac{4}{36}(9)$

$+ \frac{3}{36}(10) + \frac{2}{36}(11) + \frac{1}{36}(12) = 7$

The expected sum is 7.

25. The sample space of the red dice can be found using a table.

	0	**0**	**4**	**4**	**4**	**4**
2	20	20	24	24	24	24
3	30	30	34	34	34	34
3	30	30	34	34	34	34
9	90	90	94	94	94	94
10	100	100	104	104	104	104
11	110	110	114	114	114	114

Possible sums are: {2, 3, 6, 7, 9, 10, 11, 13, 14, 15} Probabilities are:

$P(2) = \frac{2}{36}, P(3) = \frac{4}{36}, P(6) = \frac{4}{36},$

$P(7) = \frac{8}{36}, P(9) = \frac{2}{36}, P(10) = \frac{2}{36},$

$P(11) = \frac{2}{36}, P(13) = \frac{4}{36}, P(14) = \frac{4}{36},$

$P(15) = \frac{4}{36}.$

Expectation =

$= \frac{2}{36}(2) + \frac{4}{36}(3) + \frac{4}{36}(6) + \frac{8}{36}(7)$

$+ \frac{2}{36}(9) + \frac{2}{36}(10) + \frac{2}{36}(11) + \frac{4}{36}(13)$

$+ \frac{4}{36}(14) + \frac{4}{36}(15) = 9$

The expected sum for the red dice is 9.
The sample space of the green dice can be found using a table.

	3	**3**	**3**	**3**	**3**	**3**
0	03	03	03	03	03	03
1	13	13	13	13	13	13
7	73	73	73	73	73	73
8	83	83	83	83	83	83
8	83	83	83	83	83	83
8	83	83	83	83	83	83

Possible sums are: {3, 4, 10, 11} Probabilities are:

$P(3) = \frac{6}{36}, P(4) = \frac{6}{36}, P(10) = \frac{6}{36},$

$P(11) = \frac{18}{36}$

Expectation =

$= \frac{6}{36}(3) + \frac{6}{36}(4) + \frac{6}{36}(10) + \frac{18}{36}(11)$

≈ 8.3

The expected sum for the green dice is 8.3. The red dice has a higher expected sum, so it would be the better choice.

CHAPTER 11 REVIEW EXERCISES

1. $\{11, 12, 13, 21, 22, 23, 31, 32, 33\}$

2. $\{26, 28, 62, 68, 82, 86\}$

3.

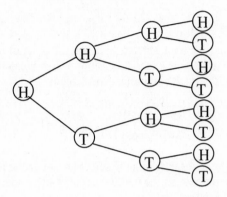

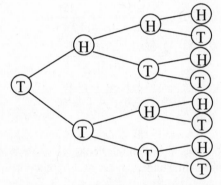

{HHHH, HHHT, HHTH, HHTT, HTHH, HTHT, HTTH, HTTT, THHH, THHT, THTH, THTT, TTHH, TTHT, TTTH, TTTT}

4.

	7	8	9
A	7A	8A	9A
B	7B	8B	9B

{7A, 8A, 9A, 7B, 8B, 9B}

5. There are 3 choices for style, so $n_1 = 3$. There are 4 choices for color, so $n_2 = 4$.
There are 6 choices for sizes, so $n_3 = 6$. By the counting principle, there are $3 \cdot 4 \cdot 6 = 72$ shoes available.

6. There are 10 numbers that can be used in each position, so n_1, n_2, n_3 and $n_4 = 10$. By the counting principle, there are $10 \cdot 10 \cdot 10 \cdot 10 = 10,000$ different combinations possible.

7. There are 5 letters that can be used in the first position, so $n_1 = 5$. There are 12 letters that can be used in the second position, so $n_2 = 12$. There are 10 digits that can be used in the third and fourth positions, so $n_3 = 10$ and $n_4 = 10$. By the counting principle, there are $5 \cdot 12 \cdot 10 \cdot 10 = 6,000$ different four-character sequences possible.

8. There are 7 binary digits, the first two of which are required to be different. There are two choices for the first digit, $n_1 = 2$, and only one choice for the second, $n_2 = 1$, since it must be different from the first. There are two choices for each of the five remaining digits, $n_3 = n_4 = n_5 = n_6 = n_7 = 2$. The counting principle then says there are $2 \cdot 1 \cdot 2 \cdot 2 \cdot 2 \cdot 2 \cdot 2 = 64$ biquinary numbers.

9. $7! = 7 \cdot 6 \cdot 5 \cdot 4 \cdot 3 \cdot 2 \cdot 1 = 5040$

10. Evaluating:
$$8! - 4! = (8 \cdot 7 \cdot 6 \cdot 5 \cdot 4 \cdot 3 \cdot 2 \cdot 1) - (4 \cdot 3 \cdot 2 \cdot 1)$$
$$= 40,320 - 24 = 40,296$$

11. $\dfrac{9!}{2!3!4!} = \dfrac{9 \cdot 8 \cdot 7 \cdot 6 \cdot 5 \cdot 4!}{2 \cdot 1 \cdot 3 \cdot 2 \cdot 1 \cdot 4!} = 1260$

12. Evaluating:
$$P(10,6) = \frac{10!}{(10-6)!} = \frac{10!}{4!}$$
$$= \frac{10 \cdot 9 \cdot 8 \cdot 7 \cdot 6 \cdot 5 \cdot 4!}{4!}$$
$$= 10 \cdot 9 \cdot 8 \cdot 7 \cdot 6 \cdot 5 = 151,200$$

13. Evaluating:
$$P(8,3) = \frac{8!}{(8-3)!} = \frac{8!}{5!} = \frac{8 \cdot 7 \cdot 6 \cdot 5!}{5!}$$
$$= 8 \cdot 7 \cdot 6 = 336$$

14. Evaluating:

$$\frac{C(6,2)\cdot C(8,3)}{C(14,5)}=\frac{\frac{6!}{2!(6-2)!}\cdot\frac{8!}{3!(8-3)!}}{\frac{14!}{5!(14-5)!}}$$

$$=\frac{\frac{6!}{2!4!}\cdot\frac{8!}{3!5!}}{\frac{14!}{5!9!}}$$

$$=\frac{\frac{6\cdot5\cdot4!}{2!4!}\cdot\frac{8\cdot7\cdot6\cdot5!}{3!5!}}{\frac{14\cdot13\cdot12\cdot11\cdot10\cdot9!}{5!9!}}$$

$$=\frac{\frac{6\cdot5}{2\cdot1}\cdot\frac{8\cdot7\cdot6}{3\cdot2\cdot1}}{\frac{14\cdot13\cdot12\cdot11\cdot10}{5\cdot4\cdot3\cdot2\cdot1}}$$

$$=\frac{15\cdot56}{2002}=\frac{840}{2002}=\frac{60}{143}$$

15. The order in which the people arrange themselves is important, so the number of ways to receive service is

$$P(7,7)=\frac{7!}{(7-7)!}=\frac{7!}{0!}=\frac{7\cdot6\cdot5\cdot4\cdot3\cdot2\cdot1}{1}$$
$$=7\cdot6\cdot5\cdot4\cdot3\cdot2\cdot1=5040$$

16. The order in which the answers are chosen is important, so the number of matches possible is

$$P(7,7)=\frac{7!}{(7-7)!}=\frac{7!}{0!}=\frac{7\cdot6\cdot5\cdot4\cdot3\cdot2\cdot1}{1}$$
$$=7\cdot6\cdot5\cdot4\cdot3\cdot2\cdot1=5040$$

17. The order in which the answers are chosen is important, so the number of matches possible is

$$P(7,5)=\frac{7!}{(7-5)!}=\frac{7!}{2!}=\frac{7\cdot6\cdot5\cdot4\cdot3\cdot2\cdot1}{2\cdot1}$$
$$=2520$$

18. We are looking for the number of permutations of the letters *letter*. With $n=6$ (number of letters), $k_1=1$ (number of l's), $k_2=2$ (number of e's), $k_3=2$ (number of t's) and $k_4=1$ (number of r's), we have

$$\frac{6!}{1!\,2!\,2!\,1!}=180$$

There are 180 different arrangements possible.

19. The order in which the heads and tails are chosen is not important, so the number of distinct arrangements is

$$C(12,4)=\frac{12!}{4!(12-4)!}=\frac{12!}{4!8!}$$
$$=\frac{12\cdot11\cdot10\cdot9\cdot8!}{4!8!}$$
$$=\frac{12\cdot11\cdot10\cdot9}{4\cdot3\cdot2\cdot1}=495.$$

20. Because each shift is different, the order of the assignments to shifts is important, so the number of different assignments possible is

$$P(5,3)=\frac{5!}{(5-3)!}=\frac{5!}{2!}=\frac{5\cdot4\cdot3\cdot2!}{2!}$$
$$=5\cdot4\cdot3=60.$$

21. The order in which the selections are made is not important, so the number of sets that can be chosen is

$$C(25,10)=\frac{25!}{10!(25-10)!}=\frac{25!}{10!15!}$$
$$=\frac{25\cdot24\cdot23\cdot22\cdot21\cdot20\cdot19\cdot18\cdot17\cdot16\cdot15!}{10!15!}$$
$$=\frac{25\cdot24\cdot23\cdot22\cdot21\cdot20\cdot19\cdot18\cdot17\cdot16}{10\cdot9\cdot8\cdot7\cdot6\cdot5\cdot4\cdot3\cdot2\cdot1}$$
$$=3,268,760.$$

22. The order in which the stocks are selected is not important, so the number of different portfolios is

$$C(11,3)=\frac{11!}{3!(11-3)!}=\frac{11!}{3!8!}=\frac{11\cdot10\cdot9\cdot8!}{3!8!}$$
$$=\frac{11\cdot10\cdot9}{3\cdot2\cdot1}=165.$$

23. The order of the choices is not important, so the number of possible ways to choose the nondefective monitors, is C(12, 3) and the number of ways to choose the defective monitors is C(3,2). Therefore, by the counting principle, there are $C(12,3)\cdot C(3,2)$ different sets.

$$C(12,3)\cdot C(3,2)=\frac{12!}{3!(12-3)!}\cdot\frac{3!}{2!(3-2)!}$$
$$=\frac{12!}{3!9!}\cdot\frac{3!}{2!1!}$$
$$=\frac{12\cdot11\cdot10\cdot9!}{3!9!}\cdot\frac{3\cdot2!}{2!1!}$$
$$=\frac{12\cdot11\cdot10}{3\cdot2\cdot1}\cdot\frac{3}{1}$$
$$=220\cdot3=660$$

24. Call the two people that refuse to sit next to each other A and B. Consider two cases.
In the first case, A sits at an end of the row. There are 2 ways A can do this. Once A is seated, there are 7 places where B could sit and not be next to A. Then there are 7! ways the remaining people can be seated, for a total of $2 \cdot 7 \cdot 7!$ ways to seat the group if A sits at an end of the row.
In the second case, A does not sit at the end of the row. There are 7 ways A can do this. Once A is seated, there are 6 places where B could sit and not be next to A. Then there are 7! ways the remaining people can be seated, for a total of $7 \cdot 6 \cdot 7!$. Summing the totals from case A and case B gives 282,240 ways.

25. There are C(4, 4) ways of choosing four aces from four aces, C(4, 4) ways of choosing four kings from four kings, and C(4, 4) ways of choosing each of the other possible cards of which there are 13 total. There are 48 choices for the fifth card in the hand. By the counting principle, the number of hands containing four of a kind is

$$13 \cdot C(4,4) \cdot 48 = 13 \cdot \frac{4!}{4!(4-4)!} \cdot 48$$
$$= 13 \cdot \frac{4!}{4!0!} \cdot 48 = 13 \cdot \frac{1}{1} \cdot 48$$
$$= 624.$$

26. The sample space is {BBBB, BBBG, BBGB, BBGG, BGBB, BGBG, BGGB, BGGG, GBBB, GBBG, GBGB, GBGG, GGBB, GGBG, GGGB, GGGG}. The elements in the event that a family will have one boys and three girls is {BGGG, GBGG, GGBG, GGGB}. Then the probability is

$$P(E) = \frac{n(E)}{n(S)} = \frac{4}{16} = \frac{1}{4}.$$

27. The sample space is {HHH, HHT, HTH, HTT, THH, THT, TTH, TTT}. The elements in the event of getting one head and two tails is {HTT, THT, TTH}. Then the probability is

$$P(E) = \frac{n(E)}{n(S)} = \frac{3}{8}.$$

28. Let E be the event that the employee is a woman.

$$P(E) = \frac{7290}{5739 + 7290} = \frac{7290}{13,029} \approx 0.56$$

29. The total number of students is 2495. Let E be the event that the student is a junior or senior. There are $483 + 445 = 928$ students that are juniors or seniors.

$$P(E) = \frac{928}{2495} \approx 0.37$$

30. The total number of students in 2495. Let E be the event that the student is not a graduate student. There are $642 + 549 + 483 + 445 = 2119$ students that are not graduate students.

$$P(E) = \frac{2119}{2495} \approx 0.85$$

31. The sample space for tossing two six-sided dice was found on page 721. The dice must be considered as distinct, so there are 36 possible outcomes. The elements in the event that the sum of the pips on the upward faces is 9 are {3 6, 4 5, 5 4, 6 3}. Then the probability is

$$P(E) = \frac{n(E)}{n(S)} = \frac{4}{36} = \frac{1}{9}.$$

32. If E is the event of tossing an 11, then E^C is the event of not tossing an 11, and

$$P(E^C) = 1 - P(E) = 1 - \frac{2}{36} = \frac{34}{36} = \frac{17}{18}.$$

33. If E is the event of tossing a sum less than 10, then E^C is the event of tossing a sum of at least ten, and

$$P(E^C) = 1 - P(E) = 1 - \frac{30}{36} = \frac{6}{36} = \frac{1}{6}.$$

34. Let A be the event of rolling an even number, and let B be the event of rolling a number less than 5. From the figure on page 721, there are 36 elements in the sample space. $n(A) = 18$, $n(B) = 6$ and $n(A$ and $B) = 4$.
$P(A$ or $B) = P(A) + P(B) - P(A$ and $B)$

$$= \frac{18}{36} + \frac{6}{36} - \frac{4}{36} = \frac{20}{36} = \frac{5}{9}$$

35. Let $B = \{$the sum is 9$\}$ and $A = \{$the sum is odd$\}$. From the table on page 721, there are four possible rolls of the dice for which the sum is 9 and the sum is odd. So $P(A \cap B) = \frac{4}{36} = \frac{1}{9}$.
There are 18 possibilities for which the sum is odd, so $P(A) = \frac{18}{36} = \frac{1}{2}$. We have

$$P(B|A) = \frac{P(A \cap B)}{P(A)} = \frac{\frac{1}{9}}{\frac{1}{2}} = \frac{2}{9}.$$

36. Let B = {the sum is 8} and A = {doubles}. From the table on page 721, there is one possible roll of the dice for which the sum is 8 and doubles. So $P(A \cap B) = \frac{1}{36}$. There are 6 possibilities for which the roll is doubles, so $P(A) = \frac{6}{36} = \frac{1}{6}$. We have

$$P(B|A) = \frac{P(A \cap B)}{P(A)} = \frac{\frac{1}{36}}{\frac{1}{6}} = \frac{1}{6}.$$

37. Let A be the event of drawing a heart, and let B be the event of drawing a black card. The sample space consists of 52 cards. $n(A) = 13$, $n(B) = 26$ and $n(A$ and $B) = 0$.
    $P(A$ or $B) = P(A) + P(B)$
    $$= \frac{13}{52} + \frac{26}{52} = \frac{39}{52} = \frac{3}{4}$$

38. Let A be the event of drawing a heart, and let B be the event of drawing a jack. The sample space consists of 52 cards. $n(A) = 13$, $n(B) = 4$ and $n(A$ and $B) = 1$.
    $P(A$ or $B) = P(A) + P(B) - P(A$ and $B)$
    $$= \frac{13}{52} + \frac{4}{52} - \frac{1}{52} = \frac{16}{52} = \frac{4}{13}$$

39. If E is the event of drawing a three, then E^C is the event of not drawing a three, and
    $$P(E^C) = 1 - P(E) = 1 - \frac{4}{52} = \frac{48}{52} = \frac{12}{13}.$$

40. Let B = {the card is red} and A = {the card is not a club}. There are 26 cards that are red and not clubs. So $P(A \cap B) = \frac{26}{52} = \frac{1}{2}$. There are 39 possibilities for cards that are not clubs, so $P(A) = \frac{39}{52} = \frac{3}{4}$. We have

$$P(B|A) = \frac{P(A \cap B)}{P(A)} = \frac{\frac{1}{2}}{\frac{3}{4}} = \frac{2}{3}.$$

41. Let E be the event of rolling a sum of 6. Refer to page 721. Because there are five ways to roll a sum of 6, there are five favorable outcomes, leaving 31 unfavorable outcomes.
    Odds in favor of $E = \frac{5}{31}$
    The odds in favor are 5 to 31.

42. Let E be the event of pulling a heart. Because there are thirteen ways to draw a heart, there are thirteen favorable outcomes, leaving 39 unfavorable outcomes.
    Odds in favor of $E = \frac{13}{39} = \frac{1}{3}$
    The odds in favor are 1 to 3.

43. Because the odds against are 4 to 5, the odds in favor are 5 to 4. Use the formula for calculating the probability of an event when the odds in favor are known.
    $$P(E) = \frac{a}{a+b} = \frac{5}{5+4} = \frac{5}{9}$$

44. Make a Punnett square.

Parents	H	h
h	Hh	hh
h	Hh	hh

To have short hair, the rodent must have two h. From the table, two of the four possible genotypes have two h's, so the probability that an offspring will have short hair is $\frac{2}{4} = \frac{1}{2}$.

45. Let A be the event of exactly one card being an ace and let B be the event of exactly one card being a face card. Then,
    $P(A$ and $B) = P(A) + P(B) - P(A$ or $B)$
    $$= 0.145 + 0.362 - 0.471$$
    $$= 0.036.$$

46. Let A = {people who like cheese flavored}, and let B = {people who like jalapeno-flavored}. Then, $n(A) = 642$, $n(B) = 487$ and $n(A$ and $B)$ =302. The total number of people represented is 1000.
    $P(A$ or $B) = P(A) + P(B) - P(A$ and $B)$
    $$= \frac{642}{1000} + \frac{487}{1000} - \frac{302}{1000} = \frac{827}{1000}$$
    If E is the event of a person liking one of the two flavors, then E^C is the event of not liking either of the two flavors, and
    $$P(E^C) = 1 - P(E) = 1 - \frac{827}{1000} = \frac{173}{1000}.$$

47. Because there are 24 chips in the box, there are
24 items in the sample space. Let E be the event
of choosing a yellow or white chip. There are 7
yellow chips and 8 white chips, so there are 7 + 8
= 15 items in E. Then the probability is

$$P(E) = \frac{n(E)}{n(S)} = \frac{15}{24} = \frac{5}{8}.$$

48. Let E be the event of getting a red chip. Because
there are 4 red chips, there are 4 favorable
outcomes, leaving 20 unfavorable outcomes.

Odds in favor of $E = \frac{4}{20} = \frac{1}{5}$

The odds in favor are 1 to 5.

49. Let $B = \{$the chip is yellow$\}$ and $A = \{$the chip is
not white$\}$. There are seven chips that are yellow
and not white. So $P(A \cap B) = \frac{7}{24}$. There are 16

chips that are not white, so $P(A) = \frac{16}{24} = \frac{2}{3}$. We

have

$$P(B|A) = \frac{P(A \cap B)}{P(A)} = \frac{\frac{7}{24}}{\frac{2}{3}} = \frac{7}{16}.$$

50. Let $E = \{$no red chips are drawn$\}$. There are 24
possibilities for the first chip drawn, 23
possibilities for the second chip drawn, 22 for
the third, 21 for the fourth and 20 for the fifth.
Thus, $n(S) = 24 \cdot 23 \cdot 22 \cdot 21 \cdot 20 = 5,100,480$.
The first time a chip is drawn, there are 20 chips
that are not red, the second time there are 19
chips that are not red, 18 chips the third time 17
chips the fourth time, and 16 chips the fifth time,
so $n(E) = 20 \cdot 19 \cdot 18 \cdot 17 \cdot 16 = 1,860,480$.

$$P(E) = \frac{1,860,480}{5,100,480} = \frac{646}{1771}.$$

51. Let $A = \{$a yellow is selected first$\}$, $B = \{$a white
is selected second$\}$ and $C = \{$a yellow is selected
third$\}$. Then
$P(A$ followed by B followed by $C)$
$= P(A) \cdot P(B|A) \cdot P(C|A$ and $B)$

$$= \frac{7}{24} \cdot \frac{8}{23} \cdot \frac{6}{22} = \frac{336}{12,144} = \frac{7}{253}.$$

52. $n(S) = 23,500$. Let E be the event that the person
voted against the proposition. Then

$$P(E) = \frac{2527 + 5370 + 712}{23,500}$$

$$= \frac{8609}{23,500} \approx 0.37.$$

53. $n(S) = 23,500$. Let $A = \{$Democrats$\}$, and let $B =$
$\{$Independents$\}$. Then, from the table, $n(A) =$
$8452 + 2527 + 894 = 11,873$, $n(B) = 1225 + 712$
$+ 686 = 2623$ and $n(A$ and $B) = 0$.
$P(A$ or $B) = P(A) + P(B) - P(A$ and $B)$

$$= \frac{11,873}{23,500} + \frac{2623}{23,500}$$

$$= \frac{14,496}{23,500} \approx 0.62$$

54. $n(S) = 23,500$. Let E be the event that the person
abstained from voting and is not a Republican.

Then $P(E) = \frac{894 + 686}{23,500} = \frac{1580}{23,500} \approx 0.067$.

55. $n(S) = 23,500$. Let $B = \{$voted for the
proposition$\}$ and $A = \{$registered Independent$\}$.
From the table, $n(A \cap B) = 1225$, $n(A) = 1225 +$
$712 + 686 = 2623$, and $n(S) = 23,500$. We have

$$P(B|A) = \frac{P(A \cap B)}{P(A)} = \frac{\frac{1225}{23,500}}{\frac{2623}{23,500}} \approx 0.47.$$

56. $n(S) = 23,500$. Let $B = \{$registered Democrat$\}$
and $A = \{$voted against the proposition$\}$. From
the table, $n(A \cap B) = 2527$ and $n(A) = 2527 +$
$5370 + 712 = 8609$. We have

$$P(B|A) = \frac{P(A \cap B)}{P(A)} = \frac{\frac{2527}{23,500}}{\frac{8609}{23,500}} \approx 0.29.$$

57. The rolls of a pair of dice are independent. Let
$A = \{6$ on the first roll$\}$, $B = \{6$ on the second
roll$\}$ and $C = \{6$ on the third roll$\}$. Then
$P(A \cap B \cap C) = P(A) \cdot P(B) \cdot P(C)$

$$= \frac{1}{6} \cdot \frac{1}{6} \cdot \frac{1}{6} = \frac{1}{216}.$$

58. Let E = {at least one 6}; then E^C = {no 6's}. Because on each toss of the die there are six possible outcomes, $n(S) = 6 \cdot 6 \cdot 6 \cdot 6 \cdot 6 = 7776$. On each toss of the die there are five numbers that are not 6's. Therefore, $n(E^C) = 5 \cdot 5 \cdot 5 \cdot 5 \cdot 5 = 3125$.

    $P(E) = 1 - P(E^C)$

    $= 1 - \dfrac{3125}{7776} = \dfrac{4651}{7776} \approx 0.60$

59. Let E = {exactly two 6's}. Because on each toss of the die there are six possible outcomes, $n(S) = 6 \cdot 6 \cdot 6 \cdot 6 \cdot 6 = 7776$. On each toss of the die there are five numbers that are not 6's and one number that is a 6. The number of possible arrangements of the two sixes is C(5, 2) = 10. Therefore, $n(E) = 10 \cdot 5 \cdot 5 \cdot 5 \cdot 1 \cdot 1 = 1250$.

    $P(E) = \dfrac{1250}{7776} \approx 0.16$

60. Let E = {at least one spade is drawn}; then E^C = {no spade is drawn}. There are 52 possibilities for each card drawn. Thus, $n(S) = 52 \cdot 52 \cdot 52 \cdot 52 = 7,311,616$. Each time a card is drawn, there are 39 cards that are not spades, so $n(E^C) = 39 \cdot 39 \cdot 39 \cdot 39 = 2,313,441$.

    $P(E) = 1 - P(E^C)$

    $= 1 - \dfrac{2,313,441}{7,311,616} = \dfrac{4,998,175}{7,311,616} = \dfrac{175}{256}$

61. Let D be the event that a dog has the disease and let T be the event that the test is positive. The probability we wish to determine is $P(D|T)$. Then

    $P(D \mid T) = \dfrac{P(D \text{ and } T)}{P(T)}$.

    A tree diagram will help us compute the needed probabilities.

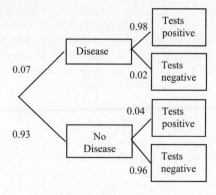

    From the diagram, $P(D$ and $T) = P(D) \cdot P(T \mid D) = (0.07)(0.98)$. To compute $P(T)$ we need to combine two branches from the diagram, one corresponding to a correct positive test result when the dog has the disease, and one corresponding to a false positive result when the dog does not have the disease: $P(T) = (0.07)(0.98) + (0.93)(0.04)$. Then

    $P(D \mid T) = \dfrac{P(D \text{ and } T)}{P(T)}$

    $= \dfrac{(0.07)(0.98)}{(0.07)(0.98) + (0.93)(0.04)} \approx 0.648$.

62. Let A = {rain the first day}, B = {rain the second day}, C = {no rain the third day} and D = {no rain the fourth day}. Then

    $P(A$ followed by B followed by C followed by $D)$

    $= P(A) \cdot P(B \mid A) \cdot P(C \mid A \text{ and } B)$

    $\qquad \cdot P(D \mid A, B \text{ and } C)$

    $= 0.15 \cdot 0.65 \cdot 0.35 \cdot 0.85 \approx 0.029$.

63. $1 - (0.988)^{12} \approx 0.135$

64. Let S_1 be the event that a 1 is rolled, in which case the player loses $3. There is 1 possible number, so $P(S_1) = \dfrac{1}{6}$. Let S_2 be the event that a 2 is rolled, in which case the player loses $2. $P(S_2) = \dfrac{1}{6}$. Let S_3 be the event that a 3 is rolled, in which case the player loses $1. There is 1 possible number, so $P(S_3) = \dfrac{1}{6}$. Let S_4 be the event that a 4 is rolled, in which case the player wins $0. $P(S_4) = \dfrac{1}{6}$. Let S_5 be the event that a 5 is rolled, in which case the player wins $1. There

is 1 possible number, so $P(S_5) = \frac{1}{6}$. Let S_6 be the event that a 6 is rolled, in which case the player wins $2. $P(S_6) = \frac{1}{6}$.

Expectation =

$P(S_1) \cdot S_1 + P(S_2) \cdot S_2 + P(S_3) \cdot S_3$
$+ P(S_4) \cdot S_4 + P(S_5) \cdot S_5 + P(S_6) \cdot S_6$

$= \frac{1}{6}(-3) + \frac{1}{6}(-2) + \frac{1}{6}(-1)$

$+ \frac{1}{6}(0) + \frac{1}{6}(1) + \frac{1}{6}(2) = -\frac{1}{2}$

Expectation is -50 cents.

65. Let S_1 be the event that both coins are tails, in which case the player wins $5. There is 1 outcome in the sample space of 4, so $P(S_1) = \frac{1}{4}$.

Let S_2 be the event that one coin is heads and one is tails, in which case the player loses $2. There are two outcomes in the sample space of 4, so $P(S_2) = \frac{2}{4} = \frac{1}{2}$. Let S_3 be the event that both coins are heads, in which case the player neither wins nor loses money. There is 1 outcome in the sample space of 4, so $P(S_3) = \frac{1}{4}$. Expectation =

$P(S_1) \cdot S_1 + P(S_2) \cdot S_2 + P(S_3) \cdot S_3$
$+ P(S_4) \cdot S_4 + P(S_5) \cdot S_5 + P(S_6) \cdot S_6$

$= \frac{1}{4}(5) + \frac{1}{2}(-2) + \frac{1}{4}(0) = 0.25$

Expectation is 25 cents.

66. Let S_1 be the event that the first prize is won, in which case the player wins $200. $P(S_1) = \frac{1}{800}$.

Let S_2 be the event that the second prize is won, in which case the player wins $75. $P(S_2) = \frac{4}{800}$.

It costs $1 to buy a ticket.
Expectation =

$P(S_1) \cdot S_1 + P(S_2) \cdot S_2 - 1$

$= \frac{1}{800}(200) + \frac{4}{800}(75) - 1 \approx$

-0.375

Expectation is -37.5 cents.

67. The sample space for tossing two six-sided dice is found on page 721. There are 36 possible outcomes. There are 3 elements in the event that

the sum of the pips on the upward faces is 4. Then the probability is

$P(E) = \frac{n(E)}{n(S)} = \frac{3}{36} = \frac{1}{12}$.

Multiply the probability by 65 and find that we can expect a sum of 4 about 5.4 times.

68. Let S_1 be the event that the person dies within one year. Then $P(S_1) = 0.000616$ and the company must pay out $40,000. Since the company charged $320 for the policy, the company's actual loss is $39,680. Let S_2 be the event that the policy holder does not die during the year of the policy. Then $P(S_2) = 0.999393$ and the company keeps the premium of $320.
Expectation = $P(S_1) \cdot S_1 + P(S_2) \cdot S_2$

$= 0.000616(-39,680) + 0.999393(320)$

$= 295.36$
Expectation is $295.36.

69. Let S_1 be the event that the person dies within one year. Then $P(S_1) = 0.001376$ and the company must pay out $25,000. Since the company charged $795 for the policy, the company's actual loss is $24,205. Let S_2 be the event that the policy holder does not die during the year of the policy. Then $P(S_2) = 0.998624$ and the company keeps the premium of $795.
Expectation = $P(S_1) \cdot S_1 + P(S_2) \cdot S_2$

$= 0.001376(-24,205) + 0.998624(795)$

$= 760.60$
Expectation is $760.60.

70. Expectation
$= 0.20(25,000) + 0.25(15,000) + 0.20(10,000) +$
$0.15(5,000) + 0.10(0) + 0.10(-5000)$
$= 5000 + 3750 + 2000 + 750 + 0 - 500$
$= 11,000$
The expected profit is $11,000.

CHAPTER 11 TEST

1. {A2, D2, G2, K2, A3, D3, G3, K3, A4, D4, G4, K4}

2. There are 3 choices for processors, so $n_1 = 3$.
There are 4 choices for disk drives, so $n_2 = 4$.
There are 3 choices for monitors, so $n_3 = 3$.
There are 2 choices for graphics cards, so $n_4 = 2$.
By the counting principle, there are
$3 \cdot 4 \cdot 3 \cdot 2 = 72$ computer systems possible.

3. The order in which the instructions is sent is important, so the number of ways to send the instructions is

$$P(10,4) = \frac{10!}{(10-4)!} = \frac{10!}{6!} = \frac{10 \cdot 9 \cdot 8 \cdot 7 \cdot 6!}{6!}$$
$$= 10 \cdot 9 \cdot 8 \cdot 7 = 5040.$$

4. The order in which the words are chosen is important, so the number of pairs possible is $P(15,10)$

$$= \frac{15!}{(15-10)!} = \frac{15!}{5!}$$
$$= \frac{15 \cdot 14 \cdot 13 \cdot 12 \cdot 11 \cdot 10 \cdot 9 \cdot 8 \cdot 7 \cdot 6 \cdot 5!}{5!}$$
$$= 15 \cdot 14 \cdot 13 \cdot 12 \cdot 11 \cdot 10 \cdot 9 \cdot 8 \cdot 7 \cdot 6$$
$$\approx 1.09 \times 10^{10}.$$

5. The order in which the computers and printers are attached is important, so the number of pairs possible is

$$P(20,4) = \frac{20!}{(20-4)!} = \frac{20!}{16!}$$
$$= \frac{20 \cdot 19 \cdot 18 \cdot 17 \cdot 16!}{16!}$$
$$= 20 \cdot 19 \cdot 18 \cdot 17 = 116,280.$$

6. Let A be the event of tossing a head, and let B be the event of rolling a 5. There are 2 outcomes possible for the coin and 6 outcomes possible for the die, so by the counting principle, there are $2 \cdot 6 = 12$ items in the sample space. $n(A) = 6$, $n(B) = 2$ and $n(A \text{ and } B) = 1$.
    $P(A \text{ or } B) = P(A) + P(B) - P(A \text{ and } B)$

$$= \frac{6}{12} + \frac{2}{12} - \frac{1}{12} = \frac{7}{12}$$

7. Let A = {brace A breaks} and B = {brace B breaks}. If the events are independent, then $P(B|A) = P(B)$.

$$P(B|A) = \frac{P(A \cap B)}{P(A)} = \frac{0.2}{0.4} = 0.5$$

    Since $P(B) = 0.5$, then A and B are independent.

8. Let A = {a heart is drawn first} and B = {a heart is drawn second}. On the first draw, there are 13 hearts in the deck of 52 cards. Therefore,

$$P(A) = \frac{13}{52} = \frac{1}{4}.$$ On the second draw, there are

only 51 cards remaining and only 12 hearts. Therefore, $P(B|A) = \frac{12}{51} = \frac{4}{17}$. We have

$$P(A \cap B) = P(A) \cdot P(B|A) = \frac{1}{4} \cdot \frac{4}{17} = \frac{1}{17}.$$

9. Let E = {no 9 is drawn}. There are 52 possibilities for the first card drawn, 51 possibilities for the second card drawn, 50 for the third and 49 for the fourth. Thus, $n(S) = 52 \cdot 51 \cdot 50 \cdot 49 = 6,497,400$. The first time a card is drawn, there are 48 cards that are not 9's, the second time there are 47 cards that are not 9's, 46 cards the third time and 45 cards the fourth time, so
    $n(E) = 48 \cdot 47 \cdot 46 \cdot 45 = 4,669,920$.

$$P(E) = \frac{4,669,920}{6,497,400} \approx 0.72.$$

10. The odds in favor are

$$\frac{P(E)}{1 - P(E)} = \frac{\frac{1}{8}}{1 - \frac{1}{8}} = \frac{\frac{1}{8}}{\frac{7}{8}} = \frac{1}{7}$$

    The odds in favor are 1 to 7.

11. Let D be the event that a person has the disease and let T be the event that the test is positive. The probability we wish to determine is $P(D|T)$. Then

$$P(D|T) = \frac{P(D \text{ and } T)}{P(T)}.$$

    A tree diagram will help us compute the needed probabilities.

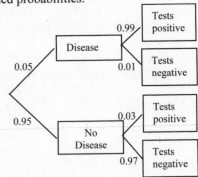

From the diagram, P(D and T) =
$P(D) \cdot P(T \mid D) = (0.05)(0.99)$. To
compute P(T) we need to combine two
branches from the diagram, one
corresponding to a correct positive test
result when the person has the disease, and
one corresponding to a false positive result
when the person does not have the disease:
P(T) = (0.05)(0.99) + (0.95)(0.03). Then

$$P(D \mid T) = \frac{P(D \text{ and } T)}{P(T)}$$

$$= \frac{(0.05)(0.99)}{(0.05)(0.99) + (0.95)(0.03)} \approx 0.635.$$

12. Let B = {person is a woman} and A =
{responded negatively}. From the table,
$n(A \cap B) = 642$, $n(A) = 1378$, and $n(S) = 2815$.
We have

$$P(B \mid A) = \frac{P(A \cap B)}{P(A)} = \frac{\frac{642}{2815}}{\frac{1378}{2815}} \approx 0.466 \,.$$

13. Make a Punnett square.

Parents	S	s
s	Ss	ss
s	Ss	ss

To have curly hair, the hamster must have two
s's. From the table, two of the four possible
genotypes have two s's, so the probability that an
offspring will have curly hair is $\frac{2}{4} = \frac{1}{2}$ or 50%.

14. The expected profit is
Expectation
= 0.18(75,000) + 0.36(50,000) +
0.31(25,000) + 0.08(0) +
0.05(-10,000) + 0.02(-20,000)
= 13,500 + 18,000 + 7750 + 0 − 500 − 400
= 38,350
The expected profit is $38,350.

Chapter 12: Statistics

EXERCISE SET 12.1

1. $\bar{x} = \dfrac{\sum x}{n} = \dfrac{2+7+5+7+14}{5} = 7$

   Ranking the numbers from smallest to largest gives 2, 5, 7, 7, 14. The middle number is 7. Thus 7 is the median.
   The number 7 occurs more often than other numbers. Thus 7 is the mode.

3. $\bar{x} = \dfrac{11+8+2+5+17+39+52+42}{8} = 22$

   Ranking the numbers from smallest to largest gives 2, 5, 8, 11, 17, 39, 42, 52. The two middle numbers are 11 and 17. The mean of 11 and 17 is 14. Thus 14 is the median. No number occurs more often than the others, so there is no mode.

5. $\bar{x} = \dfrac{\sum x}{n} = \dfrac{2.1+4.6+8.2+3.4+5.6+8.0+9.4+12.2+56.1+78.2}{10} \approx 18.8$

   Ranking the numbers from smallest to largest gives 2.1, 3.4, 4.6, 5.6, 8.0, 8.2, 9.4, 12.2, 56.1, 78.2. The two middle numbers are 8.0 and 8.2. The mean of 8.0 and 8.2 is 8.1. Thus 8.1 is the median. No number occurs more often than the others, so there is no mode.

7. $\bar{x} = \dfrac{\sum x}{n} = \dfrac{255+178+192+145+202+188+178+201}{8} \approx 192.4$

   Ranking the numbers from smallest to largest gives 145, 178, 178, 188, 192, 201, 202, 255. The two middle numbers are 188 and 192. The mean of 188 and 192 is 190. Thus 190 is the median. The number 178 occurs more often than the other numbers. Thus 178 is the mode.

9. $\bar{x} = \dfrac{\sum x}{n} = \dfrac{-12+-8+-5+-5+-3+0+4+9+21}{9} \approx 0.1$

   Ranking the numbers from smallest to largest gives –12, –8, –5, –5, –3, 0, 4, 9, 21. The middle number is -3. Thus –3 is the median. The number –5 occurs more often than the other numbers. Thus –5 is the mode.

11. a. Yes. The mean is computed by using the sum of all the data.

    b. No. The median is not affected unless the middle value, or one of the two middle values, in a data set is changed.

13. $\bar{x} = \dfrac{\sum x}{n} = \dfrac{1292}{34} = 38$

    Ranking the numbers from smallest to largest shows the two middle number to be 35 and 35. Thus 35 is the median. The numbers 33 and 35 each occur four times. Thus the mode is 33 and 35.

15. a. Answers will vary.

    b. Answers will vary.

17. Weighted Mean $= \dfrac{\sum(x \cdot w)}{\sum w}$

$= \dfrac{(96 \times 1)+(88 \times 1)+(60 \times 1)+(76 \times 1)+(85 \times 2)+(75 \times 2)+(90 \times 2)+(85 \times 2 \times 2)}{14} = \dfrac{1160}{14} \approx 82.9$

19. Weighted Mean $= \dfrac{\sum(x \cdot w)}{\sum w}$

$= \dfrac{(70 \times 10\%)+(65 \times 10\%)+(82 \times 10\%)+(94 \times 10\%)+(85 \times 10\%)+(92 \times 20\%)+(80 \times 30\%)}{15} = \dfrac{82}{1} = 82$

21. Weighted Mean $= \dfrac{\sum(x \cdot w)}{\sum w} = \dfrac{(3 \times 4)+(4 \times 3)+(1 \times 3)+(2 \times 4)}{14} = \dfrac{35}{14} \approx 2.5$

23. Mean $= \dfrac{\sum(x \cdot f)}{\sum f} = \dfrac{(2 \cdot 6)+(4 \cdot 5)+(5 \cdot 6)+(9 \cdot 3)+(10 \cdot 1)+(14 \cdot 2)+(19 \cdot 1)}{24} = \dfrac{146}{24} \approx 6.1$ points

If the data were written in a list (6 twos, 5 fours, 6 fives, 3 nines, 1 ten, 2 fourteens, 1 nineteen), the middle number would be 5. Thus 5 points is the median. 2 points and 5 points occur most. Thus 2 and 5 points are the mode.

25. Mean $= \dfrac{\sum(x \cdot f)}{\sum f} = \dfrac{(2 \cdot 1)+(4 \cdot 2)+(6 \cdot 7)+(7 \cdot 12)+(8 \cdot 10)+(9 \cdot 4)+(10 \cdot 3)}{39} = \dfrac{282}{39} \approx 7.2$

The middle score is 7. Thus 7 is the median. The score of 7 occurs more often than any other score. Thus 7 is the mode.

27. Midrange $= \dfrac{52° + 76°}{2} = 64°$

29. Midrange $= \dfrac{-56°F + 44°F}{2} = -6°F$

31. Let x be the score Ruben needs on the next test. The mean score has a weight of 6 and the new score has a weight of 1, so

$80 = \dfrac{(78 \times 6)+(x \times 1)}{7}$

$560 = 468 + x$

$92 = x$

33. a. $\bar{x} = \dfrac{0.336 + 0.213}{2} \approx 0.275$

   b. Batting average

   $\dfrac{92 + 60}{274 + 282} \approx 0.273$

c. No.

35. n = number removed $\qquad \bar{x} = 48$

$\dfrac{12\bar{x} - n}{12 - 1} = 45$

$\dfrac{12(48) - n}{11} = 45$

$576 - n = 495$

$81 = n$

37. The route is the same distance both ways, so $d_1 = d_2$.

Solving the distance/rate/time formula for time gives $t_1 = \dfrac{d_1}{r_1}; t_2 = \dfrac{d_2}{r_2}$. Since

$d_1 = d_2$, $t_2 = \dfrac{d_1}{r_2}$. The average rate for the round trip is equal to the total distance by the total time. $r = \dfrac{d_1 + d_2}{t_1 + t_2}$. Substituting the results from above and simplifying gives the desired result:

$$r = \frac{d_1 + d_2}{t_1 + t_2} = \frac{d_1 + d_1}{t_1 + t_2} = \frac{2d_1}{\dfrac{d_1}{r_1} + \dfrac{d_1}{r_2}}$$

$$= \frac{2d_1}{d_1 \left(\dfrac{1}{r_1} + \dfrac{1}{r_2} \right)} = \frac{2}{\dfrac{1}{r_1} + \dfrac{1}{r_2}} \cdot \left(\frac{r_1 r_2}{r_1 r_2} \right)$$

$$= \frac{2 r_1 r_2}{r_1 + r_2}$$

39. Yes. In the first month, Dawn's average is 0.4 and Joanne's average is 0.3973. In the second month, Dawn's average is 0.3878 and Joanne's average is 0.3875. For both months, Dawn has a total of 21 hits at 54 at-bats and an average of 0.3889, while Joanne has a total of 60 hits at 153 at-bats and an average of 0.3922. Joanne has a smaller average for the first month and the second month, but she has a larger average for both months combined.

EXERCISE SET 12.2

1. The highest temperature is $63°$ F and the lowest temperature is $-21°$ F. The range is $63°\,F - (-21°F) = 84°$ F.

3. The largest data value is 22 and the smallest data value is 1. The range is $22 - 1 = 21$.

$$\overline{x} = \frac{1 + 2 + 5 + 7 + 8 + 19 + 22}{7}$$

$$= \frac{64}{7} \approx 9.1429$$

x	$x - \overline{x}$	$(x - \overline{x})^2$
1	$1 - 9.1429$	$(-8.1429)^2 = 66.3068$
2	$2 - 9.1429$	$(-7.1429)^2 = 51.0210$
5	$5 - 9.1429$	$(-4.1429)^2 = 17.1636$
7	$7 - 9.1429$	$(-2.1429)^2 = 4.5920$
8	$8 - 9.1429$	$(-1.1429)^2 = 1.3062$
19	$19 - 9.1429$	$(9.8571)^2 = 97.1624$
22	$22 - 9.1429$	$(12.8571)^2 = 165.3050$
		402.857

$$\frac{\text{sum of the (deviations)}^2}{n-1} = \frac{402.857}{7-1}$$

$$= \frac{402.857}{6} \approx 67.1428$$

The square root of the quotient is
$s = \sqrt{67.1428} \approx 8.2$.
The variance is the square of the standard deviation.
$s^2 = (\sqrt{67.1428})^2 \approx 67.1$

5. The largest data value is 4.8 and the smallest data value is 1.5. The range is $4.8 - 1.5 = 3.3$.

$$\overline{x} = \frac{2.1 + 3.0 + 1.9 + 1.5 + 4.8}{5} = \frac{13.3}{5} = 2.66$$

x	$x - \overline{x}$	$(x - \overline{x})^2$
2.1	$2.1 - 2.66$	$(-0.56)^2 = 0.3136$
3.0	$3.0 - 2.66$	$(0.34)^2 = 0.1156$
1.9	$1.9 - 2.66$	$(-0.76)^2 = 0.5776$
1.5	$1.5 - 2.66$	$(-1.16)^2 = 1.3456$
4.8	$4.8 - 2.66$	$(2.14)^2 = 4.5796$
		6.932

$$\frac{\text{sum of (deviations)}^2}{n-1} = \frac{6.932}{5-1} = \frac{6.932}{4} = 1.733$$

The square root of the quotient is
$s = \sqrt{1.733} \approx 1.3$.

Variance is the square of the standard deviation.
Thus the variance is $s^2 = (\sqrt{1.733})^2 \approx 1.7$.

7. The largest data value is 100 and the smallest data value is 48. The range is $100 - 48 = 52$.

$$\bar{x} = \frac{48 + 91 + 87 + 93 + 59 + 68 + 92 + 100 + 81}{9} = \frac{719}{9} = 79.889$$

x	$x - \bar{x}$	$(x - \bar{x})^2$
48	$48 - 79.889$	$(-31.889)^2 = 1016.90$
91	$91 - 79.889$	$(11.111)^2 = 123.454$
87	$87 - 79.889$	$(7.111)^2 = 50.566$
93	$93 - 79.889$	$(13.11)^2 = 171.898$
59	$59 - 79.889$	$(-20.889)^2 = 436.350$
68	$68 - 79.889$	$(-11.889)^2 = 141.348$
92	$92 - 79.889$	$(12.111)^2 = 146.676$
100	$100 - 79.889$	$(20.111)^2 = 404.452$
81	$81 - 79.889$	$(1.111)^2 = 1.234$
		2492.878

$$\frac{\text{sum of the (deviations)}^2}{n-1} = \frac{2492.878}{9-1} = \frac{2492.878}{8}$$
$$= 311.610$$

The square root of the quotient is $s = \sqrt{311.610} \approx 17.7$
Variance is the square of the standard deviation. Thus the variance is $s^2 = (\sqrt{311.610})^2 \approx 311.6$.

9. All of the data values are 4. Thus, the range is $4 - 4 = 0$.

$$\bar{x} = \frac{4+4+4+4+4+4+4+4+4+4+4+4+4+4+4+4+4}{17}$$

x	$x - \bar{x}$	$(x - \bar{x})^2$
4	$4 - 4$	$(0)^2 = 0$
4	$4 - 4$	$(0)^2 = 0$
4	$4 - 4$	$(0)^2 = 0$
4	$4 - 4$	$(0)^2 = 0$
4	$4 - 4$	$(0)^2 = 0$
4	$4 - 4$	$(0)^2 = 0$
etc.	etc.	etc.
		0

$$\frac{\text{sum of the (deviations)}^2}{n-1} = \frac{0}{14-1} = \frac{0}{13} = 0.$$

The square root of the quotient is $s = \sqrt{0} = 0$.
Variance is the square of the standard deviation. Thus the variance is $s^2 = (\sqrt{0})^2 = 0$.

11. The largest data value is 11 and the smallest data value is -12. The range is $11 - (-12) = 23$.

$$\bar{x} = \frac{-8 + (-5) + (-12) + (-1) + 4 + 7 + 11}{7} = \frac{-4}{7} = -0.571$$

x	$x - \bar{x}$	$(x - \bar{x})^2$
-8	$-8 - (-0.571)$	$(-7.429)^2 = 55.190$
-5	$-5 - (-0.571)$	$(-4.429)^2 = 19.616$
-12	$-12 - (-0.571)$	$(-11.429)^2 = 130.622$
-1	$-1 - (-0.571)$	$(-0.429)^2 = 0.184$
4	$4 - (-0.571)$	$(4.571)^2 = 20.894$
7	$7 - (-0.571)$	$(7.571)^2 = 57.320$
11	$11 - (-0.571)$	$(11.571)^2 = 133.888$
		417.714

$$\frac{\text{sum of the (deviations)}^2}{n-1} = \frac{417.714}{7-1} = \frac{417.714}{6}$$
$$= 69.619$$

The square root of the quotient is $s = \sqrt{69.619} \approx 8.3$.
Variance is the square of the standard deviation. Thus the variance is
$$s^2 = (\sqrt{69.619})^2 \approx 69.6.$$

13. Opinions will vary. However, many climbers would consider rope B to be safer because of its small standard deviation.

15. The students in the college statistics course because the range of weights is greater.

17. a. Using a calculator, the mean ($\bar{x}$), to the nearest tenth, is 30.1 and the population standard deviation (σx), to the nearest tenth, is 10.1.

    b. Using a calculator, the mean ($\bar{x}$), to the nearest tenth, is 15.6 and the population standard deviation (σx), to the nearest tenth, is 6.5.

    c. Winning scores; winning scores

19. a. Using a calculator, the mean ($\bar{x}$), to the nearest tenth, is 44.9 and the population standard deviation (σx), to the nearest tenth, is 9.3.

    b. Using a calculator, the mean ($\bar{x}$), to the nearest tenth, is 42.8 and the population standard deviation (σx), to the nearest tenth, is 7.7.

    c. National League; National League

21. Using a calculator, the mean ($\bar{x}$), to the nearest tenth, is 54.8 years and the population standard deviation (σx), to the nearest tenth, is 6.2 years.

23. a. Answers will vary.

    b. The population standard deviation remains the same.

25. a. 0

    b. Yes

    c. No. Compare the standard deviations for the following two samples: 2, 4, 6, 8, 10 and 1, 3, 5, 7, 9

EXERCISE SET 12.3

1. a. $z_{85} = \dfrac{85 - 75}{11.5} \approx 0.87$

    b. $z_{95} = \dfrac{95 - 75}{11.5} \approx 1.74$

    c. $z_{50} = \dfrac{50 - 75}{11.5} \approx -2.17$

    d. $z_{75} = \dfrac{75 - 75}{11.5} = 0$

3. a. $z_{6.2} = \dfrac{6.2 - 6.8}{1.9} \approx -0.32$

    b. $z_{7.2} = \dfrac{7.2 - 6.8}{1.9} \approx 0.21$

    c. $z_{9.0} = \dfrac{9.0 - 6.8}{1.9} \approx 1.16$

    d. $z_{5.0} = \dfrac{5.0 - 6.8}{1.9} \approx -0.95$

5. a. $z_{110.5} = \dfrac{110.5 - 119.4}{13.2} \approx -0.67$

    b. Solving for x:
    $$z_x = \frac{x - 119.4}{13.2} = 2.15$$
    $$x - 119.4 \approx 28.4$$
    $$x = 147.8$$
    Her systolic blood pressure reading was 147.8 mm Hg.

7. a. $z_{214} = \dfrac{214 - 182}{44.2} \approx 0.72$

    b. Solving for x:
    $$z_x = \frac{x - 182}{44.2} = -1.58$$
    $$x - 182 = -69.836$$
    $$x \approx 112$$
    His blood cholesterol level is approximately 112 mg/dl.

9. a. $z_{65} = \dfrac{65 - 72}{8.2} \approx -0.85$

    b. $z_{102} = \dfrac{102 - 130}{18.5} \approx -1.51$

    c. $z_{605} = \dfrac{605 - 720}{116.4} \approx -0.99$
    The 65 is the highest relative score.

11. Percentile of $455 = \dfrac{4256}{7210} \cdot 100 \approx 59$
    Shaylen's score was at the 59th percentile.

13. Kevin's percentile $= \dfrac{x}{9840} \cdot 100 = 65$

$x = \dfrac{65 \cdot 9840}{100} = 6396$

6396 students scored lower than Kevin.

15. a. The median is by definition the 50th percentile. Therefore, 50% of the four-person family incomes are more than $57,500.

    b. Because $70,400 is the 88th percentile, $100 - 88 = 12\%$ of the four-person family incomes are more than $70,400.

    c. $88 - 50 = 38\%$ of the four-person family incomes are between $57,500 and $70,400.

17. 1, 3, 4, 5, 5, 7, 8, 10, 10, 10, 12, 15, 18, 26, 28, 32, 41, 85

The median (or second quartile) is the average of the 9th and 10th data values. $\dfrac{10+10}{2} = 10$.

The first quartile is the median of the 1st through 9th data values. 5.

The third quartile is the median of the 10th through 18th data values. 26.

$Q_1 = 5$, $Q_2 = 10$, $Q_3 = 26$

19.

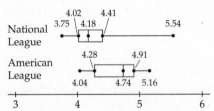

The ERAs in the National League tend to be smaller than the ERAs in the American League. Also, the range of the ERAs is larger for the National League.

21.

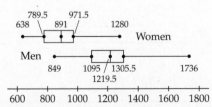

The median salary for men was approximately as high as the highest salary for women. The

lowest salary for men was approximately the median salary for women.

23. a. Using the formula:

$\dfrac{-1.101 - 0.944 - 0.157 + 0.629 + 1.573}{5} = 0$

Mean $= 0$

$(x - \bar{x})_1 = -1.101$

$(x - \bar{x})_2 = -0.944$

$(x - \bar{x})_3 = -0.157$

$(x - \bar{x})_4 = 0.629$

$(x - \bar{x})_5 = 1.573$

$(-1.101)^2 + (-0.944)^2 + (-0.157)^2$
$+ (0.629)^2 + (1.573)^2 \approx 5$

$\dfrac{5}{5} = 1$

$\sqrt{1} = 1$

The mean of the z-scores is 0, and their standard deviation is 1.

    b. Using the formula:

$\dfrac{-1.459 - 0.973 - 0.243 + 0.365 + 0.973 + 1.337}{6}$

$= 0$

Mean $= 0$

$(x - \bar{x})_1 = -1.459$

$(x - \bar{x})_2 = -0.973$

$(x - \bar{x})_3 = -0.243$

$(x - \bar{x})_4 = 0.365$

$(x - \bar{x})_5 = 0.973$

$(x - \bar{x})_6 = 1.337$

$(-1.459)^2 + (-0.973)^2 + (-0.243)^2$
$+ (0.365)^2 + (0.973)^2 + (1.337)^2 \approx 6$

$\dfrac{6}{6} = 1$

$\sqrt{1} = 1$

The mean of the z-scores is 0, and their standard deviation is 1.

c.   The mean of any set of z-scores is always 0, and their standard deviation is always 1.

25. A number is an outlier for a set of data provided that the number is less than $Q_1 - 1.5(Q_3 - Q_1)$ or greater than $Q_3 + 1.5(Q_3 - Q_1)$. 85 falls outside those boundaries; therefore, 85 is an outlier.

EXERCISE SET 12.4

1.   For right-skewing distributions, the mean is greater than the median. Therefore, the mean income of family practice physicians is greater than the median income.

3.   a.   95.4% of the data lies within 2 standard deviations of the mean.

     b. $13.6 + 2.15 + 0.15 = 15.9\%$
          15.9% of the data lie above 1 standard deviation of the mean.

     c. $34.1 + 34.1 + 13.6 = 81.8\%$
          81.8% of the data lies between 1 standard deviation below the mean and 2 standard deviations above the mean.

5.   a.   12 ounces is 2 standard deviations below the mean, and 30 ounces is 1 standard deviation above the mean. $34.1 + 34.1 + 13.6 = 81.8\%$. 81.8% of the parcels weighed between 12 and 30 ounces.

     b.   42 ounces is 3 standard deviations above the mean. Therefore, 0.15% of the parcels weighed more than 42 ounces.

7.   a.   68 miles per hour is 1 standard deviation above the mean. $(0.159)(8000) = 1272$. 1272 vehicles had a speed of more than 68 miles per hour.

     b.   40 miles per hour is 3 standard deviation below the mean. $(0.0015)(8000) = 12$. 12 vehicles had a speed of less than 40 miles per hour.

9.   When z is 1.5, A is 0.433. When z is 0, A is 0. Therefore, the area of the distribution between $z = 0$ and $z = 1.5$ is 0.433 square unit.

11. The area of the distribution between $z = 0$ and $z = -1.85$ is equal to the area of the distribution

between $z = 0$ and $z = 1.85$, due to symmetry. When z is 1.85, A is 0.468. When z is 0, A is 0. Therefore, the area of the distribution between $z = 0$ and $z = -1.85$ is 0.468 square unit.

13. When z is 1.9, A is 0.471. When z is 1, A is 0.341. $0.471 - 0.341 = 0.130$. The area of the distribution between $z = 1$ and $z = 1.9$ is 0.130 square unit.

15. The area of the distribution between $z = 0$ and $z = -1.47$ is equal to the area of the distribution between $z = 0$ and $z = 1.47$, due to symmetry. When z is 1.47, A is 0.429. When z is 0, A is 0. When z is 1.64, A is 0.450. When z is 0, A is 0. $0.429 + 0.450 = 0.879$.
     The area of the distribution between $z = -1.47$ and $z = 1.64$ is 0.879 square unit.

17. The area of the distribution between $z = 0$ and $z = 1.3$ is 0.403 square unit. The area of the distribution to the right of $z = 0$ is 0.500 square unit. $0.500 - 0.403 = 0.097$.
     The area of the distribution where $z > 1.3$ is 0.097 square unit.

19. The area of the distribution between $z = 0$ and $z = -2.22$ is 0.487 square unit. The area of the distribution to the left of $z = 0$ is 0.500 square unit. $0.500 - 0.487 = 0.013$. The area of the distribution where $z < -2.22$ is 0.013 square unit.

21. The area of the distribution between $z = 0$ and $z = -1.45$ is 0.427 square unit. The area of the distribution to the right of $z = 0$ is 0.500 square unit. $0.427 + 0.500 = 0.927$. The area of the distribution where $z \geq -1.45$ is 0.927 square unit.

23. The area of the distribution between $z = 0$ and $z = 2.71$ is 0.497 square unit. The area of the distribution to the left of $z = 0$ is 0.500 square unit. $0.497 + 0.500 = 0.997$. The area of the distribution where $z < 2.71$ is 0.997 square unit.

25. If 0.200 square unit of the distribution is to the right of z, then 0.300 square unit is between z and 0. At $A = 0.300$, $z = 0.84$.
     The z-score is 0.84.

27. If 0.184 square unit of the distribution is to the left of z, then 0.316 square unit is between z and 0. At $A = 0.316$, $z = 0.90$.
     Because z is to the left of 0, the z-score is -0.90.

29. If 0.363 square unit of the distribution is to the right of z, then 0.137 square unit is between z and 0. At $A = 0.137$, $z = 0.35$. The z-score is 0.35.

31. a. $z_{51.0} = \dfrac{51.0 - 44.8}{12.4} = 0.5$

    0.192 square unit is between $z = 0$ and $z = 0.5$. Therefore, 0.308 square unit is to the right of $z = 0.5$. 30.8% of the data is greater than 51.0.

    b. Calculating:
    $z_{47.9} = \dfrac{47.9 - 44.8}{12.4} = 0.25$

    $z_{63.4} = \dfrac{63.4 - 44.8}{12.4} = 1.5$

    0.099 square unit is between $z = 0$ and $z = 0.25$, and 0.433 square unit is between $z = 0$ and $z = 1.5$.
    $0.433 - 0.099 = 0.334$. 33.4% of the data is between 47.9 and 63.4.

33. a. $z_{404} = \dfrac{404 - 580}{160} = -1.1$

    0.364 square unit is between $z = -1.1$ and $z = 0$. Therefore, 0.136 square unit is to the left of $z = -1.1$. 13.6% of the data is less than 404.

    b. Calculating:
    $z_{460} = \dfrac{460 - 580}{160} = -0.75$

    $z_{612} = \dfrac{612 - 580}{160} = 0.2$

    0.273 square unit is between $z = -0.75$ and $z = 0$. 0.079 square unit is between $z = 0$ and $z = 0.2$. $0.273 + 0.079 = 0.352$. 35.2% of the data is between 460 and 612.

35. a. $z_{14} = \dfrac{14 - 14.5}{0.4} = -1.25$

    0.394 square unit is between $z = -1.25$ and $z = 0$. Therefore, 0.106 square unit is to the left of $z = -1.25$. 10.6% of the boxes will weigh less than 14 ounces.

    b. Calculating:
    $z_{13.5} = \dfrac{13.5 - 14.5}{0.4} = -2.5$

    $z_{15.5} = \dfrac{15.5 - 14.5}{0.4} = 2.5$

    0.494 square unit is between $z = -2.5$ and

$z = 0$, as well as $z = 0$ and $z = 2.5$.
$0.494 + 0.494 = 0.988$. 98.8% of the boxes will weigh between 13.5 and 14.5 ounces.

37. a. $z_{320} = \dfrac{320 - 350}{24} = -1.25$

    0.394 square unit is between $z = -1.25$ and $z = 0$. Therefore, 0.106 square unit is to the right of $z = -1.25$. The probability that a piece of this rope chosen at random will have a breaking point of less than 320 pounds is 0.106.

    b. Calculating:
    $z_{340} = \dfrac{340 - 350}{24} \approx -0.42$

    $z_{370} = \dfrac{370 - 350}{24} \approx 0.83$

    0.163 square unit is between $z = -0.42$ and $z = 0$. 0.297 square unit is between $z = 0$ and $z = 0.83$. $0.163 + 0.297 = 0.460$. The probability of a piece of this rope chosen at random will have a breaking point of between 340 and 370 pounds is 0.460.

39. a. $z_3 = \dfrac{3 - 2.5}{0.75} \approx 0.67$

    0.249 square unit is between $z = 0$ and $z = 0.67$. 0.500 square unit is to the left of $z = 0$. $0.249 + 0.500 = 0.749$. The probability that the time a customer spends waiting is at most 3 minutes is 0.749.

    b. $z_1 = \dfrac{1 - 2.5}{0.75} = -2$

    0.477 square unit is between $z = -2$ and $z = 0$. Therefore, 0.023 square unit is left of $z = -2$. The probability that the time a customer spends waiting is less than 1 minute is 0.023.

41. Answers will vary.

43. True.

45. True.

47. False.

49. False.

51. True.

53. a.   The two areas are identical.

b. The area of the region is the same with or without the vertical boundary at $z = 1$.

55. To bound the middle 60% of a normal distribution, $A = 0.300$. The z-value for $A = 0.300$ is 0.84. The two z-values are -0.84 and 0.84.

EXERCISE SET 12.5

1. a. Graph b, as the points are clustered closer to the least-squares line.

   b. Graph c, as the points are clustered closer to the least-squares line.

3. a.

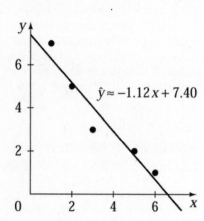

$$\hat{y} \approx -1.12x + 7.40$$

   b. $n = 5$
   $\sum x = 1 + 2 + 3 + 5 + 6 = 17$
   $\sum y = 7 + 5 + 3 + 2 + 1 = 18$
   $\sum x^2 = 1 + 4 + 9 + 25 + 36 = 75$
   $(\sum x)^2 = 17^2 = 289$
   $\sum xy = 7 + 10 + 9 + 10 + 6 = 42$

   c. Calculating:
   $$a = \frac{n(\sum xy) - (\sum x)(\sum y)}{n(\sum x^2) - (\sum x)^2}$$
   $$= \frac{5(42) - (17)(18)}{5(75) - 289} \approx -1.12$$
   $$b = \overline{y} - a\overline{x} \approx 7.40$$

   d. See the graph in part a.

   e. Yes.

   f. $\hat{y} = (-1.12)(3.4) + 7.40 \approx 3.6$

g. Calculating:
$$r = \frac{n(\sum xy) - (\sum x)(\sum y)}{\sqrt{n(\sum x^2) - (\sum x)^2} \cdot \sqrt{n(\sum y^2) - (\sum y)^2}}$$
$$= \frac{5(42) - (17)(18)}{\sqrt{5(75) - 289} \cdot \sqrt{5(88) - 324}} \approx -0.96$$

5. $n = 5$
$\sum x = 2 + 3 + 4 + 6 + 8 = 23$
$\sum y = 6 + 6 + 8 + 11 + 18 = 49$
$\sum x^2 = 4 + 9 + 16 + 36 + 64 = 129$
$\sum y^2 = 36 + 36 + 64 + 121 + 324 = 581$
$(\sum x)^2 = 23^2 = 529$
$(\sum y)^2 = 49^2 = 2401$
$\sum xy = 12 + 18 + 32 + 66 + 144 = 272$
$$a = \frac{n(\sum xy) - (\sum x)(\sum y)}{n(\sum x^2) - (\sum x)^2}$$
$$= \frac{5(272) - (23)(49)}{5(129) - 529} \approx 2.01$$
$$b = \overline{y} - a\overline{x} \approx 0.56$$
$$\hat{y} = 2.01x + 0.56$$
$$r = \frac{n(\sum xy) - (\sum x)(\sum y)}{\sqrt{n(\sum x^2) - (\sum x)^2} \cdot \sqrt{n(\sum y^2) - (\sum y)^2}}$$
$$= \frac{5(272) - (23)(49)}{\sqrt{5(129) - 529} \cdot \sqrt{5(581) - 2401}} \approx 0.96$$

7. $n = 5$
$\sum x = -3 - 1 + 0 + 2 + 5 = 3$
$\sum y = 11.8 + 9.5 + 8.6 + 8.7 + 5.4 = 44$
$\sum x^2 = 9 + 1 + 0 + 4 + 25 = 39$
$\sum y^2 = 139.24 + 90.25 + 73.96 + 75.69 + 29.16 = 408.3$
$(\sum x)^2 = 3^2 = 9$
$(\sum y)^2 = 44^2 = 1936$
$\sum xy = -35.4 - 9.5 + 0 + 17.4 + 27 = -0.5$
$$a = \frac{n(\sum xy) - (\sum x)(\sum y)}{n(\sum x^2) - (\sum x)^2}$$
$$= \frac{5(-0.5) - (3)(44)}{5(39) - 9} \approx -0.72$$
$$b = \overline{y} - a\overline{x} \approx 9.23$$
$$\hat{y} = -0.72x + 9.23$$
$$r = \frac{n(\sum xy) - (\sum x)(\sum y)}{\sqrt{n(\sum x^2) - (\sum x)^2} \cdot \sqrt{n(\sum y^2) - (\sum y)^2}}$$
$$= \frac{5(-0.5) - (3)(44)}{\sqrt{5(39) - 9} \cdot \sqrt{5(408.3) - 1936}} \approx -0.96$$

9. $n = 5$

$\sum x = 1 + 2 + 4 + 6 + 8 = 21$

$\sum y = 4.1 + 6.0 + 8.2 + 11.5 + 16.2 = 46$

$\sum x^2 = 1 + 4 + 16 + 36 + 64 = 121$

$\sum y^2 = 16.81 + 36 + 67.24 + 132.25 + 262.44 = 514.74$

$(\sum x)^2 = 21^2 = 441$

$(\sum y)^2 = 46^2 = 2116$

$\sum xy = 4.1 + 12.0 + 32.8 + 69 + 129.6 = 247.5$

$a = \dfrac{n(\sum xy) - (\sum x)(\sum y)}{n(\sum x^2) - (\sum x)^2}$

$= \dfrac{5(247.5) - (21)(46)}{5(121) - 441} \approx 1.66$

$b = \overline{y} - a\overline{x} \approx 2.25$

$\hat{y} = 1.66x + 2.25$

$r = \dfrac{n(\sum xy) - (\sum x)(\sum y)}{\sqrt{n(\sum x^2) - (\sum x)^2} \cdot \sqrt{n(\sum y^2) - (\sum y)^2}}$

$= \dfrac{5(247.5) - (21)(46)}{\sqrt{5(121) - 441} \cdot \sqrt{5(514.74) - 2116}} \approx 0.99$

11. a. $n = 6$

$\sum x = 13{,}000 + 18{,}000 + 20{,}000 + 25{,}000 + 29{,}000 + 32{,}000 = 137{,}000$

$\sum y = 46{,}100 + 44{,}600 + 43{,}300 + 41{,}975 + 40{,}975 + 39{,}750 = 256{,}700$

$\sum x^2 = (13{,}000)^2 + (18{,}000)^2 + (20{,}000)^2 + (25{,}000)^2 + (29{,}000)^2 + (32{,}000)^2 = 3{,}383{,}000{,}000$

$\sum y^2 = (46{,}100)^2 + (44{,}600)^2 + (43{,}300)^2 + (41{,}975)^2 + (40{,}975)^2 + (39{,}750)^2 = 11{,}010{,}173{,}750$

$(\sum x)^2 = (137{,}000)^2 = 18{,}769{,}000{,}000$

$(\sum y)^2 = (256{,}700)^2 = 65{,}894{,}890{,}000$

$\sum xy = (13{,}000)(46{,}100) + (18{,}000)(44{,}600) + (20{,}000)(43{,}300) + (25{,}000)(41{,}975) + (29{,}000)(40{,}975) + (32{,}000)(39{,}750) = 5{,}777{,}750{,}000$

$a = \dfrac{n(\sum xy) - (\sum x)(\sum y)}{n(\sum x^2) - (\sum x)^2}$

$= \dfrac{6(5{,}777{,}750{,}000) - (137{,}000)(256{,}700)}{6(3{,}383{,}000{,}000) - 18{,}769{,}000{,}000}$

≈ -0.328

$b = \overline{y} - a\overline{x} \approx 50{,}271$

$\hat{y} = -0.328x + 50{,}271$

b. $\hat{y} = -0.328(30{,}000) + 50{,}271 = 40{,}431$

There car will be worth \$40,431.

c. Calculating:

$r = \dfrac{n(\sum xy) - (\sum x)(\sum y)}{\sqrt{n(\sum x^2) - (\sum x)^2} \cdot \sqrt{n(\sum y^2) - (\sum y)^2}}$

$= \dfrac{6(5{,}777{,}750{,}000) - (137{,}000)(256{,}700)}{\sqrt{6(3{,}383{,}000{,}000) - 18{,}769{,}000{,}000} \cdot \sqrt{6(11{,}010{,}173{,}750) - 65{,}894{,}890{,}000}}$

≈ -0.99

d. As the number of miles the car is driven increases, the value of the car decreases.

13. a. $n = 8$

$\sum x = 0 + 1 + 2 + 3 + 4 + 5 + 6 + 7 = 28$

$\sum y = 16.3 + 16.3 + 17.1 + 18.4 + 19.9 + 20.6 + 21.2 + 23.1 = 152.9$

$\sum x^2 = 0 + 1 + 4 + 9 + 16 + 25 + 36 + 49 = 140$

$(\sum x)^2 = 28^2 = 784$

$\sum xy = 0 + 16.3 + 34.2 + 55.2 + 79.6 + 103 + 127.2 + 161.7 = 577.2$

$a = \dfrac{n(\sum xy) - (\sum x)(\sum y)}{n(\sum x^2) - (\sum x)^2}$

$= \dfrac{8(577.2) - (28)(152.9)}{8(140) - 784} \approx 1.0$

$b = \bar{y} - a\bar{x} \approx 15.6$

$\hat{y} = x + 15.6$

b. $\hat{y} = 10 + 15.6 \approx 25.6$

The predicted percentage of overweight males in 2005 is 25.6%.

15. a. $n = 6$

$\sum x = 0 + 1 + 2 + 3 + 4 + 5 = 15$

$\sum y = 69 + 86 + 109 + 128 + 141 + 159 = 692$

$\sum x^2 = 0 + 1 + 4 + 9 + 16 + 25 = 55$

$\sum y^2 = 4761 + 7396 + 11,881 + 16,384 + 19,881 + 25,281 = 85,584$

$(\sum x)^2 = 15^2 = 225$

$(\sum y)^2 = 692^2 = 478,864$

$\sum xy = 0 + 86 + 218 + 384 + 564 + 795 = 2047$

$r = \dfrac{n(\sum xy) - (\sum x)(\sum y)}{\sqrt{n(\sum x^2) - (\sum x)^2} \cdot \sqrt{n(\sum y^2) - (\sum y)^2}}$

$= \dfrac{6(2047) - (15)(692)}{\sqrt{6(55) - 225} \cdot \sqrt{6(85,584) - 478,864}}$

≈ 0.99

b. Yes, at least for the years 1998 to 2003. The correlation coefficient is very close to 1, which indicates a near perfect linear correlation.

17. a. $n = 6$

$\sum x = 415 + 442 + 462 + 476 + 513 + 530 = 2838$

$\sum y = 540 + 547 + 550 + 561 + 601 + 609 = 3408$

$\sum x^2 = 172,225 + 195,364 + 213,444 + 226,576 + 263,169 + 280,900 = 1,351,678$

$\sum y^2 = 291,600 + 299,209 + 302,500 + 314,721 + 361,201 + 370,881 = 1,940,112$

$(\sum x)^2 = (2838)^2 = 8,054,244$

$(\sum y)^2 = (3408)^2 = 11,614,464$

$\sum xy = 224,100 + 241,774 + 254,100 + 267,036 + 308,313 + 322,770 = 1,618,093$

$r = \dfrac{n(\sum xy) - (\sum x)(\sum y)}{\sqrt{n(\sum x^2) - (\sum x)^2} \cdot \sqrt{n(\sum y^2) - (\sum y)^2}}$

$= \dfrac{6(1,618,093) - (2838)(3408)}{\sqrt{6(1,351,678) - 8,054,244} \cdot \sqrt{6(1,940,112) - 11,614,464}}$

≈ 0.96

b. Yes. The median income of men rose as the median income of women rose.

19. a. $n = 3$

$\sum x = -40 + 0 + 100 = 60$

$\sum y = -40 + 32 + 212 = 204$

$\sum x^2 = 1600 + 0 + 10,000 = 11,600$

$\sum y^2 = 1600 + 1024 + 44,944 = 47,568$

$(\sum x)^2 = 60^2 = 3600$

$(\sum y)^2 = 204^2 = 41,616$

$\sum xy = 1600 + 0 + 21,200 = 22,800$

$r = \dfrac{n(\sum xy) - (\sum x)(\sum y)}{\sqrt{n(\sum x^2) - (\sum x)^2} \cdot \sqrt{n(\sum y^2) - (\sum y)^2}}$

$= \dfrac{3(22,800) - (60)(204)}{\sqrt{3(11,600) - 3600} \cdot \sqrt{3(47,568) - 41,616}}$

$= 1$

b. The data displays a perfect linear relationship.

c. $\hat{y} = 1.8x + 32$

d. $\hat{y} = 1.8(35) + 32 = 95$
    The Fahrenheit temperature that corresponds to 35°C should be 95°F.

e. As the predicted data value falls between data values that were used to obtain the least-squares line equation, this is an interpolation.

21. $n = 6$

$\sum x = 14{,}709 + 15{,}518 + 16{,}223 + 17{,}272 + 18{,}273 + 19{,}710 = 101{,}705$

$\sum y = 3247 + 3362 + 3487 + 3725 + 4081 + 4694 = 22{,}596$

$\sum x^2\; 216{,}354{,}681 + 240{,}808{,}324 + 263{,}185{,}729 + 298{,}321{,}984 + 333{,}902{,}529 + 388{,}484{,}100 = 1{,}741{,}057{,}347$

$\sum y^2 = 10{,}543{,}009 + 11{,}303{,}044 + 12{,}159{,}169 + 13{,}875{,}625 + 16{,}654{,}561 + 22{,}033{,}636 = 86{,}569{,}044$

$(\sum x)^2 = 101{,}705^2 = 10{,}343{,}907{,}025$

$(\sum y)^2 = 22{,}596^2 = 510{,}579{,}216$

$\sum xy = 47{,}760{,}123 + 52{,}171{,}516 + 56{,}569{,}601 + 64{,}338{,}200 + 74{,}572{,}113 + 92{,}518{,}740 = 387{,}930{,}293$

$$r = \frac{n(\sum xy) - (\sum x)(\sum y)}{\sqrt{n(\sum x^2) - (\sum x)^2} \cdot \sqrt{n(\sum y^2) - (\sum y)^2}}$$

$$= \frac{6(387{,}930{,}293) - (101{,}705)(22{,}596)}{\sqrt{6(1{,}741{,}057{,}347) - 10{,}343{,}907{,}025} \cdot \sqrt{6(86{,}569{,}044) - 510{,}579{,}216}}$$

≈ 0.98

Yes. The linear correlation coefficient is about 0.98.

CHAPTER 12 REVIEW EXERCISES

1. $\dfrac{12 + 17 + 14 + 12 + 8 + 19 + 21}{7} \approx 14.7$

    The mean is about 14.7. The median is the fourth ranked number, or 14. The mode is the most common number, or 12.

2. 14, as the mode is the most common number.

3. Answers will vary.

4. a. The median.

    b. The mode.

    c. The mean.

5. $(1235 + 1644 + 1576 + 1200 + 1200 + 1182 + 1212 + 1400) / 8 \approx 1331.125$
    The mean is about 1331.125 feet. The median is the average of the fourth and fifth ranked numbers, or 1223.5 feet. The mode is the most common number, or 1200 feet. The range is the difference between the highest and lowest values, or 462 feet.

6. $\dfrac{90}{2.5} = 36$. The average rate was 36 mph.

7. $[3(4) + 3(2.33) + 2(2.67) + 4(3) + 1(4)] / 13 \approx 3.10$. The student has a grade point average of about 3.10.

8. a. $z_{82} = \dfrac{82 - 72}{8} = 1.25$. Ann's z-score is 1.25.

    b. Assuming a normal distribution of scores, 89.4% of all results lie to the left of $z = 1.25$. Therefore, Ann is in the 89th percentile.

9. Mean $= \dfrac{12 + 18 + 20 + 14 + 16}{5} = 16$

    Sample variance $= \dfrac{\sum (x - \bar{x})^2}{n - 1}$

    $$\dfrac{(12-16)^2 + (18-16)^2 + (20-16)^2 + (14-16)^2 + (16-16)^2}{4}$$

    $= 10$

    The sample variance is 10 minutes^2. The sample standard deviation is 3.16 minutes.

10. $(4.08 + 4.35 + 4.42 + 4.59 + 4.69 + 5.08 + 5.39 + 5.65 + 5.80 + 6.03) / 10 = 5.008$. The mean is $5.008.

$(4.69 + 5.08) / 2 = 4.885$. The median is $4.885.

$(4.08 - 5.01)^2 + (4.35 - 5.01)^2 + (4.42 - 5.01)^2$

$+ (4.59 - 5.01)^2 + (4.69 - 5.01)^2$

$+ (5.08 - 5.01)^2 + (5.39 - 5.01)^2$

$+ (5.65 - 5.01)^2 + (5.80 - 5.01)^2$

$+ (6.03 - 5.01)^2 = 4.1508$

$\sqrt{\dfrac{4.1508}{9}} \approx 0.679$

The standard deviation is approximately $0.679.

11. a. The second student's mean is 5 points higher than the first student's mean.

   b. The two standard deviations are the same.

12. a. $z_{72} = \dfrac{72 - 81}{5.2} \approx -1.73$

   b. $z_{84} = \dfrac{84 - 81}{5.2} \approx 0.58$

13. Drawing the plot:

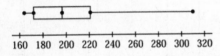

14. a. 8 students scored at least an 84 on the test.

   b. 40 students took the test.

   c. The uniform class width is 8.

15. a. $14 + 6 + 10 + 8 + 2 = 40$. 40% of the states paid an average teacher salary of at least $44,000.

   b. $14 + 14 + 6 = 34$. The probability that a state selected at random paid an average teacher salary of at least $41,000 but less than $50,000 is 0.34.

16. $n = 5$

$\sum x = 0 + 5 + 10 + 12 + 13 = 40$
$\sum y = 5 + 29 + 60 + 68 + 70 = 232$
$\sum x^2 = 0 + 25 + 100 + 144 + 169 = 438$
$\sum y^2 = 25 + 841 + 3600 + 4624 + 4900 = 13{,}990$
$(\sum x)^2 = 40^2 = 1600$

$(\sum y)^2 = 232^2 = 53{,}824$
$\sum xy = 0 + 145 + 600 + 816 + 910 = 2471$

$r = \dfrac{n(\sum xy) - (\sum x)(\sum y)}{\sqrt{n(\sum x^2) - (\sum x)^2} \cdot \sqrt{n(\sum y^2) - (\sum y)^2}}$

$= \dfrac{5(2471) - (40)(232)}{\sqrt{5(438) - 1600} \cdot \sqrt{5(13{,}990) - 53{,}824}}$

≈ 0.997

Yes. $r \approx 0.997$, which is greater than 0.9.

17. a. $n = 4$

$\sum x = 0 + 1 + 2 + 3 = 6$
$\sum y = 11 + 18.5 + 30.5 + 50 = 110$
$\sum x^2 = 0 + 1 + 4 + 9 = 14$
$(\sum x)^2 = 6^2 = 36$
$\sum xy = 0 + 18.5 + 61 + 150 = 229.5$

$a = \dfrac{n(\sum xy) - (\sum x)(\sum y)}{n(\sum x^2) - (\sum x)^2}$

$= \dfrac{4(229.5) - (6)(110)}{4(14) - 36} \approx 12.9$

$b = \overline{y} - a\overline{x} \approx 8.15$

$\hat{y} = 12.9x + 8.15$

   b. $n = 4$

$\sum x = 0 + 1 + 2 + 3 = 6$
$\sum y = 71 + 66 + 63 + 57 = 257$
$\sum x^2 = 0 + 1 + 4 + 9 = 14$
$(\sum x)^2 = 6^2 = 36$
$\sum xy = 0 + 66 + 126 + 171 = 363$

$a = \dfrac{n(\sum xy) - (\sum x)(\sum y)}{n(\sum x^2) - (\sum x)^2}$

$= \dfrac{4(363) - (6)(257)}{4(14) - 36} \approx -4.5$

$b = \overline{y} - a\overline{x} \approx 71$

$\hat{y} = -4.5x + 71$

   c. greater than since:

$12.9(5) + 8.15 = 72.65$

$-4.5(5) + 71 = 48.5$

18. No. No information is given about how the scores are distributed below the mean.

19. a. $z_8 = \dfrac{8 - 6.5}{1} = 1.5$

0.433 square unit lies between $z = 1.5$ and $z = 0$. 0.500 square unit lies left of $z = 0$. $0.433 + 0.500 = 0.933$. The probability that a customer spends at most 8 minutes waiting is 0.933.

b. $z_6 = \dfrac{6-6.5}{1} = -0.5$

0.191 square unit lies between $z = -0.5$ and $z = 0$. Therefore, 0.309 square unit lies to the left of $z = -0.5$. The probability that a customer spends less than 6 minutes waiting is 0.309.

20. a. $z_{49.6} = \dfrac{49.6-50}{0.5} = -0.8$

0.288 square units lie between $z = -0.8$ and $z = 0$. Therefore, 0.212 square units lies to the left of $z = -0.8$. 21.2% of the sacks will weigh less than 49.6 pounds.

b. 95.4% of the sacks will weigh between 49 and 51 pounds, as 49 and 51 are two standard deviations below and above the mean, respectively.

21. a. $z_{7.25} = \dfrac{7.25-6.5}{0.5} = 1.5$

0.433 square unit lies between $z = 0$ and $z = 1.5$. Therefore, 0.067 square unit lies to the right of $z = 1.5$. 6.7% of the company's telephones will last at least 7.25 years.

b. Calculating:

$z_{5.8} = \dfrac{5.8-6.5}{0.5} = -1.4$

$z_{6.8} = \dfrac{6.8-6.5}{0.5} = 0.6$

0.419 square unit lies between $z = -1.4$ and $z = 0$. 0.226 square unit lies between $z = 0$ and $z = 0.6$. $0.419 + 0.226 = 0.645$. 64.5% of the company's telephones will last between 5.8 and 6.8 years.

c. $z_{6.9} = \dfrac{6.9-6.5}{0.5} = 0.8$

0.288 square unit lies between $z = 0$ and $z = 0.8$. 0.500 square unit lies to the left of $z = 0$. $0.288 + 0.500 = 0.788$. 78.8% of the company's phones will last less than 6.9 years.

22. $(91.4 + 92.6 + 94.5 + 94.6 + 94.3 + 91.5) / 6 =$ 93.15 million miles

23. a.

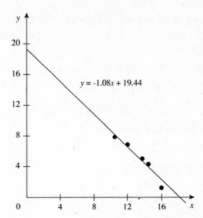

b. $n = 5$
$\sum x = 10 + 12 + 14 + 15 + 16 = 67$
$\sum y = 8 + 7 + 5 + 4 + 1 = 25$
$\sum x^2 = 100 + 144 + 196 + 225 + 256 = 921$
$(\sum x)^2 = 67^2 = 4489$
$\sum xy = 80 + 84 + 70 + 60 + 16 = 310$

c. Calculating:
$$a = \dfrac{n(\sum xy)-(\sum x)(\sum y)}{n(\sum x^2)-(\sum x)^2}$$
$$= \dfrac{5(310)-(67)(25)}{5(921)-4489} \approx -1.08$$
$$b = \bar{y} - a\bar{x} \approx 19.44$$
$$\hat{y} = -1.08x + 19.44$$

d. See the graph in part a.

e. Yes.

f. $\hat{y} = -1.08(8) + 19.44 \approx 10.8$

g. $\sum y^2 = 64 + 49 + 25 + 16 + 1 = 155$
$(\sum y)^2 = 25^2 = 625$
$$r = \dfrac{n(\sum xy)-(\sum x)(\sum y)}{\sqrt{n(\sum x^2)-(\sum x)^2}\cdot\sqrt{n(\sum y^2)-(\sum y)^2}}$$
$$= \dfrac{5(310)-(67)(25)}{\sqrt{5(921)-4489}\cdot\sqrt{5(155)-625}} \approx -0.95$$

24. a. $n = 5$
$\sum x = 80 + 100 + 110 + 150 + 170 = 610$
$\sum y = 6.2 + 7.4 + 8.3 + 11.1 + 12.7 = 45.7$
$\sum x^2 = 6400 + 10{,}000 + 12{,}100 + 22{,}500 + 28{,}900 = 79{,}900$
$\sum y^2 = 38.44 + 54.76 + 68.89 + 123.21 + 161.29 = 446.59$

$(\sum x)^2 = 610^2 = 372,100$
$(\sum y)^2 = (45.7)^2 = 2088.49$
$\sum xy = 496 + 740 + 913 + 1665 + 2159 = 5973$

$r = \dfrac{n(\sum xy) - (\sum x)(\sum y)}{\sqrt{n(\sum x^2) - (\sum x)^2} \cdot \sqrt{n(\sum y^2) - (\sum y)^2}} =$

$\dfrac{5(5973) - (610)(45.7)}{\sqrt{5(79,900) - 372,100} \cdot \sqrt{5(446.59) - 2088.49}}$

≈ 1.00

b. Calculating:

$a = \dfrac{n(\sum xy) - (\sum x)(\sum y)}{n(\sum x^2) - (\sum x)^2}$

$= \dfrac{5(5973) - (610)(45.7)}{5(79,900) - 372,100} \approx 0.07$

$b = \overline{y} - a\overline{x} \approx 0.29$

$\hat{y} = 0.07x + 0.29$

c. $\hat{y} = 0.07(195) + 0.29 = 13.94$
A weight of 195 pounds should stretch the spring 13.94 inches.

25. a. $n = 12$
$\sum x = 10.5 + 12.9 + 15 + 20 + 60 + 75 + 110 + 156 + 163 + 175 + 200 + 250 = 1247.4$
$\sum y = 0.20 + 0.24 + 0.27 + 0.36 + 1.09 + 1.42 + 2.01 + 2.68 + 2.87 + 3.10 + 3.64 + 4.61 = 22.49$
$\sum x^2 = 110.25 + 166.41 + 225 + 400 + 3600 + 5625 + 12,100 + 24,336 + 26,569 + 30,625 + 40,000 + 62,500 = 206,256.66$
$\sum y^2 = 0.04 + 0.0576 + 0.0729 + 0.1296 + 1.1881 + 2.0164 + 4.0401 + 7.1824 + 8.2369 + 9.61 + 13.2496 + 21.2521 = 67.0757$
$(\sum x)^2 = (1247.4)^2 = 1,556,006.76$
$(\sum y)^2 = (22.49)^2 = 505.8001$
$\sum xy = 2.1 + 3.096 + 4.05 + 7.2 + 65.4 + 106.5 + 221.1 + 418.08 + 467.81 + 542.5 + 728 + 1152.5 = 3718.336$

$a = \dfrac{n(\sum xy) - (\sum x)(\sum y)}{n(\sum x^2) - (\sum x)^2}$

$= \dfrac{12(3718.336) - (1247.4)(22.49)}{12(206,256.66) - 1,556,006.76} \approx 0.018$

$b = \overline{y} - a\overline{x} \approx 0.0005$

$\hat{y} = 0.018x + 0.0005$

b. Calculating:

$r = \dfrac{n(\sum xy) - (\sum x)(\sum y)}{\sqrt{n(\sum x^2) - (\sum x)^2} \cdot \sqrt{n(\sum y^2) - (\sum y)^2}}$

$= \dfrac{12(3718.336) - (1247.4)(22.49)}{\sqrt{12(206,256.66) - 1,556,006.76} \cdot \sqrt{12(67.0757) - 505.8001}}$

≈ 0.999

Yes, a linear model of the data is a reasonable model, as r is very close to 1.

c. $\hat{y} = 0.018(100) + 0.0005 \approx 1.80$
The expected download time is 1.80 seconds.

26. a. $n = 7$
$\sum x = 5 + 10 + 15 + 20 + 25 + 30 + 34 = 139$
$\sum y = 161.2 + 248.8 + 322.2 + 391.4 + 456.5 + 515.8 + 562.0 = 2657.9$
$\sum x^2 = 25 + 100 + 225 + 400 + 625 + 900 + 1156 = 3431$
$\sum y^2 = 25,985.4 + 61,901.4 + 103,813 + 153,194 + 208,392 + 266,050 + 315,844 = 1,135,180$
$(\sum x)^2 = 139^2 = 19,321$
$(\sum y)^2 = (2657.9)^2 = 7,064,430$
$\sum xy = 806 + 2488 + 4833 + 7828 + 11,412.5 + 15,474 + 19,108 = 61,949.5$

$a = \dfrac{n(\sum xy) - (\sum x)(\sum y)}{n(\sum x^2) - (\sum x)^2}$

$= \dfrac{7(61,949.5) - (139)(2657.9)}{7(3431) - 19,321} \approx 13.67$

$b = \overline{y} - a\overline{x} \approx 108.24$

$\hat{y} = 13.67x + 108.24$

$r = \dfrac{n(\sum xy) - (\sum x)(\sum y)}{\sqrt{n(\sum x^2) - (\sum x)^2} \cdot \sqrt{n(\sum y^2) - (\sum y)^2}}$

$= \dfrac{7(61,949.5) - (139)(2657.9)}{\sqrt{7(3431) - 19,321} \cdot \sqrt{7(1,135,180) - 7,064,430}}$

≈ 0.998

b. $\hat{y} = 13.67(37) + 108.24 \approx 614.0$
The CPI for 2007 will be 614.0.

CHAPTER 12 TEST

1. $\dfrac{3 + 7 + 11 + 12 + 7 + 9 + 15}{7} \approx 9.1$

The mean is 9.1. The median is the fourth rank number, or 9. The mode is the most common number, or 7.

2. $\dfrac{0.5(86)+0.5(50)+90+75+84+2(88)}{6} \approx 82.2$

Pam's course average is 82.2.

3. The range is the difference between the largest and smallest numbers, or 24.

$\text{Mean} = \dfrac{7+11+12+15+22+31}{6} \approx 16.3$

$[(7-16.3)^2 + (11-16.3)^2 + (12-16.3)^2$

$+(15-16.3)^2 + (22-16.3)^2$

$+(31-16.3)^2]/5 \approx 76.7$

The sample variance is 76.7.

The standard deviation is 8.76.

4. a. $z_{77} = \dfrac{77-65}{10.2} \approx 1.18$

    b. $z_{60} = \dfrac{60-65}{10.2} \approx -0.49$

5. The plot is shown below:

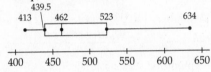

6. a. $14 + 11 + 8 = 33$

    33% of the movie attendees are at least 40 years old.

    b. $12 + 9 + 19 = 40$

    40% of the movie attendees are at least 21 but less than 40 years old.

7. a. 54 ounces is 2 standard deviations away from 34 ounces. Therefore, 47.7% of the parcels weighed between 34 and 54 ounces.

    b. 24 ounces is 1 standard deviation away from 34 ounces. Therefore, $13.6 + 2.15 + 0.15 = 15.9\%$ of the parcels weighed less than 24 ounces.

8. a. $z_{17} = \dfrac{17-18}{0.8} = -1.25$

    0.394 square unit lies between $z = -1.25$ and $z = 0$. Therefore, 0.106 square unit lies to the left of $z = -1.25$. 10.6% of the boxes will weigh less than 17 ounces.

b. Calculating:

$z_{18.4} = \dfrac{18.4-18}{0.8} = 0.5$

$z_{19} = \dfrac{19-18}{0.8} = 1.25$

0.191 square unit lies between $z = 0$ and $z = 0.5$. 0.394 square unit lies between $z = 0$ and $z = 1.25$. Therefore, 0.203 square unit lies between $z = 0.5$ and $z = 1.25$. 20.3% of the boxes weigh between 18.4 and 19 ounces.

9. $n = 4$

$\sum x = 0 + 1 + 2 + 3 = 6$

$\sum y = 71 + 66 + 63 + 57 = 257$

$\sum x^2 = 0 + 1 + 4 + 9 = 14$

$\sum y^2 = 5041 + 4356 + 3969 + 3249 = 16{,}615$

$(\sum x)^2 = 6^2 = 36$

$(\sum y)^2 = 257^2 = 66{,}049$

$\sum xy = 0 + 66 + 126 + 171 = 363$

$r = \dfrac{n(\sum xy)-(\sum x)(\sum y)}{\sqrt{n(\sum x^2)-(\sum x)^2} \cdot \sqrt{n(\sum y^2)-(\sum y)^2}}$

$= \dfrac{4(363)-(6)(257)}{\sqrt{4(14)-36} \cdot \sqrt{4(16{,}615)-66{,}049}} \approx -0.99$

Yes, there is a linear relationship.

10. a. $n = 8$

$\sum x = 93.2 + 92.3 + 91.9 + 89.5 + 89.6 + 90.5 + 91.9 + 91.7 = 730.6$

$\sum y = 28 + 26 + 39 + 56 + 56 + 36 + 32 + 32 = 305$

$\sum x^2 = 8686.24 + 8519.29 + 8445.61 + 8010.25 + 8028.16 + 8190.25 + 8445.61 + 8408.89 = 66{,}734.30$

$(\sum x)^2 = (730.6)^2 = 533{,}776.36$

$\sum xy = 2609.6 + 2399.8 + 3584.1 + 5012 + 5017.6 + 3258 + 2940.8 + 2934.4 = 27{,}756.3$

$a = \dfrac{n(\sum xy)-(\sum x)(\sum y)}{n(\sum x^2)-(\sum x)^2}$

$= \dfrac{8(27{,}756.3)-(730.6)(305)}{8(66{,}734.30)-533{,}776.36} \approx -7.98$

$b = \bar{y} - a\bar{x} \approx 767.12$

$\hat{y} = -7.98x + 767.12$

b. $\hat{y} = -7.98(89) + 767.12 \approx 57$

57 calories are expected in a soup that is 89% water.

Chapter 13: Apportionment and Voting

EXERCISE SET 13.1

1. To calculate the standard divisor, divide the total population p by the number of items to apportion n.

3. The standard quota for a state is the whole number part of the quotient of the state's population divided by the standard divisor.

5. a. $\dfrac{1547-1215}{1215}=0.273$

   b. $\dfrac{1498-1195}{1195}=0.254$

   c. Salinas

7.

	Summer Hill Average	Seaside Average	Abs. Unfair
Summer Hill receives assoc.	$\dfrac{5289}{587+1}$ $=8.99$	$\dfrac{6215}{614}$ $=10.12$	1.13
Seaside receives assoc.	$\dfrac{5289}{587}$ $=9.01$	$\dfrac{6215}{614+1}$ $=10.11$	1.10

   Relative unfairness if Summer Hill receives the new associate: $\dfrac{1.13}{8.99}=0.126$

   Relative unfairness if Seaside receives the new associate: $\dfrac{1.10}{10.11}=0.109$

   Using the apportionment principle, Seaside Mall should get the new associate.

9. a. $\dfrac{p}{n}=\dfrac{281,424,177}{435}\approx 646,952$

   There is one representative for every 646,952 citizens in the U.S.

b. Underrepresented; the average constituency is greater than the standard divisor.

c. Over-represented; the average constituency is less than the standard divisor.

11. a. $\dfrac{p}{n}=\dfrac{1781}{48}\approx 37.10$

   There is one new nurse for every 37.10 beds.

   b.

Hospital	Quot.	Q	SQ	beds
Sharp	$\dfrac{242}{37.1}$	6.52	6	7
Palomar	$\dfrac{356}{37.1}$	9.60	9	10
Tri-City	$\dfrac{308}{37.1}$	8.30	8	8
Del Raye	$\dfrac{190}{37.1}$	5.12	5	5
R. Verde	$\dfrac{275}{37.1}$	7.41	7	7
Bel Aire	$\dfrac{410}{37.1}$	11.05	11	11

   c. The modified standard divisor is 34.5.

Hospital	Quot.	Q	SQ	beds
Sharp	$\dfrac{242}{34.5}$	7.01	7	7
Palomar	$\dfrac{356}{34.5}$	10.32	10	10
Tri-City	$\dfrac{308}{34.5}$	8.93	8	8
Del Raye	$\dfrac{190}{34.5}$	5.51	5	5
R. Verde	$\dfrac{275}{34.5}$	7.97	7	7
Bel Aire	$\dfrac{410}{34.5}$	11.88	11	11

   d. They are identical.

13. The population paradox occurs when the population of one state is increasing faster than that of another state, yet the first state still loses a representative.

15. The Balinski-Young Impossibility Theorem states that any apportionment method will either violate the quota rule or will produce paradoxes such as the Alabama paradox.

17. a. The standard divisor is 6.57.

Resort	A	B	C	D
Guest rooms	23	256	182	301
@ 115	4	39	27	45
@ 116	3	39	28	46

Yes. Resort A will lose a TV.

b. The standard divisor is 6.51.

@ 117	4	39	28	46

No. No resort loses a TV.

c. The standard divisor is 6.46.

@ 118	3	40	28	47

Yes. Resort A will lose a TV.

19. a. The standard divisor is 61.9.

Office	Boston	Chicago
Employees	151	1210
Vice presidents	2	20

b. The standard divisor is 62.3.

Office	Boston	Chicago	SF
Employees	151	1210	135
Quota	2.424	19.422	2.167
VPs	3	19	2

Yes. Chicago lost a vice president and Boston gained one.

21. a.

	avg. constit. in 5th grade	avg. constit. in 6th grade
Add to 5th grade	$\dfrac{604}{(19+1)}$ $= 30.2$	$\dfrac{698}{21}$ ≈ 33.2
Add to 6th grade	$\dfrac{604}{19}$ ≈ 31.8	$\dfrac{698}{(21+1)}$ ≈ 31.7

The absolute unfairness of the apportionment if the 5th grade gets the new teacher is $33.2 - 30.2$; the absolute unfairness of the apportionment if the sixth grade gets the new teacher is $31.8 - 31.7$. The relative unfairness of the apportionment for each grade equals the absolute unfairness of the apportionment for the grade divided by the average constituency if the grade gets the new teacher. If the fifth grade gets the new teacher, the relative unfairness of the apportionment is $(33.2 - 30.2)/30.2 \approx 0.0993$. If the sixth grade gets the new teacher, the relative unfairness of the apportionment is $(31.8 - 31.7)/31.7 \approx 0.0032$. Because the smaller relative unfairness results from adding the teacher to the sixth grade, sixth grade should get the teacher.

b. The Huntington-Hill number for the fifth grade is $\dfrac{(S_5)^2}{t(t+1)} = \dfrac{(604)^2}{(19)(20)} \approx 960$. The Huntington-Hill number for the sixth grade is $\dfrac{(698)^2}{(21)(22)} \approx 1054.6$. The grade with the greater Huntington-Hill number, which is grade 6, receives the new teacher. This is the same result as using the relative unfairness of the apportionment.

23.

School	Computers	Students	H-H number
Rose	26	625	556
Lincoln	22	532	559
Midway	26	620	548
Valley	31	754	573

Because Valley School has the greatest Huntington-Hill number, it should be assigned the new computer.

25. a. The modified standard divisor is 88.

Div.	Quot.	Q	SQ	PCs
Lib. Arts	3455/88	39.26	39	39
Bus.	5780/88	65.68	66	66
Hum.	1896/88	21.55	22	22
Sci.	4678/88	53.16	53	53

This is the same result as the Hamilton method.

b. Using the Jefferson method, the humanities division gets one fewer computer and the sciences division gets one more computer, compared with the Webster method.

27. The Jefferson and Webster methods

29. The Huntington-Hill method

31. Answers will vary.

EXERCISE SET 13.2

1. A majority means that a choice receives more than 50% of the votes. A plurality means that the choice with the most votes wins. It is possible to have a plurality without a majority if there are more than two choices.

3. If there are n choices in an election, each voter ranks the choices by giving n points to the voter's first choice, $n-1$ points to the voter's second choice, and so on, with the voter's least favorite choice receiving 1 point. The choice with the most points is the winner.

5. In the pairwise comparison voting method, each choice is compared one-on-one with each of the other choices. A choice receives 1 point for a win, 0.5 points for a tie, and 0 points for a loss. The choice with the greatest number of points is the winner.

7. No; no. By Arrow's Impossibility Theorem, no possible voting system involving three or more choices satisfies the fairness criteria.

9. a. Al Gore

   b. No, since 50,999,897 is less than 50% of the total votes cast: 50,456,002 + 50,999,897 + 2,882,955 = 104,338,854

   c. George Bush

11. a. 35

    b. 18

    c. Scooby Doo-13 votes

13. Theater:
    $(15+7) \cdot 3 + (8+7) \cdot 2 + 13 = 109$

DVD: $(13+7) \cdot 3 + 7 \cdot 2 + (15+8) = 97$

Pay-per-view: $8 \cdot 3 + (13+15) \cdot 2 + (7+7) = 94$

This group of consumers prefers to go to the theater.

15. Mickey Mouse: $(10 \cdot 3) + (11 \cdot 2) + 14 = 66$

    Bugs Bunny: $(12 \cdot 3) + (14 \cdot 2) + 9 = 73$

    Scooby Doo: $(13 \cdot 3) + (10 \cdot 2) + 12 = 71$

    The students prefer Bugs Bunny.

17. Lee: $(41 \cdot 4) + (81 \cdot 3) + (80 \cdot 2) + 31 = 598$

    Brewer: $(84 \cdot 4) + 0 + (86 \cdot 2) + 63 = 571$

    Garcia: $(36 \cdot 4) + (121 \cdot 3) + (31 \cdot 2) + 45 = 614$

    Turley: $(72 \cdot 4) + (31 \cdot 3) + (36 \cdot 2) + 94 = 547$

    Elaine Garcia should be class president.

19. Red and white has the least first place votes (none) and is eliminated. The preference schedule is now:

green/yellow	3	1	3	1
red/blue	2	3	2	3
blue/white	1	2	1	2
votes	4	2	5	4

Red and blue now has no first-place votes, and is eliminated. The preference schedule is now:

green/yellow	2	1	2	1
blue/white	1	2	1	2
votes	4	2	5	4

Blue and white has 9 first-place votes, and green and yellow has 6. The team chooses blue and white uniforms.

21. Garcia has the least first-place votes (36) and is eliminated. The preference schedule is now:

Lee	1	2	1	2	3	2
Brewer	3	1	2	3	1	3
Turley	2	3	3	1	2	1
votes	36	53	41	27	31	45

Turley now has the least first-place votes (72) and is eliminated. The preference schedule becomes:

Lee	1	2	1	1	2	1
Brewer	2	1	2	2	1	2
votes	36	53	41	27	31	45

Lee has 149 first-place votes; Brewer has 84 first-place votes. Raymond Lee should be class president.

23. a. "Buy new computers for the club," which has 19 first-place votes.

    b. Neither "throw an end-of-year party" nor "donate to charity" received any first-place votes, and are eliminated first. Of the remaining options, "establish a scholarship" received the fewest first-place votes (8) and is also eliminated. The final preference schedule is now:

convention	1	1	2	1	2
computers	2	2	1	2	1
votes	8	5	12	9	7

"Travel to a convention" has 22 first-place votes; "Buy new computers" has 19 first-place votes. The money should be spent to pay for several members to travel to a convention.

c. Scholarship:
$(8 \cdot 5) + (5 \cdot 4) + (21 \cdot 3) + (7 \cdot 2) + 0 = 137$
Convention:
$(14 \cdot 5) + (20 \cdot 4) + 0 + 0 + 7 = 157$
Computers:
$(19 \cdot 5) + 0 + (13 \cdot 3) + (9 \cdot 2) + 0 = 152$
Party: $0 + (16 \cdot 4) + 0 + (8 \cdot 2) + 17 = 97$
Charity: $0 + 0 + (7 \cdot 3) + (17 \cdot 2) + 17 = 72$
The money should be spent to pay for several members to travel to a convention.

d. Answers will vary.

25.

versus	Star Wars	Empire	Jedi	Menace
Star Wars	--	Star Wars	Jedi	Star Wars
Empire Strikes Back	--	--	Jedi	Menace
Return of the Jedi	--	--	--	Jedi
Phantom Menace	--	--	--	--

Total votes = 2532
Star Wars is preferred over Empire Strikes Back on 429 + 1137 + 384 = 1950 ballots, and Empire Strikes Back is preferred over Star Wars on 2532 − 1950 = 582 ballots. Star Wars wins the match. Star Wars is preferred on 813 ballots over Return of the Jedi, so Jedi is preferred on 1719 ballots, and Jedi wins the matchup. Star Wars is preferred on all ballots over the Phantom Menace. Empire Strikes Back is preferred over Return of the Jedi on 966 ballots, so Jedi is preferred over Empire on 1566 ballots, and is the winner of that matchup. Empire Strikes Back is also preferred over Phantom Menace on 966 ballots, so Menace wins that matchup. Finally, Return of the Jedi is favored on all ballots over the Phantom Menace.
Return of the Jedi has three wins, Star Wars has two, and the Phantom Menace has one. Website visitors prefer Return of the Jedi.

27.

versus	Bulldog	Panther	Hornet	Bobcat
Bulldog	--	Panther	Bulldog	Bobcat
Panther	--	--	Panther	Bobcat
Hornet	--	--	--	Hornet
Bobcat	--	--	--	--

Total votes = 3150
Bulldog has 390 ballots versus 2760 for Panther. Bulldog has 1701 ballots versus 1449 for Hornet. Bulldog has 390 ballots versus 2760 for Bobcat. Panther has 2235 ballots versus 915 for Hornet.
Panther has 1449 ballots versus 1701 for Bobcat. Hornet has 1839 ballots versus 1311 for Bobcat. Panther and Bobcat have two wins, and Hornet and Bulldog have one each. The vote is a tie between Panther and Bobcat.

29.

versus	red/white	green/yellow	red/blue	blue/white
red/white	--	red/white	red/white	blue/white
green/yellow	--	--	red/blue	blue/white
red/blue	--	--	--	blue/white
blue/white	--	--	--	--

Total votes = 15
Red/white has 9 ballots versus 6 for green/yellow; red/white has 10 ballots versus 5 for red/blue; red/white has 4 ballots versus 11 for blue/white; green/yellow has 6 ballots versus 9 for red/blue; green/yellow has 6 ballots versus 9 for blue/white; red/blue has 0 ballots versus 15 for blue/white.
Blue and white wins three matchups, red and white wins two matchups, and red and blue wins one matchup. The players prefer blue and white uniforms.

31.

versus	Mickey	Bugs	Scooby
Mickey	--	Bugs	Scooby
Bugs	--	--	Bugs
Scooby	--	--	--

No. Using plurality voting, Scooby Doo won the election, but Bugs Bunny wins both head-to-head contests.

33.

versus	Scholarship	Convention	Computer	Party	Charity
Scholarship	--	convention	scholarship	scholarship	scholarship
Convention	--	--	convention	convention	convention
Computers	--	--	--	computer	computer
Party	--	--	--	--	party
Charity	--	--	--	--	--

Yes. Using the Borda Count method, the members voted to pay to travel to a convention. Traveling to a convention also won all of its head-to-head contests, so the Condorcet criterion is satisfied.

35. No. Using the Borda Count method, Elaine Garcia won the election, although no candidate received a majority of the first-place votes.

37.
a. Lorenz: $(2691 \cdot 3) + 0 + 2653 = 10,726$
Beasley:
$(2416 \cdot 3) + (237 \cdot 2) + 2691 = 10,413$
Hyde: $(237 \cdot 3) + (5107 \cdot 2) + 0 = 10,925$
Stephen Hyde wins the election.

b. Stephen Hyde received the fewest number of first-place votes.

c. John Lorenz beats Marcia Beasley by $2691 - 2653$, and beats Stephen Hyde by the same vote, so Lorenz wins all head-to-head comparisons.

d. The candidate that wins all the head-to-head matches does not win the election.

e. Lorenz gets $(2691 \cdot 2) + 2653 = 8035$ points, while Hyde gets
$(2653 \cdot 2) + 2691 = 7997$ points.

f. The candidate winning the original election (Stephen Hyde) did not remain the winner in a recount in which a losing candidate withdrew from the race.

39. Cynthia: $0 + (16 \cdot 4) + (27 \cdot 3) + (22 \cdot 2) + 16 = 205$
Andrew:
$(16 \cdot 5) + (22 \cdot 4) + (10 \cdot 3) + (6 \cdot 2) + 27 = 237$

Jen: $(10 \cdot 5) + (43 \cdot 4) + (6 \cdot 3) + 0 + 22 = 262$

Hector: $(28 \cdot 5) + 0 + 0 + (43 \cdot 2) + 10 = 236$

Medin: $(27 \cdot 5) + 0 + (38 \cdot 3) + (10 \cdot 2) + 6 = 275$

Medin is the president, Jen is the vice-president, Andrew is the secretary, and Hector is the treasurer.

41. Chang only wins the Borda Count method, with 77 points. He loses a plurality contest, loses plurality with elimination, and ties the pairwise comparison with Huck.

EXERCISE SET 13.3

1. a. 6

   b. 4

   c. 3

   d. 6

   e. No

   f. A and C

   g. $2^4 - 1 = 15$

   h. 6

3.

Winning coalition	Critical voters
{A, B}	A, B
{A, C}	A, C
{A, B, C}	A

$BPI(A) = \dfrac{3}{5} = 0.6$

$BPI(B) = BPI(C) = \dfrac{1}{5} = 0.2$

5.

Winning coalition	Critical voters
{A, B}	A, B
{A, B, C}	A, B
{A, B, D}	A, B
{A, C, D}	A, C, D
{A, B, C, D}	A

$BPI(A) = \dfrac{5}{10} = 0.5$

$BPI(B) = \dfrac{3}{10} = 0.3; BPI(C) = \dfrac{1}{10} = 0.1$
$= BPI(D)$

7.

Winning coalitions	Critical voters
{A, B}	A, B
{A, B, C}	A, B
{A, B, D}	A, B
{A, B, E}	A, B
{A, C, D}	A, C, D
{A, C, E}	A, C, E
{B, C, D}	B, C, D
{A, B, C, D}	none
{A, B, C, E}	A
{A, B, D, E}	A, B
{A, C, D, E}	A, C
{B, C, D, E}	B, C, D
{A, B, C, D, E}	none

$BPI(A) = \dfrac{9}{25} = 0.36; BPI(B) = \dfrac{7}{25} = 0.28$

$BPI(C) = \dfrac{5}{25} = 0.20; BPI(D) = \dfrac{3}{25} = 0.12$

$BPI(E) = \dfrac{1}{25} = 0.04$

9. No coalition without voter A can reach a quota. Voter A has a Banzhaf power index of 1; all the other voters have a BPI of 0.

11.

Winning Coalition	Critical Voters
{A, B}	A, B
{A, C}	A, C
{A, B, C}	A
{A, B, D}	A, B
{A, B, E}	A, B
{A, C, D}	A, C
{A, C, E}	A, C
{A, D, E}	A, D, E
{B, C, D}	B, C, D
{A, B, C, D}	none
{A, B, C, E}	A
{A, B, D, E}	A
{A, C, D, E}	A
{B, C, D, E}	B, D, C
{A, B, C, D, E}	none

$$BPI(A) = \frac{11}{25} = 0.44$$

$$BPI(B) = BPI(C) = \frac{5}{25} = 0.20$$

$$BPI(D) = \frac{3}{25} = 0.12$$

$$BPI(E) = \frac{1}{25} = 0.04$$

13. a. In Exercise 9, voter A never needs any of the other voters in any winning coalition, and is thus the dictator.

    b. Veto power means that one or more voters are critical in every winning coalition. In exercises 3 and 5, voter A is critical in each winning coalition. In exercise 6, there is only one coalition, so each voter has veto power. In exercise 9, voter A is a dictator and thus has veto power. In exercise 12, voter A is critical in all winning coalitions.

    c. None

    d. In exercise 6, although voters have different weights, the only winning coalition comprises all the voters, so in practice each voter has one vote. In exercise 8, each voter has one vote.

15. The winning coalitions are {D, T}, {D,P}, and {T, P}, with each voter critical to each. The director, teacher, and principal each have a Banzhaf power index of 0.33.

17. a. {12: 1, 1, 1, 1, 1, 1, 1, 1, 1, 1, 1, 1,}

    b. Yes.

    c. Yes, every juror has a veto, as the verdict must be unanimous.

    d. Because this is a veto power system, divide the voter power of 1 by the quota of 12.

19. A is the dictator, and all of B, C, D, and E are dummies.

21. None.

23. a. The winning coalitions are {A, B}, {A, C}, {A, B, C}. A is critical to all three (and has

veto power), and B and C are each critical in their own coalition with A. The BPI for A is 0.6, and for B and C the BPI is 0.2

    b. With respect to the formation of winning coalitions, the coaches are the same.

25. a. There are four winning coalitions: {A, B}, {A, C}, {B, C}, and {A, B, C}. The critical voters are, respectively, A and B, A and C, B and C, and none of them. Each voter has a BPI of 0.33.

    b. Despite the varied weights, this is a majority system. Any two of the three voters are needed for a quota.

27. a. 11 and 14

    b. 15 and 16

    c. No. D is the dummy for $q = 11$, but not for $q = 12$, for instance. The existence of a dummy depends on the combinations of voter weight as well as the quota.

29. Yes. Voters with the same weight can be arithmetically substituted for each other, and when doing this, the arithmetic of winning coalitions cannot change. Thus, the BPI values remain the same.

31. In the weighted system, all voters have equal weight and the quota is equal to 50% of the sum of the weights rounded up to the next whole number, or a strict majority.

CHAPTER 13 REVIEW EXERCISES

1. a. Relative unfairness if High Desert receives the new controller: $\frac{326 - 297}{297} \approx 0.098$

    b. Relative unfairness if Eastlake receives the new controller: $\frac{302 - 253}{253} \approx 0.194$

    c. Using the apportionment principle, High Desert Airport should get the new controller.

2.

	Morena Valley Average	West Keyes Average	Abs. Unfair
Morena Valley receives professor	$\dfrac{1437}{38+1}$ $= 36.85$	$\dfrac{1504}{46}$ $= 32.70$	4.15
West Keyes receives professor	$\dfrac{1437}{38}$ $= 37.82$	$\dfrac{1504}{46+1}$ $= 32.00$	5.82

Relative unfairness if Morena Valley receives the new professor: $\dfrac{4.15}{36.85} \approx 0.113$

Relative unfairness if Eastlake receives the new controller: $\dfrac{5.82}{32.00} \approx 0.182$

Morena Valley should receive the new professor.

3.   a.   The standard divisor is $\dfrac{9609}{50} = 192.18$.

Div.	Enrol.	Quota	Proj.
Health	1280	6.66	7
Bus.	3425	17.82	18
Eng.	1968	10.24	10
Sci.	2936	15.28	15

Health division gets 7 projectors, the Business division gets 18, the Engineering division gets 10, and the Science division gets 15.

  b.   The modified standard divisor is 183.4.

Div.	Enrol.	Quota	Proj.
Health	1280	6.98	6
Bus.	3425	18.68	18
Eng.	1968	10.73	10
Sci.	2936	16.01	16

  c.   A modified standard divisor of 191 yields the following apportionment.

Div.	Enrol.	Quota	Proj.
Health	1280	6.70	7
Bus.	3425	17.93	18
Eng.	1968	10.30	10
Sci.	2936	15.37	15

4.   a.   The standard divisor is $\dfrac{110}{35} = 3.143$.

Airport	Agts.	Quota	Empl.
Newark	28	8.91	9
Cleve.	19	6.05	6
Chi.	34	10.82	11
Phila.	13	4.14	4
Det.	16	5.09	5

Newark gets 9 security employees, Cleveland 6, Chicago 11, Philadelphia 4, and Detroit 5.

  b.   The modified standard divisor is 3.05.

Airport	Agts.	Quota	Empl.
Newark	28	9.18	9
Cleve.	19	6.23	6
Chicago	34	11.15	11
Phila.	13	4.26	4
Detroit	16	5.25	5

  c.   The same modified standard divisor of 3.05 yields the Webster apportionment.

Airport	Agts.	Quota	Empl.
Newark	28	9.18	9
Cleve.	19	6.23	6
Chi.	34	11.15	11
Phila.	13	4.26	4
Det.	16	5.25	5

5.   a.   The standard divisor is 13.79. Office A has a quota is 1.38, B has a quota of 14.14, C has a quota of 22.34, and D has a quota of 29.15. Office A gets the next printer, and no office loses a printer. The Alabama paradox does not occur.

  b.   The standard divisor is 13.59, and the quotas are as follows: A:1.40; B: 14.35; C: 22.66; and D: 29.58. Offices C and D get the new printers, and Office A falls back to 1 printer. This is an example of the Alabama paradox.

6. a.   The standard divisor is 20.96.

Center	Quota	Autos
A	1.48	2
B	5.15	5
C	3.34	3
D	15.70	16
E	2.34	2

Regional center D gets the additional automobile. There is no Alabama paradox, as no center loses an automobile.

b.   The standard divisor is 20.24.

Center	Quota	Autos
A	1.53	2
B	5.34	5
C	3.46	4
D	16.25	16
E	2.42	2

Regional center C gets the additional automobile. There is no Alabama paradox, as no center loses an automobile.

7. a.   The standard divisor is 151.4. Los Angeles gets 9 servers, and Newark gets 2.

b.   The standard divisor is now 148. LA gets 10, and Newark and KC get 1 each. Yes, this is the new states' paradox, since Newark loses a server without having lost people.

8. a.   The standard divisor is 1408. The apportionment is A: 10, B: 3, C: 21.

b.   The standard divisor is 1428. The apportionment is A: 11, B: 2, C: 21. Yes, this is the population paradox. Although the population of region B has grown at a higher rate than the population region A, region B still loses an inspector in reapportionment.

9. Yes. See exercise 6b above.

10. Yes. It fulfills the quota rule, but, by the Balinski-Young Impossibility Theorem, produces paradoxes such as the new states paradox.

11. a.   The Huntington-Hill number is given by

$$\frac{(P_A)^2}{a(a+1)},$$ where P_A is the population, and a

is the current apportionment.

Bldg	A	B	C
Guard	25	43	18
Empl.	414	705	293
H-H	263.7	262.7	251.0

Building A has the greatest Huntington-Hill number, and is assigned the next guard.

b.

Bldg	A	B	C
Guard	26	43	18
Empl.	414	705	293
H-H	244.2	262.7	251.0

Building B now has the greatest Huntington-Hill number, and is assigned the 88th guard.

12. a.   Cynthia: $(3\cdot11)+(2\cdot97)+(1\cdot112)=339$

Hannah: $(3\cdot112)+(1\cdot97)+(1\cdot11)=444$

Shannon = $(3\cdot97)+(2\cdot112)+(2\cdot11)=537$

Shannon is the homecoming queen.

b.

versus	Cynthia	Hannah	Shannon
Cynthia	--	Hannah	Shannon
Hannah	--	--	Hannah
Shannon	--	--	--

Hannah wins all head-to-head contests.

c.   The candidate who wins all head-to-head contests should win the election of all candidates, but while Hannah wins all head-to-head contests, Shannon wins the overall election.

d.   Hannah, with 112 votes, wins both the plurality and the majority of votes.

e.   Hannah won the majority of first-place votes, but Shannon won the overall election.

13. a.   The resulting preference schedule:

Hannah	1	2	2
Shannon	2	1	1
votes	112	97	11

Hannah wins the election, 112 to 108.

b. The winner of the original election, Shannon, did not win the recount after a losing candidate withdrew.

14. The montonicity criterion has been violated. If a candidate wins an election, then, if the only change in voters' preferences is that supporters of another candidate change their votes to support the winner, then the result must remain the same. Margaret won the scholarship, but, when voters changed their preference from Terry to Margaret, Jean won the scholarship.

15. a. 18   b. 18

     c. Yes   d. A and C

     e. 15   f. 6

16. a. 35   b. 35

     c. Yes   d. A

     e. 31   f. 10

17. The winning coalitions are {A, B}, with A and B critical members; {A, C}, with A and C critical members; and {A, B, C}, with A as the critical member. $BPI(A) = \dfrac{3}{5} = 0.6$

$BPI(B) = BPI(C) = \dfrac{1}{5} = 0.2$

18. In a one-person, one-vote system, each voter has the same power as any other. As there are five voters in this system, each one has a BPI of $\dfrac{1}{5} = 0.2$.

19.

Winning coalitions	Critical voters
{A, B}	A, B
{A, C}	A, C
{A, B, C}	A
{A, B, D}	A, B
{A, C, D}	A, C
{B, C, D}	B, C, D
{A, B, C, D}	none

$BPI(A) \approx \dfrac{5}{12} = 0.42$

$BPI(B), BPI(C) = \dfrac{3}{12} = 0.25$

$BPI(D) = \dfrac{1}{12} \approx 0.08$

20.

Winning Coalition	Critical Voters
{A, B}	A, B
{A, C}	A, C
{A, B, C}	A
{A, B, D}	A, B
{A, B,E}	A, B
{A, C, D}	A, C
{A, C, E,}	A, C
{A, D, E}	A, D, E
{A, B, C, D}	A
{A, B, C, E}	A
{A, B, D, E}	A
{A, C, D, E}	A
{A, B, C, D, E}	A

$BPI(A) \approx 0.62; BPI(B) = BPI(C) \approx 0.14;$
$BPI(D) = BPI(E) \approx 0.05$

21. A is the dictator, and B, C, D, and E are dummies.

22. There is no dictator, and D is a dummy.

23.

Winning Coalition	Critical Voters
{A, B}	A, B
{A, C}	A, C
{A, D}	A, D
{A, B, C}	A
{A, B, D}	A
{A, C, D}	A
{A, B, C, D}	A
{A, E}	A, E
{B, C, D, E}	B, C, D, E

$BPI(A) = 0.5; BPI(B) = BPI(C) = BPI(D) = BPI(E) = 0.125$

24. a. Ortega, with 8 votes.

     b. No, he has 8 of 23 votes.

c. Kelly: $(2\cdot8)+(3\cdot5)+(3\cdot4)+(4\cdot6)=67$
Ortega: $(4\cdot8)+(2\cdot5)+(1\cdot4)+(2\cdot6)=58$
Nisbet: $(3\cdot8)+(1\cdot5)+(4\ \cdot4)+(3\cdot6)=63$
Toyama: $(1\cdot8)+(4\cdot5)+(2\cdot4)+(1\cdot6)=42$
Crystal Kelly wins the essay contest.

25. a. Vail 23, Aspen 22, Powderhorn 18
The club chooses Vail.

b. Aspen:
$(5\cdot14)+(5\cdot8)+(3\cdot11)+(4\cdot18)$
$+(3\cdot12)=251$
Copper Mtn.:
$(1\cdot14)+(2\cdot8)+(4\cdot11)+(2\cdot18)$
$+(2\cdot12)=134$
Powderhorn:
$(3\cdot14)+(4\cdot8)+(1\cdot11)+(5\cdot18)$
$+(1\cdot12)=187$
Telluride:
$(2\cdot14)+(1\cdot8)+(2\cdot11)+(1\cdot18)$
$+(4\cdot12)=124$
Vail:
$(4\cdot14)+(3\cdot8)+(5\cdot11)+(3\cdot18)$
$+(5\cdot12)=249$

Using the Borda Count method, the club chooses Aspen.

26. Schneider has no first-place votes, and is eliminated. The preference schedule is now:

Reynolds	2	2	1	3
Hernandez	1	3	3	2
Kim	3	1	2	1
votes	132	214	93	119

Reynolds now has the fewest first-place votes, and is eliminated. The preference schedule is now:

Hernandez	1	2	2	2
Kim	2	1	1	1
votes	132	214	93	119

Kim wins the election, 426 to 132.

27. Twix has no first-place votes, and is eliminated. The preference schedule is now:

Crunch	1	3	3	2	2
Snickers	2	1	2	3	1
Milky Way	3	2	1	1	3
votes	15	38	27	16	22

Nestle Crunch has the fewest first-place votes, and is eliminated. The preference schedule is now:

Snickers	1	1	2	2	1
Milky Way	2	2	1	1	2
votes	15	38	27	16	22

Snickers is the favorite candy bar, 75 to 43.

28.

versus	GR	LH	AK	JS
GR	--	GR	AK	GR
LH	--	--	AK	JS
AK	--	--	--	AK
JS	--	--	--	--

Kim wins 3 matches to 2 for Reynolds and 1 for Schneider.

29.

versus	Crunch	Snick.	Milky	Twix
Crunch	--	Snick.	Milky	Twix
Snick.	--	--	Snick.	Snick.
Milky	--	--	--	Milky
Twix	--	--	--	--

Snickers is the favorite candy bar, with 3 matches compared to 2 for Milky Way and 1 for Twix.

CHAPTER 13 TEST

1.

	Spring Valley Average	Summerville Average	Abs. Unfair
Spring Valley receives carrier	$\dfrac{67,530}{158+1}$ $=424.72$	$\dfrac{53,950}{129}$ $=418.22$	6.5
Summerville receives carrier	$\dfrac{67,530}{158}$ $=427.41$	$\dfrac{53,950}{129+1}$ $=415$	12.41

Relative unfairness if Spring Valley receives the

new carrier: $\dfrac{6.5}{424.72} \approx 0.015$

Relative unfairness if Summerville receives the

new carrier: $\dfrac{12.41}{415} \approx 0.030$

Spring Valley should receive the new carrier.

2. a. The standard divisor is 58.6.

Div.	Empl.	Quota	Comp.
Sales	1008	17.20	17
Adver.	234	3.99	4
Serv.	625	10.67	11
Manu.	3114	53.14	53

Sales get 17 computers, advertising 4, service 11, and manufacturing 53.

   b. The modified standard divisor is 57.0.

Div.	Empl.	Quota	Comp.
Sales	1008	17.68	17
Adver.	234	4.11	4
Serv.	625	10.96	10
Manu.	3114	54.63	54

No, the quota rule is not violated; each division gets either the standard quota, or one more, computers.

3. a. The formula for the Huntington-Hill number

is $\dfrac{\left(P_A\right)^2}{a(a+1)}$.

The Huntington-Hill number for Cedar Falls equals 77,792 and the Huntington-Hill number for Lake View equals 70,290.

   b. Cedar Falls has the higher Huntington-Hill number, and should be assigned the new counselor.

4. a. 33   b. 26

   c. No.   d. A and C

   e. $2^5 - 1 = 31$ f. 10

5. a. Aquafina, which has 39 votes to 31 for Arrowhead and 30 for Evian.

   b. No, Aquafina has 39 of 100 votes and needs 51 for a majority.

   c. Arrowhead:

$(2 \cdot 22) + (1 \cdot 17) + (3 \cdot 31) + (1 \cdot 11)$

$\quad + (2 \cdot 19) = 203$

Evian:

$(1 \cdot 22) + (2 \cdot 17) + (2 \cdot 31) + (3 \cdot 11)$

$\quad + (3 \cdot 19) = 208$

Aquafina:

$(3 \cdot 22) + (3 \cdot 17) + (1 \cdot 31) + (2 \cdot 11)$

$\quad + (1 \cdot 19) = 189$

Evian is the favored brand of water, using the Borda Count method.

6.

versus	NY	Dal.	LA	Atl.
NY	--	NY	NY	NY
Dal.	--	--	LA	Atl.
LA	--	--	--	LA
Atl.	--	--	--	--

New York wins 3 matches, LA wins 2, and Atlanta wins 1. New York should be selected.

7. a. Evening had the fewest first-place votes and is eliminated. The preference schedule is now:

Noon	2	2	2	2	1
PM	1	3	1	3	2
votes	12	16	9	5	13

Noon has the fewest first-place votes and is eliminated. The preference schedule is now:

AM	2	1	2	1	2
PM	1	2	1	2	1
votes	12	16	9	5	13

Afternoon has 34 first-place votes and morning has 21. She should schedule the session for the afternoon.

b. Morning:

$(1 \cdot 12) + (4 \cdot 16) + (1 \cdot 9) + (3 \cdot 5)$

$+ (2 \cdot 13) = 126$

Noon:

$(2 \cdot 12) + (3 \cdot 16) + (3 \cdot 9) + (2 \cdot 5)$

$+ (4 \cdot 13) = 161$

Afternoon:

$(4 \cdot 12) + (2 \cdot 16) + (4 \cdot 9) + (1 \cdot 5)$

$+ (3 \cdot 13) = 160$

Evening:

$(3 \cdot 12) + (1 \cdot 16) + (2 \cdot 9) + (4 \cdot 5)$

$+ (1 \cdot 13) = 103$

Using the Borda Count method, the best time for the session is noon.

8. a. Proposal A, with 40 votes compared to 9 for Proposal B and 39 for Proposal C.

b. Proposal B, which wins 48 votes compared to 40 for Proposal A.

c. Eliminating a losing choice changed the outcome of the election.

d.

versus	A	B	C
A	--	B	C
B	--	--	B
C	--	--	--

Proposal B wins all head-to-head contests.

e. Proposal B won all head-to-head contests but lost the vote when all the choices were on the ballot.

9.

Winning Coalition	Critical Voters
{A, B}	A, B
{A, C}	A, C
{A, B, C}	A
{A, B, D}	A, B
{A, C, D}	A, C
{B, C, D}	B, C, D
{A, B, C, D}	none

$BPI(A) = \dfrac{5}{12} \approx 0.42$

$BPI(B) = BPI(C) = \dfrac{3}{12} = 0.25$

$BPI(D) = \dfrac{1}{12} \approx 0.08$

10.

Winning Coalition	Critical Voters
{A, B}	A, B
{A, C}	A, C
{A, B, C}	A
{A, D}	A, D
{B, C, D}	B, C, D

$BPI(A) = \dfrac{4}{10} = 0.4$

$BPI(B) = BPI(C) = BPI(D) = \dfrac{2}{10} = 0.2$

Appendix: The Metric System of Measurement

EXERCISES

1. In the metric system, the meter is the basic unit of length, the liter is the basic unit of liquid measure, and the gram is the basic volume of weight.

3. kilometer

5. centimeter

7. gram

9. meter

11. gram

13. milliliter

15. gram

17. millimeter

19. milligram

21. gram

23. kiloliter

25. milliliter

27. a. See the table below.

   b. When multiplying by a number greater than one, the decimal point in the number is moved x places to the right, where x equals the magnitude's exponent. When multiplying by a number less than one, the decimal point in the number is moved x places to the left, where x equals the absolute value of the magnitude's exponent.

Metric system prefix	Symbol	Magnitude	Means multiply the basic unit by:
tera-	T	10^{12}	1,000,000,000,000
giga-	G	10^{9}	1,000,000,000
mega-	M	10^{6}	1,000,000
kilo-	k	10^{3}	1000
hecto-	h	10^{2}	100
deca-	da	10^{1}	10
deci-	d	10^{-1}	0.1
centi-	c	10^{-2}	0.01
milli-	m	10^{-3}	0.001
micro-	μ	10^{-6}	0.000 001
nano-	n	10^{-9}	0.000 000 001
pico-	p	10^{-12}	0.000 000 000 001

29. 910

31. 1.856

33. 7.285

35. 8 000

37. 0.034

39. 29.7

41. 7.530

43. 9 200

45. 36

47. 2 350

49. 83

51. 0.716

53. 6.302

55. 458

57. 9.2

59. $10 \text{ carats} \cdot \dfrac{200 \text{ mg}}{1 \text{ carat}} \cdot \dfrac{1 \text{ g}}{1\ 000 \text{ mg}} = 2 \text{ g}$

   The precious stone weighs 2 grams.

61. $\dfrac{800 \cdot 30}{1\ 000} = 24$. 24 liters of chlorine are used.

63. $\dfrac{3\,780}{230} \approx 16$. There are 16 servings in the

    container.

65. $\dfrac{2\,000}{500} = 4$. The patient should take 4 tablets per

    day.

67. Comparing:

    $\dfrac{19.80}{12 \cdot 1\,000} = 0.001\,65$

    $\dfrac{14.50}{24 \cdot 340} \approx 0.001\,78$

    The 12 one-liter bottles cost less.

69. $\dfrac{150\,000\,000 \cdot 1\,000}{300\,000\,000} = 500$. It takes 500 seconds

    for light from the sun to reach Earth.

71. The operator made \$658 per kl. The total
    amount from selling the gasoline is
    $0.658 \cdot 85 = \$55{,}930$. The profit is
    $55{,}930 - 38{,}500 = \$17{,}430$

73. 10 kg is 10,000 gr. The number of bags the store

    can make is $\dfrac{10\,000}{200} = 50$. The store spends

    $75 + 50(0.06) = \$78$. The profit is
    $50(2.89) - 78 = \$66.50$.